ANALYTIC
GEOMETRY

ANALYTIC GEOMETRY
THIRD EDITION

DOUGLAS F. RIDDLE

St. Joseph's University

Wadsworth Publishing Company
Belmont, California

A Division of Wadsworth, Inc.

ISBN 0-534-01030-X

Mathematics Editor: Richard Jones
Production: Greg Hubit Bookworks
Cover art: *Circular Forms* (1912) by Robert Delaunay. © A.D.A.G.P.

Printed in the United States of America

6 7 8 9 10—86

Library of Congress Cataloging in Publication Data

Riddle, Douglas F.
  Analytic geometry.

  First-2nd eds. published as: Analytic geometry with vectors.
  Includes index.
  1. Geometry, Analytic. I. Title.
QA551.R48 1982   516.3   81-14695
ISBN 0-534-01030-X   AACR2

CONTENTS

PREFACE   ix

1

PLANE ANALYTIC GEOMETRY   1

1.1   The Cartesian Plane   1
1.2   Distance Formula   4
1.3   Point-of-Division Formulas   10
1.4   Inclination and Slope   16
1.5   Parallel and Perpendicular Lines   21
1.6   Angle from One Line to Another   24
1.7   Graphs and Points of Intersection   30
1.8   An Equation of a Locus   33

2

VECTORS IN THE PLANE   38

2.1   Directed Line Segments and Vectors   38
2.2   The Dot Product   48
2.3   Applications of Vectors   55

v

3

THE LINE   62

3.1   Point-Slope and Two-Point Forms   62
3.2   Slope-Intercept and Intercept Forms   68
3.3   Distance from a Point to a Line   74
3.4   Families of Lines   81
3.5   Fitting a Line to Empirical Data   88

4

THE CIRCLE   95

4.1   The Standard Form for an Equation of a Circle   95
4.2   Conditions to Determine a Circle   101
4.3   Families of Circles   108

5

CONIC SECTIONS   115

5.1   Introduction   115
5.2   The Parabola   115
5.3   The Ellipse   123
5.4   The Hyperbola   131
5.5   Reflection Properties of Conics   138
5.6   Conics and a Right Circular Cone   147

6

TRANSFORMATION OF COORDINATES   152

6.1   Translation of Conic Sections   152

6.2   Translation of General Equations   163
6.3   Rotation   168
6.4   The General Equation of Second Degree   173

7

CURVE SKETCHING   181

7.1   Intercepts and Asymptotes   181
7.2   Symmetry, Sketching   190
7.3   Radicals and the Domain of the Equation   198
7.4   Direct Sketching of Conics   203

8

TRANSCENDENTAL CURVES   208

8.1   Trigonometric Functions   208
8.2   Inverse Trigonometric Functions   212
8.3   Exponential and Logarithmic Functions   216
8.4   Hyperbolic Functions   220

9

POLAR COORDINATES AND PARAMETRIC EQUATIONS   224

9.1   Polar Coordinates   224
9.2   Graphs in Polar Coordinates   226
9.3   Points of Intersection   232
9.4   Relationships between Rectangular and Polar Coordinates   235
9.5   Conics in Polar Coordinates   239
9.6   Parametric Equations   245
9.7   Parametric Equations of a Locus   252

10

SOLID ANALYTIC GEOMETRY   261

10.1   Introduction: The Distance and Point-of-Division Formulas   261
10.2   Vectors in Space   267
10.3   Direction Angles, Cosines, and Numbers   273
10.4   The Line   279
10.5   The Cross Product   287
10.6   The Plane   296
10.7   Distance between a Point and a Plane or Line; Angles between
       Lines or Planes   303
10.8   Cylinders and Spheres   310
10.9   Quadric Surfaces   314
10.10 Cylindrical and Spherical Coordinates   320

TABLES   329

Table 1   Common Logarithms   330
Table 2   Trigonometric Functions   332
Table 3   Squares, Square Roots, and Prime Factors   333

ANSWERS TO ODD-NUMBERED PROBLEMS   335

INDEX   401

PREFACE

Analytic Geometry, Third Edition, designed for students with a reasonably sound background in algebra and trigonometry, contains more than enough material for a three-semester-hour or five-quarter-hour course in analytic geometry.

As in the previous editions, my aim has been to write a text that students find to be interesting, clear and readable, and that instructors find will enhance their presentations of the material. Several changes have been made in this edition to further this aim. Although the vector approach of the previous editions has been retained, it has been supplemented with a nonvector approach to the same material. Thus the instructor has the option of taking either approach without having to depart from the text.

Although most of the problems have remained from the second edition, there have been several changes with regard to them. Of the few changes in the problems themselves, most have been to simplify problems with tediously difficult arithmetic. Review problems have been added to each chapter; students find these to be helpful in studying for examinations. In addition, most problem sets have been divided into three sections labeled A, B, and C. Section A consists of routine problems that every student can be expected to master. Those of section B are less routine but still not a great challenge. The average student can be expected to master most of them. The C problems present something of a challenge; only the better students can be expected to work them. Of course, any such system of classification must be very subjective; it should be considered to be a rough guide only. Finally, another change with regard to the problems is that answers are provided to *all* odd-numbered problems. In some cases (for example, when a problem asks the student to prove something analytically), the answer given is a complete solution of the problem. While some students are certainly going to look up the solution rather than working such problems, it is hoped that the majority will look upon them as an aid to check their own work and to help them with solutions to the even-numbered problems.

Other improvements are: many more worked examples; more explanation and improved graphs in solid analytic geometry; several results that had previously not stood out on the page have been put into theorem form (Theorems 9.2, 10.34, and 10.35) to emphasize their importance; the introduction to the chapter on conic sections has been completely rewritten and expanded to give a smoother transition into that material; the chapters on polar coordinates and parametric equations have been combined; and the section on translation has been divided into two sections, one devoted entirely to the translation of conic sections, and the other to more general translations.

Finally, since there is clearly more material here than can be covered in a three-semester-hour or five-quarter-hour course, we are presented with the problem of what to omit. There are several sections that are clearly intended as enrichment material that might be assigned to some of the better students. Included in this category are Sections 3.5, 5.5, and 5.6. The first of these, Section 3.5, is intended to give a preview of an application that the student is likely to encounter in other courses in chemistry, physics, or engineering. Another group of topics includes material that is covered elsewhere—either in this course or another. For example, Section 1.8 helps to tie together the two main problems of analytic geometry; but the material is covered in Chapters 3, 4, and 5. Similarly, the graphs of the transcendental curves, while important, are covered in courses in trigonometry and calculus. Finally, there is material that is simply less important. Again any such judgment is quite subjective, and the following should be taken as a rough guideline only. The following suggested coverage is for a semester or quarter of 45 class days. It allows for 6 days for examinations and review and 2 days for the inevitable loss of time at the beginning and end of the semester or quarter. This leaves 37 days to cover the material listed below.

Chapter	Sections
1	1–7
2	1–3
3	1–3
4	1–2
5	1–4
6	1, 3, 4
7	1–4
9	1–4, 6
10	1–6, 8–10

My thanks go to William E. Bradkin, American River College; Clifton F. Gary, Oscar Rose Junior College; Larry F. Heath, University of Texas at Arlington; and Harold Huneke, University of Oklahoma at Norman, for their many helpful suggestions.

1

PLANE ANALYTIC GEOMETRY

1.1

THE CARTESIAN PLANE

Analytic geometry provides a bridge between algebra and geometry that makes it possible for geometric problems to be solved algebraically (or analytically). It also allows us to solve algebraic problems geometrically, but the former is far more important, especially when numbers are assigned to essentially geometric concepts. Consider, for instance, the length of a line segment or the angle between two lines. Even if the lines and points in question are accurately known, the number representing the length of a segment or the angle between two lines can be determined only approximately by measurement. Algebraic methods provide an exact determination of the number.

The association between the algebra and geometry is made by assigning numbers to points. Suppose we look at this assignment of numbers to the points on a line. First of all, we select a pair of points, O and P, on the line, as shown in Figure 1.1. The point O, which we call the origin, is assigned the number zero, and the point P is assigned the number one. Using $\overline{OP}$ as our unit of length,* we assign numbers to all other points on the line in the following way: Q on the P side of the origin is assigned the positive number x if and only if its distance from the origin is x. A point Q on the opposite side of the origin is assigned the negative number $-x$ if and only if its distance from the origin is x units. In this way every point on the line is assigned a real number, and for each real number there corresponds a point on the line.

Thus, a **scale** is established on the line, which we now call a **coordinate line**. The number representing a given point is called the **coordinate** of that point, and the point is called the **graph** of the number.

Just as points on a line (a one-dimensional space) are represented by single numbers, points in a plane (a two-dimensional space) can be represented by pairs

---

*We shall use the notation AB for the line segment joining the points A and B, and $\overline{AB}$ for its length.

1

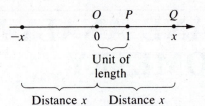

Figure 1.1

of numbers. Later we shall see that points in a three-dimensional space can be represented by triples of numbers.

In order to represent points in a plane by pairs of numbers, we select two intersecting lines and establish a scale on each line, as shown in Figure 1.2. The point of intersection is the origin. These two lines, called the axes, are distinguished by identifying symbols (usually by the letters x and y). For a given point P in the plane, there corresponds a point P_x on the x axis. It is the point of intersection of the x axis and the line containing P and parallel to the y axis. (If P is on the y axis, this line coincides with the y axis.) Similarly, there exists a point P_y on the y axis which is the point of intersection of the y axis and the line through P that parallels (or is) the x axis. The coordinates of these two points on the axes are the **coordinates** of P. If a is the coordinate of P_x and b is the coordinate of P_y, then the point P is represented by (a, b). In this example, a is called the **x coordinate**, or **abscissa**, of P and b is the **y coordinate**, or **ordinate**, of P.

In a coordinate plane, the following conventions normally apply:

1. The axes are taken to be perpendicular to each other.

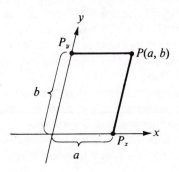

Figure 1.2

2. The x axis is a horizontal line with the positive coordinates to the right of the origin, and the y axis is a vertical line with the positive coordinates above the origin.

3. The same scale is used on both axes.

These conventions, of course, need not be followed when others are more convenient. We shall violate the third rather frequently when considering figures that would be very difficult to sketch if we insisted upon using the same scale on both axes. In such cases, we shall feel free to use different scales, remembering that we have distorted the figure in the process. Unless a departure from convention is specifically stated or is obvious from the context, we shall always follow the first two conventions.

We can now identify the coordinates of the points in Figure 1.3. Note that all points on the x axis have the y coordinate zero, while those on the y axis have the x coordinate zero. The origin has both coordinates zero, since it is on both axes.

The axes separate the plane into four regions, called **quadrants**. It is convenient to identify them by the numbers shown in Figure 1.4. The points on the axes are not in any quadrant.

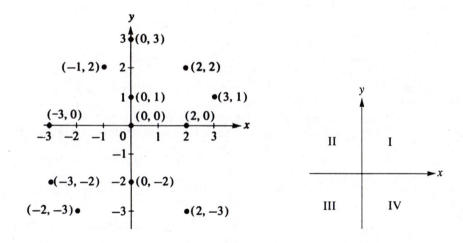

Figure 1.3   Figure 1.4

The coordinates of a point determined in this way are sometimes called cartesian coordinates, after the French mathematician and philosopher René Descartes. In the appendix of a book published in 1637, Descartes gave the first description of analytic geometry. From it came the developments that eventually led to the invention of the calculus.

1.2

DISTANCE FORMULA

Suppose we consider the distance between two points on a coordinate line. Let P_1 and P_2 be two points on a line, and let them have coordinates x_1 and x_2, respectively. If P_1 and P_2 are both to the right of the origin, with P_2 farther right than

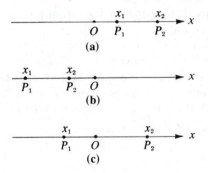

Figure 1.5

P_1 (as in Figure 1.5a), then

$$\overline{P_1P_2} = \overline{OP_2} - \overline{OP_1} = x_2 - x_1.$$

Expressing the distance between two points is only slightly more complicated if the origin is to the right of one or both of the points. In Figure 1.5b,

$$\overline{P_1P_2} = \overline{P_1O} - \overline{P_2O} = -x_1 - (-x_2) = x_2 - x_1,$$

and in Figure 1.5c,

$$\overline{P_1P_2} = \overline{P_1O} + \overline{OP_2} = -x_1 + x_2 = x_2 - x_1.$$

Thus, we see that $\overline{P_1P_2} = x_2 - x_1$ in all three of these cases in which P_2 is to the right of P_1. If P_2 were to the left of P_1, then

$$\overline{P_1P_2} = x_1 - x_2,$$

as can be easily verified. Thus, $\overline{P_1P_2}$ can always be represented as the larger coordinate minus the smaller. Since $x_2 - x_1$ and $x_1 - x_2$ differ only in that one is the negative of the other and since distance is always nonnegative, we see that $\overline{P_1P_2}$ is the difference that is nonnegative. Thus,

$$\overline{P_1P_2} = |x_2 - x_1|.$$

This form is especially convenient when the relative positions of P_1 and P_2 are unknown. However, since absolute values are sometimes rather bothersome, we will avoid them whenever the relative positions of P_1 and P_2 are known.

Let us now turn our attention to the more difficult problem of finding the distance between two points in the plane. Suppose we are interested in the distance between $P_1 = (x_1, y_1)$ and $P_2 = (x_2, y_2)$ (see Figure 1.6). A vertical line is drawn

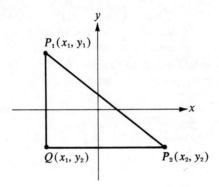

Figure 1.6

through P_1 and a horizontal line through P_2 intersecting at a point $Q = (x_1, y_2)$. Assuming P_1 and P_2 are not on the same horizontal or vertical line, $P_1 P_2 Q$ forms a right triangle with the right angle at Q. Now we can use the Theorem of Pythagoras to determine the length of $P_1 P_2$. By the previous discussion,

$$\overline{QP_2} = |x_2 - x_1| \qquad \text{and} \qquad \overline{P_1 Q} = |y_2 - y_1|$$

(the absolute values are retained here, since we want the resulting formula to hold for *any* choice of P_1 and P_2, not merely for the one shown in Figure 1.6). Now by the Pythagorean theorem,

$$\overline{P_1 P_2} = \sqrt{|x_2 - x_1|^2 + |y_2 - y_1|^2}.$$

But, since $|x_2 - x_1|^2 = (x_2 - x_1)^2 = (x_1 - x_2)^2$, the absolute values may be dropped and we have

$$\overline{P_1 P_2} = \sqrt{(x_2 - x_1)^2 + (y_2 - y_1)^2}.$$

Thus we have proved the following theorem.

Theorem 1.1 *The distance between two points $P_1 = (x_1, y_1)$ and $P_2 = (x_2, y_2)$ is*

$$\overline{P_1 P_2} = \sqrt{(x_2 - x_1)^2 + (y_2 - y_1)^2}.$$

In deriving this formula, we assumed that P_1 and P_2 are not on the same horizontal or vertical line; however, the formula holds even in these cases. For example, if P_1 and P_2 are on the same horizontal line, then $y_1 = y_2$ and $y_2 - y_1 = 0$. Thus,

$$\overline{P_1 P_2} = \sqrt{(x_2 - x_1)^2} = |x_2 - x_1|.$$

Note that $\sqrt{(x_2 - x_1)^2}$ is *not always* $x_2 - x_1$. Since the symbol $\sqrt{}$ indicates the nonnegative square root, we see that if $x_2 - x_1$ is negative, then $\sqrt{(x_2 - x_1)^2}$ is not equal to $x_2 - x_1$ but, rather, equals $|x_2 - x_1|$. Suppose, for example, that $x_2 - x_1 = -5$. Then $\sqrt{(x_2 - x_1)^2} = \sqrt{(-5)^2} = \sqrt{25} = 5 = |x_2 - x_1|$.

Theorem 1.1 depends upon the convention that the axes are perpendicular. If this convention is not followed, Theorem 1.1 cannot be used, but another more general formula based on the law of cosines can be derived. However, we shall not derive it here, since the convention of using perpendicular axes is so widely observed.

Example 1   Find the distance between $P_1 = (1, 4)$ and $P_2 = (-3, 2)$.

Solution $\overline{P_1 P_2} = \sqrt{(-3 - 1)^2 + (2 - 4)^2} = 2\sqrt{5}$

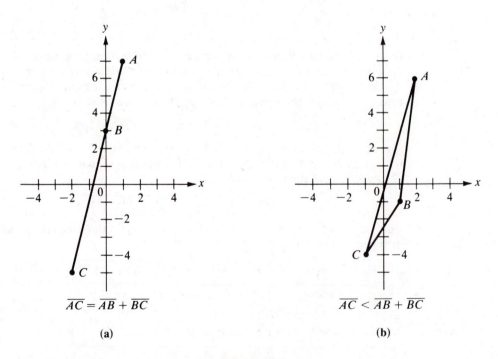

$\overline{AC} = \overline{AB} + \overline{BC}$ $\overline{AC} < \overline{AB} + \overline{BC}$

(a)   (b)

Figure 1.7

Example 2 Determine whether $A = (1, 7)$, $B = (0, 3)$, and $C = (-2, -5)$ are collinear.

Solution
$$\overline{AB} = \sqrt{(0 - 1)^2 + (3 - 7)^2} = \sqrt{17}$$
$$\overline{BC} = \sqrt{(-2 - 0)^2 + (-5 - 3)^2} = \sqrt{68} = 2\sqrt{17}$$
$$\overline{AC} = \sqrt{(-2 - 1)^2 + (-5 - 7)^2} = \sqrt{153} = 3\sqrt{17}$$

Since $\overline{AC} = \overline{AB} + \overline{BC}$, the three points must be collinear (Figure 1.7a). If they were not, they would form a triangle and any one side would be less than the sum of the other two (Figure 1.7b).

Example 3 Show that $(1, 2)$, $(4, 7)$, $(-6, 13)$, and $(-9, 8)$ are the vertices of a rectangle.

Solution The points are plotted in Figure 1.8. Let us check lengths.

$$\overline{P_1 P_2} = \sqrt{(4 - 1)^2 + (7 - 2)^2} = \sqrt{34}$$
$$\overline{P_3 P_4} = \sqrt{(-9 + 6)^2 + (8 - 13)^2} = \sqrt{34}$$
$$\overline{P_2 P_3} = \sqrt{(-6 - 4)^2 + (13 - 7)^2} = \sqrt{136}$$
$$\overline{P_4 P_1} = \sqrt{(1 + 9)^2 + (2 - 8)^2} = \sqrt{136}$$

Although $\overline{P_1 P_2} = \overline{P_3 P_4}$ and $\overline{P_2 P_3} = \overline{P_4 P_1}$, we are not justified in saying that we have a rectangle; we can merely conclude that we have a parallelogram. But if the diagonals of a parallelogram are equal, then the parallelogram is a rectangle (see Problem 30). Let us then consider the lengths of the diagonals.

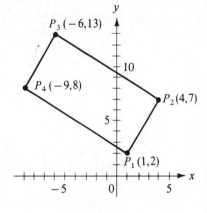

$P_3(-6, 13)$

$P_4(-9, 8)$

$P_2(4, 7)$

$P_1(1, 2)$

Figure 1.8

$$\overline{P_1 P_3} = \sqrt{(-6 - 1)^2 + (13 - 2)^2} = \sqrt{170}$$
$$\overline{P_2 P_4} = \sqrt{(-9 - 4)^2 + (8 - 7)^2} = \sqrt{170}$$

Since the parallelogram has equal diagonals, we may conclude that it is a rectangle.

A second method of verifying that the parallelogram is a rectangle is to establish that one of the interior angles is a right angle. We can do this by the converse of the Pythagorean Theorem, which states that if $\overline{AB}^2 + \overline{BC}^2 = \overline{AC}^2$, then ABC is a right triangle with right angle at B. Thus, instead of finding the lengths of both diagonals, we find only one, say $\overline{P_1 P_3} = \sqrt{170}$. It is then a simple matter to verify that $\overline{P_1 P_3}^2 = \overline{P_1 P_2}^2 + \overline{P_2 P_3}^2$, showing that the angle at P_2 is $90°$.

When we use the methods of analytic geometry to prove geometric theorems, such proofs are called analytic proofs. When carrying out analytic proofs, we should recall that a plane does not come fully equipped with coordinate axes—they are imposed upon the plane to make the transition from geometry to algebra. Thus, we are free to place the axes in any position we choose in relation to the given figure. We place them in a way that makes the algebra as simple as possible.

Example 4   Prove analytically that the diagonals of a rectangle are equal.

Solution   First we place the axes in a convenient position. Let us put the x axis on one side of the rectangle and the y axis on another, as illustrated in Figure 1.9. Since we have a rectangle, the coordinates of B and D determine those of C.

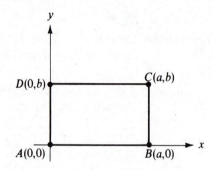

$$\overline{AC} = \sqrt{(a - 0)^2 + (b - 0)^2} = \sqrt{a^2 + b^2}$$
$$\overline{BD} = \sqrt{(0 - a)^2 + (b - 0)^2} = \sqrt{a^2 + b^2}$$

Since $\overline{AC} = \overline{BD}$, the theorem is proved.

Figure 1.9

PROBLEMS

A   *In Problems 1–8, find the distance between the given points.*

1. $(1, -3), (2, 5)$ **2.** $(4, 13), (-1, 5)$ **3.** $(3, -2), (3, -4)$

4. $(-5, 1), (0, -10)$ **5.** $(1/2, 3/2), (-5/2, 2)$ **6.** $(2/3, 1/3), (-4/3, 4/3)$

7. $(\sqrt{2}, 1), (2\sqrt{2}, 3)$ **8.** $(\sqrt{3}, -\sqrt{2}), (-3\sqrt{3}, \sqrt{2})$

In Problems 9–14, determine whether the three given points are collinear.

9. $(2, 1), (4, 3), (-1, -2)$

10. $(3, 2), (4, 6), (0, -8)$

11. $(-2, 3), (7, -2), (2, 5)$

12. $(2, -1), (-1, 4), (5, -6)$

13. $(1, -1), (3, 3), (0, -3)$

14. $(1, \sqrt{2}), (4, 3\sqrt{2}), (10, 6\sqrt{2})$

In Problems 15–18, *determine whether the three given points are the vertices of a right triangle.*

15. $(0, 2), (-2, 4), (1, 3)$

16. $(-1, 3), (4, 6), (-3, 1)$

17. $(9, 6), (-5, 4), (7, 10)$

18. $(9, -2), (8, 0), (-6, -7)$

B *In Problems* 19–22, *find the unknown quantity.*

19. $P_1 = (1, 5), \quad P_2 = (x, 2), \quad \overline{P_1 P_2} = 5$

20. $P_1 = (-3, y), \quad P_2 = (9, 2), \quad \overline{P_1 P_2} = 13$

21. $P_1 = (x, x), \quad P_2 = (1, 4), \quad \overline{P_1 P_2} = \sqrt{5}$

22. $P_1 = (x, 2x), \quad P_2 = (2x, 1), \quad \overline{P_1 P_2} = \sqrt{2}$

23. Show that $(5, 2)$ is on the perpendicular bisector of the segment AB where $A = (1, 3)$ and $B = (4, -2)$.

24. Show that $(-2, 4), (2, 0), (2, 8)$, and $(6, 4)$ are the vertices of a square.

25. Show that $(1, 1), (4, 1), (3, -2)$, and $(0, -2)$ are the vertices of a parallelogram.

26. Find all possible values for y so that $(5, 8), (-4, 11)$, and $(2, y)$ are the vertices of a right triangle.

27. Determine whether each of the following points is inside, on, or outside the circle with center $(-2, 3)$ and radius 5: $(1, 7), \ (-3, 8), \ (2, 0), \ (-5, 7), \ (0, -1), \ (-5, -1),$ $(-6, 6), (4, 2)$.

C **28.** Find the center and radius of the circle circumscribed about the triangle with vertices $(5, 1), (6, 0)$, and $(-1, -7)$.

29. Show that a triangle with vertices $(x_1, y_1), (x_2, y_2)$, and (x_3, y_3) has area

$$\frac{1}{2} | x_1 y_2 + x_2 y_3 + x_3 y_1 - x_1 y_3 - x_2 y_1 - x_3 y_2 |$$

$$= \left| \frac{1}{2} \begin{vmatrix} x_1 & y_1 & 1 \\ x_2 & y_2 & 1 \\ x_3 & y_3 & 1 \end{vmatrix} \right|.$$

[*Hint:* Consider the rectangle with sides parallel to the coordinate axes and containing the vertices of the triangle.]

30. Prove analytically that if the diagonals of a parallelogram are equal, then the parallelogram is a rectangle. [*Hint:* Place the axes as shown in Figure 1.10 (page 10) and show that $\overline{AC} = \overline{BD}$ implies that A is the origin.]

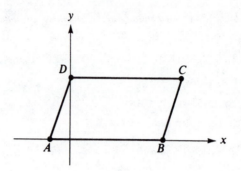

Figure 1.10

31. Prove analytically that the sum of the lengths of two sides of a triangle is greater than the length of the third side.

1.3

POINT-OF-DIVISION FORMULAS

Suppose we want to find the point which is some fraction of the way from A to B. Is it possible to express the coordinates of the point we want in terms of the coordinates of A and B? Let $A = (x_1, y_1)$ and $B = (x_2, y_2)$ be given and let $P = (x, y)$ be the point we are seeking. If we let

$$r = \frac{\overline{AP}}{\overline{AB}}$$

(see Figure 1.11), then P is 1/3 of the way from A to B when $r = 1/3$, P is 4/5 of the way from A to B when $r = 4/5$, and so on. Thus we generalize the problem to one in which x and y are to be expressed in terms of x_1, y_1, x_2, y_2, and r. The problem can be simplified considerably by working with the x's and y's separately.

If A, B, and P are projected onto the x axis (see Figure 1.11) to give the points A_x, B_x, and P_x, respectively, we have, from elementary geometry,

$$r = \frac{\overline{AP}}{\overline{AB}} = \frac{\overline{A_x P_x}}{\overline{A_x B_x}} = \frac{x - x_1}{x_2 - x_1}.$$

Solving for x gives

$$x = x_1 + r(x_2 - x_1).$$

By projecting onto the y axis, we have

$$y = y_1 + r(y_2 - y_1).$$

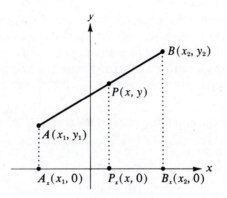

Figure 1.11

These two results, known as point-of-division formulas, are stated in the following theorem.

Theorem 1.2   *If $A = (x_1, y_1)$, $B = (x_2, y_2)$, and P is a point such that $r = \overline{AP}/\overline{AB}$, then the coordinates of P are*

$$x = x_1 + r(x_2 - x_1) \qquad \text{and} \qquad y = y_1 + r(y_2 - y_1).$$

Figure 1.12 gives a geometric interpretation of the terms in the point-of-division formula for x. A careful examination of this figure will help you to understand and

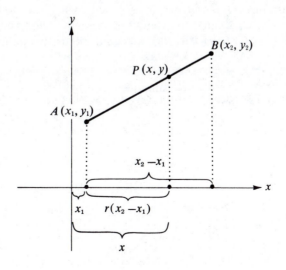

Figure 1.12

remember the formulas. Note that if B is to the left of A, $x_2 - x_1$ is negative, but it has the same absolute value as the distance between the projections of A and B on the x axis. You might sketch this and compare your sketch with Figure 1.12. Of course, a similar figure can be used to interpret the y terms of the point-of-division formula.

Example 1 Find the point one-third of the way from $A = (2, 5)$ to $B = (8, -1)$.

Solution

$$r = \frac{\overline{AP}}{\overline{AB}} = \frac{1}{3}$$

$$x = x_1 + r(x_2 - x_1)$$

$$= 2 + \frac{1}{3}(8 - 2)$$

$$= 4$$

$$y = y_1 + r(y_2 - y_1)$$

$$= 5 + \frac{1}{3}(-1 - 5)$$

$$= 3$$

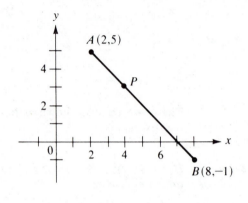

Figure 1.13

Thus the desired point is $(4, 3)$ (see Figure 1.13).

So far we have tacitly assumed that r is between 0 and 1. If r is either 0 or 1, the point-of-division formulas would give us $P = A$ or $P = B$, respectively, a result that $r = \overline{AP}/\overline{AB}$ would lead us to expect. Similarly, if $r > 1$, then $r = \overline{AP}/\overline{AB}$ indicates $\overline{AP} > \overline{AB}$, which is exactly what the point-of-division formulas give. Thus if we wanted to extend the segment AB beyond B to a point P which is r times as far from A as B is, we could still use the point-of-division formulas.

Example 2 If the segment AB, where $A = (-3, 1)$ and $B = (2, 5)$, is extended beyond B to a point P twice as far from A as B is (see Figure 1.14), find P.

Solution

$$r = \frac{\overline{AP}}{\overline{AB}} = 2$$

$$x = x_1 + r(x_2 - x_1) \qquad y = y_1 + r(y_2 - y_1)$$

$$= -3 + 2[2 - (-3)] \qquad = 1 + 2(5 - 1)$$

$$= 7 \qquad\qquad\qquad = 9$$

Thus $P = (7, 9)$.

While negative values of r do not make sense in $r = \overline{AP}/\overline{AB}$, we find that their use in the point-of-division formulas has the effect of extending the segment AB in the reverse direction—that is, from B through A to P. Suppose, for example, that $r = -2$. Then $\overline{AP}$ is twice $\overline{AB}$, and P and B are on opposite sides of A. However, we can get the same result by reversing the roles of A and B and using a positive value of r.

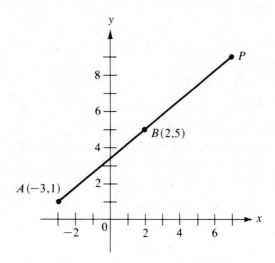

Figure 1.14

Example 3   Given the segment AB, where $A = (-3, 1)$ and $B = (2, 5)$, is extended beyond A to a point P twice as far from B as A is (see Figure 1.15); find P.

Solution   Suppose we use a negative value of r. Since $\overline{AP} = \overline{AB}$ with A between P and B,

$$r = -\frac{\overline{AP}}{\overline{AB}} = -1.$$

Therefore

$$x = x_1 + r(x_2 - x_1)$$
$$= -3 - 1[2 - (-3)]$$
$$= -8$$

$$y = y_1 + r(y_2 - y_1)$$
$$= 1 - 1(5 - 1)$$
$$= -3$$

Thus $P = (-8, -3)$.

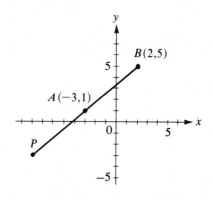

Figure 1.15

Reversing the roles of A and B, we have

$$r = \frac{\overline{BP}}{\overline{BA}} = 2, \quad B = (x_1, y_1) = (2, 5), \quad \text{and} \quad A = (x_2, y_2) = (-3, 1).$$

$$
\begin{aligned}
x &= x_1 + r(x_2 - x_1) & y &= y_1 + r(y_2 - y_1) \\
&= 2 + 2(-3 - 2) & &= 5 + 2(1 - 5) \\
&= -8 & &= -3
\end{aligned}
$$

and $P = (-8, -3)$, as before.

One very important special case of the point-of-division formulas arises when $r = 1/2$, which gives the midpoint of the segment AB. Using the point-of-division formulas, we have the following theorem.

Theorem 1.3   *If P is the midpoint of AB, then the coordinates of P are*

$$x = \frac{x_1 + x_2}{2} \quad and \quad y = \frac{y_1 + y_2}{2}.$$

Thus, to find the midpoint of a segment AB, we merely average both the x and y coordinates of the given points. A moment of thought will reveal the reasonableness of this; the average of two grades is halfway between them, the average of two temperatures is halfway between them, and so forth.

Example 4   Find the midpoint of the segment AB, where $A = (1, 5)$ and $B = (-3, -1)$.

Solution

$$
\begin{aligned}
x &= \frac{x_1 + x_2}{2} & y &= \frac{y_1 + y_2}{2} \\
&= \frac{1 - 3}{2} & &= \frac{5 - 1}{2} \\
&= -1 & &= 2
\end{aligned}
$$

Thus $P = (-1, 2)$.

Example 5   Prove analytically that the segment joining the midpoints of two sides of a triangle is parallel to the third side and one-half its length.

Solution Let us place the axes as indicated in Figure 1.16 and let D and E be the midpoints of AC and BC, respectively. By the midpoint formula, $D = (a/2, c/2)$ and $E = (b/2, c/2)$. Since D and E have identical y coordinates, DE is horizontal and therefor parallel to AB. Finally $\overline{DE} = b/2 - a/2 = (b - a)/2$ and $\overline{AB} = b - a$; thus $\overline{DE} = \overline{AB}/2$.

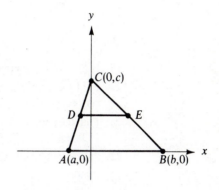

Figure 1.16

PROBLEMS

A *In Problems* 1–6, *find the point P such that* $\overline{AP}/\overline{AB} = r$.

1. $A = (3, 4)$, $B = (7, 0)$, $r = 1/4$ **2.** $A = (4, -2)$, $B = (-2, -5)$, $r = 2/3$

3. $A = (5, -1)$, $B = (-4, -5)$, $r = 1/5$ **4.** $A = (2, 4)$, $B = (-5, 2)$, $r = 2/5$

5. $A = (-4, 1)$, $B = (3, 8)$, $r = 3$ **6.** $A = (-6, 2)$, $B = (4, 4)$, $r = 5/2$

In Problems 7–10, *find the midpoint of the segment AB.*

7. $A = (5, -2)$, $B = (-1, 4)$ **8.** $A = (-3, 3)$, $B = (1, 5)$

9. $A = (4, -1)$, $B = (3, 3)$ **10.** $A = (-1, 4)$, $B = (0, 2)$

B **11.** If $A = (3, 5)$, $P = (6, 2)$, and $\overline{AP}/\overline{AB} = 1/3$, find B.

12. If $P = (4, 7)$, $B = (2, -1)$, and $\overline{AP}/\overline{AB} = 2/5$, find A.

13. If $P = (2, -5)$, $B = (4, -3)$, and $\overline{AP}/\overline{AB} = 1/2$, find A.

14. If $A = (3, 3)$, $P = (5, 2)$, and $\overline{AP}/\overline{AB} = 3/5$, find B.

In Problems 15–18, *find the point P between A and B such that AB is divided in the given ratio.*

15. $A = (5, -3)$, $B = (-1, 6)$, $\overline{AP}/\overline{PB} = 1/2$

16. $A = (-1, -3)$, $B = (-8, 11)$, $\overline{AP}/\overline{PB} = 3/4$

17. $A = (2, -1), B = (4, 5), \overline{AP}/\overline{PB} = 2/3$

18. $A = (5, 8), B = (2, -1), \overline{AP}/\overline{PB} = 5/1$

19. If $P = (4, -1)$ is the midpoint of the segment AB, where $A = (2, 5)$, find B.

20. Find the center and radius of the circle circumscribed about the right triangle with vertices $(1, 1), (1, 4)$, and $(7, 4)$.

21. Find the point of intersection of the medians of the triangle with vertices $(5, 2), (0, 4)$, and $(-1, -1)$. (See Problem 26.)

22. Prove analytically that the diagonals of a parallelogram bisect each other.

23. Find the point of intersection of the diagonals of the parallelogram with vertices $(1, 1)$, $(4, 1), (3, -2)$, and $(0, -2)$.

24. Prove analytically that the midpoint of the hypotenuse of a right triangle is equidistant from the three vertices.

25. Prove analytically that the vertex and the midpoints of the three sides of an isosceles triangle are the vertices of a rhombus.

26. Prove analytically that the medians of a triangle are concurrent at a point two-thirds of the way from each vertex to the midpoint of the opposite side.

27. The point $(1, 4)$ is at a distance 5 from the midpoint of the segment joining $(3, -2)$ and $(x, 4)$. Find x.

C **28.** The midpoints of the sides of a triangle are $(-1, 3), (1, -2)$, and $(5, -3)$. Find the vertices.

29. Three vertices of a parallelogram are $(2, 5), (-7, 1)$, and $(4, -6)$. Find the fourth vertex. [*Hint:* There is more than one solution. Sketch all possible parallelograms using the three given vertices.]

30. Show that if a triangle has vertices $(x_1, y_1), (x_2, y_2)$, and (x_3, y_3), then the point of intersection of its medians is

$$\left(\frac{x_1 + x_2 + x_3}{3}, \frac{y_1 + y_2 + y_3}{3} \right)$$

31. Prove analytically that the sum of the squares of the four sides of a parallelogram is equal to the sum of the squares of the two diagonals.

1.4

INCLINATION AND SLOPE

An important concept in the description of a line and one that is used quite extensively throughout calculus has to do with the inclination of a line. First let us recall the convention from trigonometry which states that angles measured in the counterclockwise direction are positive, while those measured in the clockwise direction are negative. Thus we have the following definition.

Definition   *The **inclination** of a line that intersects the x axis is the measure of the smallest nonnegative angle which the line makes with the positive end of the x axis. The inclination of a line parallel to the x axis is 0.*

We shall use the symbol θ to represent an inclination. The inclination of a line is always less than 180°, or π radians, and every line has an inclination. Thus, for any line,

$$0° \leq \theta < 180° \qquad \text{or} \qquad 0 \leq \theta < \pi.$$

Figure 1.17 shows several lines with their inclinations. Note that the angular measure is given in both degrees and radians. Although there is no reason to show preference for one over the other at this time, radian measure is the preferred way of representing an angle in more advanced courses.

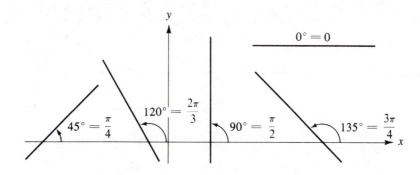

Figure 1.17

While the inclination of a line may seem like a simple representation, we cannot, in general, find a simple relationship between the inclination of a line and the coordinates of points on it without resorting to tables of trigonometric functions. Thus, we consider another expression related to the inclination—namely, the slope of a line.

Definition   *The **slope** m of a line is the tangent of the inclination; thus,*

$$m = \tan \theta.$$

While it is possible for two different angles to have the same tangent, it is not possible for lines having two different inclinations to have the same slope. The reason for this is the restriction on the inclination, $0° \leq \theta < 180°$. Nevertheless, one minor problem does arise from the use of slope since the tangent of 90° is not defined. Thus vertical lines have inclination 90° but no slope. *Do not confuse "no slope" with "zero slope."* A horizontal line definitely has a slope and that slope is the number 0, but there is no number at all (not even 0) which is the slope of a vertical line. Some might object to this nonexistence of tan 90° by saying that it is "infinity," or " ∞ ." However, infinity is not a number. Also, while the symbol ∞ is quite useful in calculus when dealing with limits, its use in algebra or in an algebraic development of trigonometry leads to trouble.

While the nonexistence of the slope of certain lines is somewhat bothersome, it is more than counterbalanced by the simple relationship between the slope and the coordinates of a pair of points on the line. Recall that if θ is as shown in either of

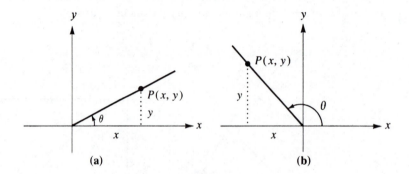

(a)   (b)

Figure 1.18

the two positions in Figure 1.18, then

$$\tan \theta = \frac{y}{x}.$$

Unfortunately, the lines with which we are dealing are not always so conveniently placed. Suppose we have a line with a pair of points, $P_1 = (x_1, y_1)$ and $P_2 = (x_2, y_2)$, on it (see Figure 1.19). If we place a pair of axes parallel to the old axes, with P_1 as the new origin, then the coordinates of P_2 with respect to this new coordinate system are $x = x_2 - x_1$ and $y = y_2 - y_1$. Now θ is situated in a position that allows us to use the definition of tan θ and state the following theorem.

Theorem 1.4   *A line through $P_1 = (x_1, y_1)$ and $P_2 = (x_2, y_2)$, where $x_1 \neq x_2$, has slope*

$$m = \frac{y_2 - y_1}{x_2 - x_1} = \frac{y_1 - y_2}{x_1 - x_2}.$$

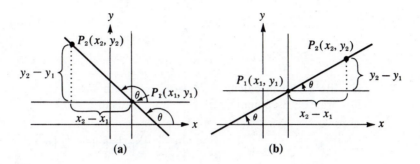

Figure 1.19

Note the agreement of the subscripts in the numerator and denominator. The slope of the line joining two points is the difference of the y coordinates divided by the difference of the x coordinates *taken in the same order*. The numerator and denominator are signed (or directed) vertical and horizontal distances.

One description of the slope of a line is that it is the vertical rise of the line divided by the horizontal run, or simply, rise over run. When the "run" is to the left, it is negative, giving a negative slope. This description of the slope as rise over run is one that is often used to describe the pitch of a roof or the grade of a road. They are often given as percentages, where a 1% grade corresponds to a slope of 0.01.

Example 1 Find the slope of the line containing $P_1 = (1, 5)$ and $P_2 = (7, -7)$.

Solution
$$m = \frac{y_2 - y_1}{x_2 - x_1} = \frac{-7 - 5}{7 - 1} = \frac{-12}{6} = -2$$

Since
$$m = \tan \theta = -2,$$
$$\theta = \arctan(-2) = 117°.$$

The line through P_1 and P_2 is shown in Figure 1.20. A slope of -2 means that, as we move a unit distance to the right on the line, we move down (because of the minus) a distance 2.

Example 2 Graph the line through $(2, 1)$ with slope 3/2.

Solution Remember that the numerator and denominator of the slope represent vertical and horizontal distances. Starting at the point $(2, 1)$, we proceed horizontally a distance 2 in the positive direction (to the right) and vertically a distance 3, again in the positive

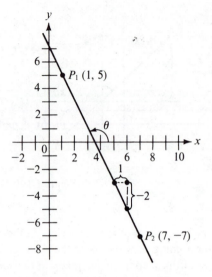

Figure 1.20

direction (upward). This takes us to the point (4, 4). The desired line is the line joining the given point (2, 1) and the point (4, 4) as shown in Figure 1.21.

Note that if the slope had been $-3/2$, then, putting the minus with the numerator, the only difference would be that the vertical distance would be in the negative direction or downward.

Since a vertical line has no slope, Theorem 1.4 does not hold in that case; however, $x_1 = x_2$ for any pair of points on a vertical line, and the right-hand side of the slope formula is also nonexistent. Thus there is no slope when the right-hand side of the slope formula does not exist.

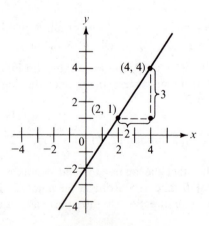

Figure 1.21

1.5

PARALLEL AND PERPENDICULAR LINES

If two nonvertical lines are parallel, they must have the same inclination and, thus, the same slope (see Figure 1.22). If two parallel lines are vertical, then neither one has slope. Similarly, if $m_1 = m_2$ or if neither line has slope, then the two lines are parallel. Thus, two lines are parallel if and only if $m_1 = m_2$ or neither line has slope.

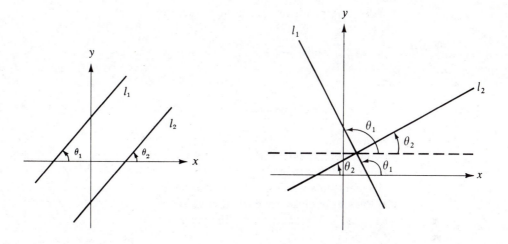

Figure 1.22   Figure 1.23

If two nonvertical lines l_1 and l_2 with the respective inclinations θ_1 and θ_2 are perpendicular (see Figure 1.23), then (assuming l_1 to be the line with the larger inclination)

$$\theta_1 - \theta_2 = 90°,$$

and

$$\theta_1 = \theta_2 + 90°.$$

Thus

$$\tan \theta_1 = \tan (\theta_2 + 90°) = -\cot \theta_2 = -\frac{1}{\tan \theta_2}$$

or

$$m_1 = -\frac{1}{m_2}.$$

On the other hand, if $m_1 = 1/m_2$, the argument can be traced backward to show that the difference of the inclinations is 90° and the lines are perpendicular. Therefore we have the following theorem.

Theorem 1.5 *The lines l_1 and l_2 with slopes m_1 and m_2, respectively, are*

(a) parallel or coincident if and only if $m_1 = m_2$,
(b) perpendicular if and only if $m_1 m_2 = -1$.

Example 1 Find the slopes of l_1 containing $(1, 5)$ and $(3, 8)$ and l_2 containing $(-4, 1)$ and $(0, 7)$; determine whether l_1 and l_2 are parallel, coincident, perpendicular, or none of these.

Solution
$$m_1 = \frac{8 - 5}{3 - 1} = \frac{3}{2}$$

and

$$m_2 = \frac{7 - 1}{0 + 4} = \frac{6}{4} = \frac{3}{2}$$

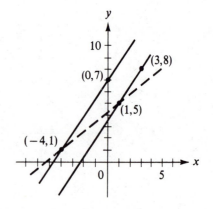

Figure 1.24

We now know that l_1 and l_2 are either parallel or coincident. While it is clear from Figure 1.24 that they are parallel rather than coincident, this would not be so obvious if the lines were closer together. Let us show it analytically. We begin by finding the slope of the line l_3 joining $(1, 5)$ on l_1 and $(-4, 1)$ on l_2.

$$m_3 = \frac{5 - 1}{1 + 4} = \frac{4}{5}$$

If l_1 and l_2 were coincident, then l_3 would be coincident to both of them; therefore it would have the same slope as l_1 and l_2. Since its slope is 4/5 rather than 3/2, l_1 and l_2 are not coincident; they are parallel.

PROBLEMS

A *In Problems 1–8 find the slope (if any) and the inclination of the line through the given points.*

1. $(2, 3), (5, 8)$ **2.** $(-1, 4), (4, 2)$ **3.** $(-2, -2), (4, 2)$

4. $(3, -5), (1, -1)$

5. $(-4, 2), (-4, 5)$

6. $(2, 3), (-4, 3)$

7. $(a, a), (b, b)$

8. $(a, a), (-a, 2a)$

In Problems 9–14, graph the line through the given point and having the given slope.

9. $(5, -2), m = 2$

10. $(-2, 4), m = 3/4$

11. $(3, 1), m = -1/3$

12. $(0, 4), m = -3$

13. $(4, 2), m = 0$

14. $(-3, 1)$, no slope

In Problems 15–22, find the slopes of the lines through the two pairs of points; then determine whether the lines are parallel, coincident, perpendicular, or none of these.

15. $(1, -2), (-2, -11);$ $(2, 8), (0, 2)$

16. $(1, 5), (-2, -7);$ $(7, -1), (3, 0)$

17. $(1, 5), (-1, -1);$ $(0, 3), (2, 7)$

18. $(1, 3), (-1, -1);$ $(0, 2), (4, -2)$

19. $(1, 1), (4, -1);$ $(-2, 3), (7, -3)$

20. $(1, -4), (6, 1);$ $(2, 3), (-1, 6)$

21. $(1, 2), (3, 2);$ $(4, 1), (4, -2)$

22. $(1, 5), (1, 1);$ $(-2, 2), (-2, 4)$

23. Find the pitch (slope) of the roof shown in Figure 1.25.

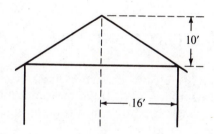

Figure 1.25

B **24.** Find the percentage grade of the section of road shown in Figure 1.26.

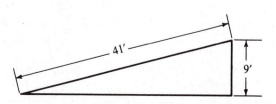

Figure 1.26

25. If the line through $(x, 5)$ and $(4, 3)$ is parallel to a line with slope 3, find x.

26. If the line through $(x, 5)$ and $(4, 3)$ is perpendicular to a line with slope 3, find x.

27. If the line through $(x, 1)$ and $(0, y)$ is coincident with the line through $(1, 4)$ and $(2, -3)$, find x and y.

28. If the line through $(-2, 4)$ and $(1, y)$ is perpendicular to one through $(-2, 4)$ and $(x, 2)$, find a relationship between x and y.

29. If the line through $(x, 4)$ and $(3, 7)$ is parallel to one through $(x, -1)$ and $(5, 1)$, find x.

30. Show by means of slopes that $(1, 1)$, $(4, 1)$, $(3, -2)$, and $(0, -2)$, are the vertices of a parallelogram.

31. Show by means of slopes that $(-2, 4)$, $(2, 0)$, $(6, 4)$, and $(2, 8)$ are the vertices of a square.

32. A certain section of railroad roadbed rises 100 feet per mile of track. What is the percentage grade for this section of track?

33. Prove analytically that the diagonals of a square intersect at right angles.

34. Prove analytically that one median of an isosceles triangle is an altitude.

35. Prove analytically that the diagonals of a rhombus intersect at right angles.

C **36.** Prove analytically that the medians of an equilateral triangle are altitudes.

1.6

ANGLE FROM ONE LINE TO ANOTHER

If l_1 and l_2 are two intersecting lines, then an angle from l_1 to l_2 is any angle measured from l_1 to l_2. If the measurement is in the counterclockwise direction, then the angle is positive; if it is in the clockwise direction, the angle is negative.

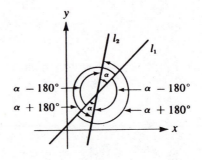

Figure 1.27

While there are many angles from l_1 to l_2, all are related (see Figure 1.27) in that, if α is one of them, all can be expressed in the form

$$\alpha + n \cdot 180°,$$

where n is an integer (positive, negative, or zero). Since any two of these angles differ from each other by a multiple of $180°$, they all have the same tangent.

Theorem 1.6 *If l_1 and l_2 are nonperpendicular lines with slopes m_1 and m_2, respectively, and α is any angle from l_1 to l_2, then*

$$\tan \alpha = \frac{m_2 - m_1}{1 + m_1 m_2}.$$

Proof Figure 1.28 shows that

$$\alpha = \theta_2 - \theta_1$$

for one of the angles α from l_1 to l_2. Thus

$$\tan \alpha = \frac{\tan \theta_2 - \tan \theta_1}{1 + \tan \theta_1 \tan \theta_2}.$$

But, since $m_1 = \tan \theta_1$ and $m_2 = \tan \theta_2$, we have

$$\tan \alpha = \frac{m_2 - m_1}{1 + m_1 m_2}.$$

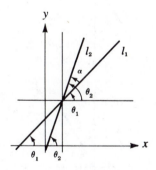

Figure 1.28

We have assumed in this argument that l_1 and l_2 intersect. If they do not, then $m_1 = m_2$. Using $m_1 = m_2$ in Theorem 1.6, we find that $\tan \alpha = 0$ and $\alpha = 0°$. Thus we shall use the convention that $\alpha = 0°$ if l_1 and l_2 are parallel. This is in agreement with the convention that $m = 0$ for horizontal lines.

The trigonometric identity used in this proof is, of course, true only when $\tan \alpha$ and $(m_2 - m_1)/(1 + m_1 m_2)$ both exist. $\tan \alpha$ does not exist if $\alpha = 90°$, but then $m_2 = -1/m_1$ and $1 + m_1 m_2 = 0$, which gives the one case in which $(m_2 - m_1)/(1 + m_1 m_2)$ does not exist. Thus Theorem 1.6 holds for all values of α except $\alpha = 90°$, for which case neither side of the equation exists.

Definition   *The **angle** from l_1 to l_2 is the smallest nonnegative angle from l_1 to l_2.*

Example 1   If l_1 and l_2 have slopes $m_1 = 3$ and $m_2 = -2$, respectively, find the angle from l_1 to l_2.

Solution
$$\tan \alpha = \frac{m_2 - m_1}{1 + m_1 m_2} = \frac{-2 - 3}{1 + 3(-2)} = 1$$

Thus $\alpha = \arctan 1 = 45°$.

Theorem 1.6 assumes that both l_1 and l_2 have slopes. If one of the two lines is vertical, it has no slope; but it does have an inclination of 90°. If the inclination of the other line is θ, then we have one of the four situations shown in Figure 1.29. From (a)

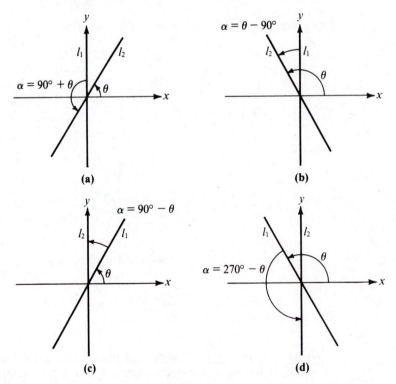

Figure 1.29

and (b) of this figure we see that when l_1 is vertical, the angle α from l_1 to l_2 is either $\theta + 90°$ or $\theta - 90°$. Since these two values differ by 180°, which is the period of the tangent, it follows that

$$\tan \alpha = \tan (\theta + 90°) = \tan (\theta - 90°)$$

$$= -\cot \theta = -\frac{1}{\tan \theta} = -\frac{1}{m},$$

where m is the slope of l_2.

Similarly, when l_2 is vertical, we see from (c) and (d) of Figure 1.29 that $\alpha = 90° - \theta$ or $\alpha = 270° - \theta$. Again they differ by 180°; again their tangents are equal. Thus when l_2 is vertical,

$$\tan \alpha = \tan (90° - \theta) = \cot \theta = \frac{1}{\tan \theta} = \frac{1}{m}.$$

Example 2 Find the angle from l_1 to l_2, where l_1 is a vertical line and l_2 has slope 1/2.

Solution The inclinition of l_1 is 90°; the inclination of l_2 is θ, where $\tan \theta = 1/2$ (see Figure 1.30). Since we want the angle α from l_1 to l_2, we see that one angle from l_1 to l_2 is

$$\alpha' = \theta + 90°$$

and

$$\tan \alpha = \tan \alpha' = \tan (\theta + 90°)$$
$$= -\cot \theta$$
$$= -\frac{1}{\tan \theta} = -\frac{1}{1/2} = -2.$$

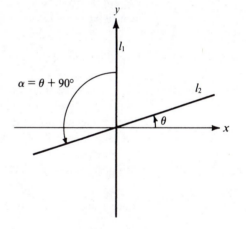

Figure 1.30

Thus the angle we want is the smallest positive angle whose tangent is -2. Using tables or a pocket calculator, we see that α is approximately 117°.

Example 3 Find the slope of the line bisecting the angle from l_1, with slope 7, to l_2, with slope 1.

Solution Let m be the slope of the desired line. Since $\alpha_1 = \alpha_2$ (see Figure 1.31), we have

$$\tan \alpha_1 = \tan \alpha_2$$

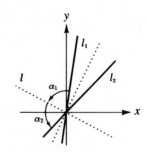

Figure 1.31

and

$$\frac{m - m_1}{1 + m_1 m} = \frac{m_2 - m}{1 + m_2 m}$$

$$\frac{m - 7}{1 + 7m} = \frac{1 - m}{1 + m}$$

$$(m - 7)(1 + m) = (1 + 7m)(1 - m)$$

$$8m^2 - 12m - 8 = 0$$

$$m = -1/2 \quad \text{or} \quad m = 2$$

We have two answers, but obviously we want only one. Which one? Since one of them is the negative reciprocal of the other, they represent slopes of perpendicular lines, one of which is the bisector of the angle from l_1 to l_2, while the other bisects the angle from l_2 to l_1. An inspection of Figure 1.31 shows that the answer we want is $m = -1/2$.

Example 4   Find the slope of the line bisecting the angle from l_1, with slope 2, to l_2, with no slope.

Solution   Let m and θ be the slope and inclination, respectively, of the desired line. Since l_2 has no slope, it is a vertical line (see Figure 1.32). Thus $\alpha_2 = 90° - \theta$ and

$$\tan \alpha_2 = \cot \theta = \frac{1}{\tan \theta} = \frac{1}{m}.$$

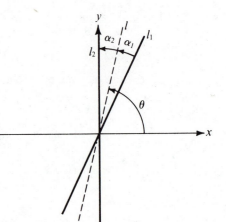

Figure 1.32

Since l and l_1 have slopes m and $m_1 = 2$,

$$\tan \alpha_1 = \frac{m - m_1}{1 + mm_1} = \frac{m - 2}{1 + 2m}.$$

Finally, the equality of α_1 and α_2 gives

$$\tan \alpha_1 = \tan \alpha_2$$

$$\frac{m - 2}{1 + 2m} = \frac{1}{m}$$

$$m^2 - 2m = 1 + 2m$$

$$m^2 - 4m - 1 = 0$$

$$m = \frac{4 \pm \sqrt{16 + 4}}{2}$$

$$= 2 \pm \sqrt{5}$$

Again we have two answers, representing the slopes of the bisectors of the angle from l_2 to l_1 as well as that from l_1 to l_2. It is easily seen from Figure 1.32 that the one we want is $m = 2 + \sqrt{5}$.

PROBLEMS

A   *In Problems 1–8, find the angle from l_1 to l_2 with slopes m_1 and m_2, respectively.*

1. $m_1 = -2, m_2 = 3$ **2.** $m_1 = 1, m_2 = 4$ **3.** $m_1 = -3, m_2 = 2$

4. $m_1 = 5, m_2 = -1$ **5.** $m_1 = 10, m_2$ does not exist   **6.** $m_1 = 0, m_2 = -1$

7. $m_1 = 2/3, m_2$ does not exist

8. m_1 does not exist, $m_2 = -2$

In Problems 9–16, find the angle from l_1 to l_2, where l_1 and l_2 contain the points indicated.

9. $l_1 : (1, 4), (3, -1);$ $l_2 : (3, 2), (5 - 1)$ **10.** $l_1 : (2, 5), (-3, 10);$ $l_2 : (-1, -3), (3, 3)$

11. $l_1 : (4, 5), (1, 1);$ $l_2 : (3, -3), (0, 4)$ **12.** $l_1 : (1, 1), (0, 5);$ $l_2 : (4, 3), (-1, 2)$

13. $l_1 : (3, 4), (3, -1); l_2 : (2, 5), (-1, 2)$ **14.** $l_1 : (-1, 2), (-1, -1); l_2 : (-3, 4), (1, 0)$

15. $l_1 : (5, 1), (3, -3); l_2 : (5, 1), (5, -3)$ **16.** $l_1 : (3, -4), (2, 3); l_2 : (2, -3), (2, 4)$

B   *In Problems 17–24, find the slope of the line bisecting the angle from l_1 to l_2 with slope m_1 and m_2, respectively.*

17. $m_1 = 3, m_2 = -2$ **18.** $m_1 = 1, m_2 = -7$ **19.** $m_1 = 2, m_2 = 3$

20. $m_1 = -1, m_2 = 2$ **21.** $m_1 = -3, m_2 = 5$ **22.** $m_1 = 2, m_2 = 0$

23. $m_1 = 3/4, m_2$ does not exist

24. m_1 does not exist, $m_2 = 1$

25. Find the interior angles of the triangle with vertices $A = (1, 5)$, $B = (3, -1)$, and $C = (-1, -1)$.

26. Find the interior angles of the triangle with vertices $A = (3, 2)$, $B = (4, 5)$, and $C = (-1, -1)$.

27. Find the slope of the line l_1 such that the angle from l_1 to l_2 is Arctan $2/3$, where l_2 contains $(2, 1)$ and $(-4, -5)$.

28. Find the slope of the line l_1 such that the angle from l_1 to l_2 is $45°$, where the slope of l_2 is -2.

29. Find the slope of the line l_2 such that the tangent of the angle from l_1 to l_2 is $-1/2$, where l_1 is a vertical line.

30. A line with slope 1 bisects the angle from l_1 to l_2, where l_2 has slope 2. What is the slope of l_1?

31. A line with slope $3 + \sqrt{10}$ bisects the angle from l_1 to l_2, where l_1 has slope 1. What is the slope of l_2?

32. Show by means of angles that $A = (1, 0)$, $B = (4, 4)$, and $C = (8, 1)$ are the vertices of an isosceles triangle.

C **33.** Lines l_1 and l_2 have slopes m and $1/m$, respectively. What is the slope of the line bisecting the angle from l_1 to l_2?

34. If lines l_1 and l_2 have slopes 1 and m, respectively, what is the slope of the line bisecting the angle from l_1 to l_2?

1.7

GRAPHS AND POINTS OF INTERSECTION

The graph of an equation in two variables x and y is simply the set of all points (x, y) in the plane whose coordinates satisfy the given equation. The determination of the graph of an equation is one of the principal problems of analytic geometry. Although we shall consider other methods in Chapter 7, we consider only point-by-point plotting here. To do this, we assign a value to either x or y, substitute the assigned value into the given equation, and solve for the other.

Example 1   Graph $2x + 3y = 6$.

Solution

x	y
-6	6
-3	4
0	2
3	0
6	-2

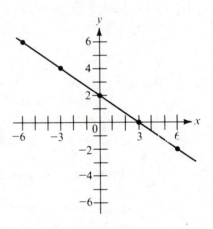

Figure 1.33

Example 2   Graph $x^2 + y^2 = 25$.

Solution

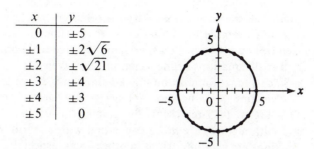

x	y
0	± 5
± 1	$\pm 2\sqrt{6}$
± 2	$\pm \sqrt{21}$
± 3	± 4
± 4	± 3
± 5	0

Figure 1.34

One obvious question that arises is how many points must be plotted before drawing the graph. There is no specific answer—plot as many as are needed to get a reasonable idea of the appearance of the graph.

Since each point of a graph satisfies the given equation, a point of intersection of two graphs is simply a point that satisfies both equations. Thus, any such point can be found by solving the two equations simultaneously.

Example 3   Find all points of intersection of $x^2 + y^2 = 25$ and $x + y = 2$.

Solution   Solving the second equation for y and substituting into the first, we have

$$x^2 + (2 - x)^2 = 25$$
$$2x^2 - 4x - 21 = 0$$
$$x = \frac{4 \pm \sqrt{16 + 168}}{4} = \frac{2 \pm \sqrt{46}}{2} = 1 \pm \frac{1}{2}\sqrt{46}$$
$$y = 2 - x = 1 \mp \frac{1}{2}\sqrt{46}$$

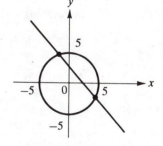

Thus, the two points of intersection are

$$\left(1 + \frac{1}{2}\sqrt{46},\ 1 - \frac{1}{2}\sqrt{46}\right)$$

Figure 1.35

and

$$\left(1 - \frac{1}{2}\sqrt{46},\ 1 + \frac{1}{2}\sqrt{46}\right).$$

A **function** is an association, or pairing, of the numbers of one set, called the **domain** of the function, with those of another (not necessarily different) set, called the **range**, such that no number in the domain is associated with more than one number in the range. A function is often given as an equation giving y in terms of x. For example, the equation $y = x^2$ defines y as a function of x, since it pairs each real number x with a single nonnegative real number y which is the square of x. Not all equations define functions, however.

Since a function must be single-valued, it is a simple matter to identify the graph of a function of x—no vertical line can interset the graph in more than one point. Thus Figure 1.33 represents the graph of a function of x, while Figure 1.34 does not. Although we are more concerned with equations than functions in this text, functions are very important in other areas, especially calculus.

PROBLEMS

A   *Plot the graphs of the equations in Problems 1–12. Indicate which give y as a function of x.*

1. $y = 2x - 1$ **2.** $x + 2y = 3$ **3.** $3x - 5y = 2$

4. $4x + 2y = 3$ **5.** $y = x^2 + 1$ **6.** $y = 2x - x^2$

7. $y = x^2 - x - 2$ **8.** $y = x^2 - 4x + 3$ **9.** $x^2 + y^2 = 1$

10. $x^2 + y^2 = 16$ **11.** $y = x^3$ **12.** $y = x^3 - 2x^2$

In Problems 13–16, find the points of intersection and sketch the graphs of the equations.

13. $3x - 5y = 2$ **14.** $2x + y = -1$
    $4x + 2y = 1$ $3x + 2y = 0$

15. $3x + 2y = -1$ **16.** $3x - 2y = 10$
    $x - 4y = -12$ $6x + 6y = -1$

B   *Plot the graphs of the equations in Problems 17–24. Indicate which give y as a function of x.*

17. $x^2 - y^2 = 4$ **18.** $x^2 - y^2 = -4$ **19.** $4x^2 + y^2 = 4$

20. $x^2 - 4y^2 = 4$ **21.** $y = \dfrac{x}{x + 1}$ **22.** $y = \dfrac{x + 1}{x}$

23. $y^2 = x^3$ **24.** $\sqrt{x} + \sqrt{y} = 4$

In Problems 25–32, find the points of intersection and sketch the graphs of the equations.

25. $x - y + 2 = 0$ **26.** $x - 2y = -1$ **27.** $x + y = 1$
    $y = x^2$ $y^2 = x + 4$ $x^2 + y^2 = 5$

28. $2x + y = 1$ **29.** $y = x^2$ **30.** $y = x^2$
    $x^2 + y^2 = 2$ $x = y^2$ $x^2 + y^2 = 2$

31. $x + 2y = 3$ **32.** $2x + 3y = 6$
    $x^2 + y^2 = 4$ $x^2 + y^2 = 16$

C   **33.** Plot the graph of $x^2 + y^2 = 0$.

34. Plot the graph of $x^2 + y^2 = -1$.

35. Graph $y = x$, $y = x^3$, and $y = x^5$, using the same axes. What do you think the graph of $y = x^{99}$ looks like? Verify by plotting a few points with the aid of a pocket calculator.

36. Graph $y = x^2$, $y = x^4$, and $y = x^6$, using the same axes. What do you think the graph of $y = x^{100}$ looks like? Verify by plotting a few points with the aid of a pocket calculator.

37. Graph $y = x$, $y = -x$, $y = |x|$, and $y = \sqrt{x^2}$, noting the similarities and differences. Recall that the definition of $|x|$ is given by

$$|x| = \begin{cases} x & \text{if } x \geq 0, \\ -x & \text{if } x < 0. \end{cases}$$

Also note that, while there are normally two square roots of a number, the symbol $\sqrt{}$ is used to represent the nonnegative square root.

38. Find the points of intersection and sketch the graphs of

(a) $x^2 + y^2 = 4$ **(b)** $y = \sqrt{4 - (x - 2)^2}$
    $(x - 2)^2 + y^2 = 4$ $x = \sqrt{4 - y^2}$

Note the similarities and differences.

1.8

AN EQUATION OF A LOCUS

In the last section we considered one of the two principal problems of analytic geometry—finding the graph of an equation. Let us now consider the other—finding an equation of a locus. In other words, given a description of a curve, we want to find an equation representing that curve. Since an equation of a curve is a relationship satisfied by the x and y coordinates of each point on the curve (but by no other point), we need merely consider an arbitrary point (x, y) on the curve and give the description of the curve in terms of x and y. Let us consider some examples.

Example 1   Find an equation for the set of all points in the xy plane which are equidistant from $(1, 3)$ and $(-2, 5)$.

Solution   Let (x, y) be one such point (see Figure 1.36). Then

$$\sqrt{(x - 1)^2 + (y - 3)^2}$$
$$= \sqrt{(x + 2)^2 + (y - 5)^2}$$
$$(x - 1)^2 + (y - 3)^2$$
$$= (x + 2)^2 + (y - 5)^2$$

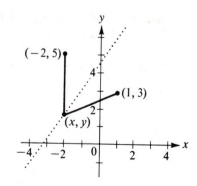

Figure 1.36

$$x^2 - 2x + 1 + y^2 - 6y + 9 = x^2 + 4x + 4 + y^2 - 10y + 25$$
$$6x - 4y + 19 = 0$$

Note in the above example that the first equation given is an equation of the desired curve. Of course, we want the equation in as simple a form as possible. In carrying out the simplification, however, we must be sure that the final equation is equivalent to the original. One way to be sure of this is to simplify only by reversible steps.

It might also be noted in the example that the first step in simplifying was squaring both sides of the original equation. This is not normally a reversible operation. (If $x = 5$, then $x^2 = 25$; but if $x^2 = 25$, then $x = \pm 5$.) However, since we are dealing only with positive distances, we need only consider positive square roots in reversing the operation. Thus it is reversible. The remaining operations are clearly reversible; the first and last equations are equivalent.

There is a second way to verify that the first and last equations are equivalent. Clearly, any point (x, y) that satisfies the first equation must satisfy the subsequent ones. The only question in doubt is: Does any point satisfying the last equation also satisfy the first? In order to answer this question, let us consider a point (x, y) which satisfies the last equation. Solving for y, we have

$$y = \frac{6x + 19}{4}.$$

Let us now substitute this expression for y into both sides of the first equation. Since both sides simplify to

$$\frac{\sqrt{52x^2 + 52x + 65}}{4},$$

we see that the first equation is satisfied by any such point (x, y). In this way, we see that the first and last equations are equivalent.

Example 2 Find an equation for the set of all points (x, y) such that the sum of its distances from $(3, 0)$ and $(-3, 0)$ is 8 (see Figure 1.37).

Solution
$$\sqrt{(x - 3)^2 + y^2} + \sqrt{(x + 3)^2 + y^2} = 8$$
$$\sqrt{(x - 3)^2 + y^2} = 8 - \sqrt{(x + 3)^2 + y^2}$$
$$x^2 - 6x + 9 + y^2 = 64 - 16\sqrt{(x + 3)^2 + y^2} + x^2$$
$$+ 6x + 9 + y^2$$
$$16\sqrt{(x + 3)^2 + y^2} = 64 + 12x$$
$$4\sqrt{(x + 3)^2 + y^2} = 16 + 3x$$
$$16x^2 + 96x + 144 + 16y^2 = 256 + 96x + 9x^2$$
$$7x^2 + 16y^2 = 112$$

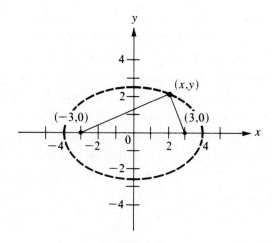

Figure 1.37

Again, the steps are reversible. Although we squared both sides twice, both sides of the equation had to be positive in each instance.

PROBLEMS

Find an equation for the set of all points (x, y) satisfying the given conditions.

A
   1. It is equidistant from $(5, 8)$ and $(-2, 4)$.
   2. It is equidistant from $(2, 3)$ and $(-4, 1)$.
   3. Its distance from $(5, 8)$ is 3.
   4. Its distance from $(3, 1)$ is 4.

B
   5. It is on the line having slope 2 and containing the point $(3, -2)$.
   6. It is on the line having slope -1 and containing the point $(-2, 5)$.
   7. It is on the line containing $(3, -2)$ and $(5, 3)$.
   8. It is on the line containing $(4, 2)$ and $(2, -1)$.
   9. Its distance from $(0, 0)$ is three times its distance from $(4, 0)$.
   10. Its distance from $(2, 5)$ is twice its distance from $(-3, 1)$.
   11. It is the vertex of a right triangle with hypotenuse the segment from $(2, 5)$ to $(-1, 4)$.
   12. It is the vertex of a right triangle with hypotenuse the segment from $(3, -2)$ to $(1, 4)$.
   13. It is equidistant from $(4, 0)$ and the y axis.
   14. It is equidistant from $(2, 3)$ and the x axis.

15. It is twice as far from $(3, 0)$ as it is from the y axis.

16. It is twice as far from the y axis as it is from $(3, 0)$.

17. The sum of its distances from $(0, 4)$ and $(0, -4)$ is 10.

18. The sum of its distances from $(4, -2)$ and $(2, 5)$ is 10.

19. The sum of the squares of its distances from $(3, 0)$ and $(-3, 0)$ is 50.

20. The sum of the squares of its distances from $(4, -2)$ and $(2, 5)$ is 20.

21. The difference of its distances from $(3, 0)$ and $(-3, 0)$ is 2.

22. The difference of its distances from $(4, 2)$ and $(1, -3)$ is 4.

23. The product of its distances from the coordinate axes is 4.

24. The product of its distances from $(4, 0)$ and the y axis is 4.

REVIEW PROBLEMS

A   **1.** Find the point of intersection of $2x + y = 5$ and $x - 3y = 7$.

2. Use distances to determine whether or not the three points $(1, 5)$, $(-2, -1)$, and $(4, 10)$ are collinear. Check your work by using slopes.

3. Find the lengths of the medians of the triangle with vertices $(-3, 4)$, $(5, 5)$, and $(3, -2)$.

4. Use distances to determine whether or not the points $(1, 6)$, $(5, 3)$, and $(3, 1)$ are the vertices of a right triangle. Check your work by using slopes.

5. Determine x so that $(x, 1)$ is on the line joining $(0, 4)$ and $(4, -2)$.

6. Line l_1 contains the points $(4, 7)$ and $(2, 3)$, while l_2 contains $(5, 6)$ and $(-3, 4)$. Are l_1 and l_2 parallel, perpendicular, coincident, or none of these?

7. Find the slopes of the altitudes of the triangle with vertices $(-2, 4)$, $(3, 3)$, and $(-5, -2)$.

B   **8.** Find the points of trisection of the segment joining $(2, -5)$ and $(-3, 7)$.

9. Find an equation of the perpendicular bisector of the segment joining $(5, -3)$ and $(-1, 1)$.

[*Hint:* What is the relationship between a point on the desired line and the two given points?]

10. Find the points of intersection of $x - 7y + 2 = 0$ and $x^2 + y^2 - 4x + 6y - 12 = 0$. Sketch.

11. The point $(5, -2)$ is at a distance $\sqrt{13}$ from the midpoint of the segment joining $(5, y)$ and $(-1, 1)$. Find y.

12. Prove analytically that the lines joining the midpoints of adjacent sides of a quadrilateral form a parallelogram.

13. Find the point of intersection of the medians of the triangle with vertices $(4, -3)$, $(-2, 1)$, and $(0, 5)$. (See Problem 26, page 16.)

14. Find an equation for the set of all points (x, y) such that it is equidistant from $(0, 1)$ and the x axis.

C **15.** Find the center of the circle circumscribed about the triangle with vertices $(-1, 1)$, $(6, 2)$, and $(7, -5)$.

16. Two vertices of an equilateral triangle are $(a, -a)$ and $(-a, a)$. Find the third.

17. A square has all its vertices in the first quadrant and one of its sides joins $(3, 1)$ and $(6, 3)$. Find the other two vertices.

18. A parallelogram has three vertices $(3, 4)$, $(6, 3)$, and $(1, 0)$ and the fourth vertex in the first quadrant. Find the fourth vertex.

19. Find an equation for the set of all points (x, y) such that the angle from the x axis to the line joining (x, y) and the origin equals y.

2

VECTORS IN THE PLANE

2.1

DIRECTED LINE SEGMENTS AND VECTORS

Since quantities such as force, velocity, and acceleration have direction as well as magnitude, it is convenient to represent them geometrically. To do so we use the concept of vectors, which have both magnitude and direction. Not only are vectors important in physics and engineering, but their use can considerably simplify geometric problems, especially in solid analytic geometry. One reason vectors are so useful is the wide range of interpretations they may be given. Since we are interested mainly in the geometric applications, vectors will be introduced geometrically by means of directed line segments.

Suppose A and B are points (not necessarily different) in space. The directed line segment from A to B is represented by $\overrightarrow{AB}$; B is called the **head** and A the **tail** of this segment. Two directed line segments $\overrightarrow{AB}$ and $\overrightarrow{CD}$ are **equivalent**, $\overrightarrow{AB} \equiv \overrightarrow{CD}$, (1) if both are of length zero or (2) if both have the same positive length, both lie on the same or parallel lines, and both are directed in the same way (see Figure 2.1, in which $\overrightarrow{AB} \equiv \overrightarrow{CD}$ and $\overrightarrow{EF} \equiv \overrightarrow{GH}$). With this information, we can easily prove the following theorem.

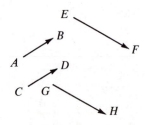

Figure 2.1

Theorem 2.1   (a) $\overrightarrow{AB} \equiv \overrightarrow{AB}$ for any directed line segment $\overrightarrow{AB}$.
(b)   If $\overrightarrow{AB} \equiv \overrightarrow{CD}$, then $\overrightarrow{CD} \equiv \overrightarrow{AB}$.
(c)   If $\overrightarrow{AB} \equiv \overrightarrow{CD}$ and $\overrightarrow{CD} \equiv \overrightarrow{EF}$, then $\overrightarrow{AB} \equiv \overrightarrow{EF}$.

Now let us choose an arbitrary directed line segment $\overrightarrow{AB}$. Let M_1 be the set of all directed line segments equivalent to $\overrightarrow{AB}$. Now let us choose another segment $\overrightarrow{CD}$ not in M_1 and let M_2 be the set of all directed line segments equivalent to $\overrightarrow{CD}$. Proceeding in this way, we can partition the set of all directed line segments into a collection of subsets, no two of which have any element in common. These subsets are what we call **vectors**. Thus a vector is a certain set of mutually equivalent directed line segments.

Definition   *The set of all directed line segments equivalent to a given directed line segment is a* **vector v**. *Any member of that set is a* **representative** *of* **v**. *The set of all directed line segments equivalent to one of length zero is called the* **zero vector, 0**.

It might be noted that a vector has magnitude (length) and direction, but not position. Any representative of a given vector has not only magnitude and direction but also position. Let us now consider how vectors may be combined.

Definition   *Suppose* **u** *and* **v** *are vectors. Let* $\overrightarrow{AB}$ *be a representative of* **u**. *Let* $\overrightarrow{BC}$ *be that representative of* **v** *with tail at B. The* **sum u + v** *of* **u** *and* **v** *is the vector* **w**, *having* $\overrightarrow{AC}$ *as a representative.*

This is represented geometrically in Figure 2.2. Since the sum of two vectors is given in terms of representatives of those vectors, the question remains, "Is the sum well defined—that is, is it independent of the representatives used?" Theorem 2.1 and the congruence of triangles easily show that the sum is well defined.

It might be noted that this definition is equivalent to the well-known parallelogram law for the addition of vectors (see Figure 2.2). Let us observe that the figures given here represent vectors graphically by means of representative directed line segments. In Figure 2.2, the vector **u** is represented by two equivalent directed line segments, both of which are labeled **u**.

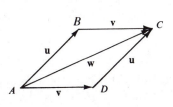

Figure 2.2

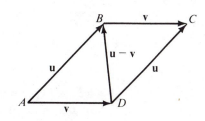

Figure 2.3

We can now use the sum of two vectors in order to define the difference.

Definition   *If* **u** *and* **v** *are vectors, then* **u** − **v** *is the vector* **w** *such that* **u** = **v** + **w**.

Geometrically, **u** − **v** can again be given by a parallelogram law; but **u** − **v** is represented by the other diagonal. This is shown in Figure 2.3. Note that the direction on **u** − **v** is from the head of **v** to the head of **u**. It is easily verified that **u** − **v** is a vector **w** such that **u** = **v** + **w**.

When dealing with vectors, we use the term *scalar* for a real number. A scalar has magnitude but not direction. In order to distinguish a scalar from a vector, we use **boldface Roman type** for vector symbols and *lightface italic* for scalars. This use of boldface type for vectors is generally followed in printed works; however in hand-written work, a vector is usually written with an arrow over it.

Definition   *If* **v** *is a vector, then* |**v**| *is the length of any representative of* **v**. *It is called the* **absolute value**, *or* **length**, *of* **v**.

Note that the absolute value of a vector is not a vector, but a scalar.

Definition   *If* k *is a scalar and* **v** *a vector, then* k**v** *is a vector whose length is* |k| |**v**| *and whose direction is the same as or opposite to the direction of* **v**, *according to whether* k *is positive or negative. It is called a* **scalar multiple** *of* **v**.

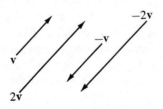

Figure 2.4

Figure 2.4 gives several examples of scalar multiples of the vector **v**.

Let us take note of the fact that we are not adding and multiplying ordinary numbers; thus it is not obvious that the rules of ordinary arithmetic hold—they must be proved from the given definitions.

Theorem 2.2   *The following properties hold for arbitrary vectors* **u**, **v**, *and* **w** *and arbitrary scalars a and b.*

(a)	$\mathbf{u} + \mathbf{v} = \mathbf{v} + \mathbf{u}$	*(f)*	$0\mathbf{v} = \mathbf{0}$
(b)	$\mathbf{u} + (\mathbf{v} + \mathbf{w}) = (\mathbf{u} + \mathbf{v}) + \mathbf{w}$	*(g)*	$a\mathbf{0} = \mathbf{0}$
(c)	$(ab)\mathbf{v} = a(b\mathbf{v})$	*(h)*	$\lvert a\mathbf{v} \rvert = \lvert a \rvert \, \lvert \mathbf{v} \rvert$
(d)	$(a + b)\mathbf{v} = a\mathbf{v} + b\mathbf{v}$	*(i)*	$\lvert \mathbf{u} + \mathbf{v} \rvert \leq \lvert \mathbf{u} \rvert + \lvert \mathbf{v} \rvert$
(e)	$\mathbf{v} + \mathbf{0} = \mathbf{v}$	*(j)*	$a(\mathbf{u} + \mathbf{v}) = a\mathbf{u} + a\mathbf{v}$

The proofs of these properties are quite simple. For example, Figure 2.2 can be used to prove (a). From triangle ABC and the definition of the sum of two vectors, it follows that $\overrightarrow{AC}$ is a representative of $\mathbf{u} + \mathbf{v}$. But triangle ADC shows that $\overrightarrow{AC}$ is also a representative of $\mathbf{v} + \mathbf{u}$. Thus $\mathbf{u} + \mathbf{v} = \mathbf{v} + \mathbf{u}$.

The remaining proofs are left for the student (see Problem 44). Be careful not to confuse the scalar 0 with the zero vector $\mathbf{0}$.

If we form the scalar multiple of the nonzero vector $\mathbf{v}$ and the scalar $1/\lvert \mathbf{v} \rvert$, the result is easily seen to be the unit vector (that is, the vector of length 1) in the direction of $\mathbf{v}$. It is usually written

$$\frac{\mathbf{v}}{\lvert \mathbf{v} \rvert}.$$

Of special interest are the unit vectors along the axes.

Definition   *If* $O = (0, 0)$, $X = (1, 0)$, *and* $Y = (0, 1)$, *then the vectors represented by* $\overrightarrow{OX}$ *and* $\overrightarrow{OY}$ *are denoted by* **i** *and* **j**, *respectively, and are called* **basis vectors**.

Theorem 2.3   *Every vector in the xy plane can be written in the form*

$$a\mathbf{i} + b\mathbf{j}$$

in one and only one way. The numbers a and b are called the **components** *of the vector.*

Proof   Suppose we have a vector $\mathbf{v}$ in the plane. Let us consider the representative of $\mathbf{v}$ with its tail at the origin O (see Figure 2.5). The head is at $P = (a, b)$. Let us project P onto both axes, giving points $A = (a, 0)$ and $B = (0, b)$. Since $\overrightarrow{OA}$ represents a vector of length $\lvert a \rvert$ that is either in the direction of $\mathbf{i}$ or in the opposite direction, depending upon whether a is positive or negative, it represents $a\mathbf{i}$. Similarly $\overrightarrow{OB}$ represents $b\mathbf{j}$. It is clear that $\mathbf{v} = a\mathbf{i} + b\mathbf{j}$.

Since the point P can be represented in rectangular coordinates by a pair (a, b) of numbers in one and only one way, the vector $\mathbf{v}$ has one and only one representation in component form.

It might be noted that this implies that a pair of numbers, a and b, defines a vector in the plane. Thus the vector $a\mathbf{i} + b\mathbf{j}$ is sometimes represented by the ordered pair (a, b). In fact, some authors define a vector as an ordered pair of numbers and derive

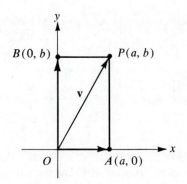

Figure 2.5

their geometric properties from this. It might be argued that the point (a, b) and the vector (a, b) are easily confused. However, since the vector (a, b) is represented by a directed line segment from the origin to the point (a, b), we may use the point as an interpretation of the vector. One of the reasons that vectors are so useful is their wide range of interpretations.

Theorem 2.4   *If $\overrightarrow{AB}$, where $A = (x_1, y_1)$ and $B = (x_2, y_2)$, represents a vector $\mathbf{v}$ in the xy plane, then $\mathbf{v} = (x_2 - x_1)\mathbf{i} + (y_2 - y_1)\mathbf{j}$.*

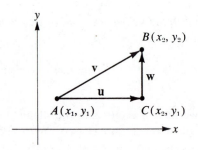

Figure 2.6

Proof   Let C be the point (x_2, y_1) (see Figure 2.6). Then

$$\mathbf{v} = \mathbf{u} + \mathbf{w},$$

where $\mathbf{u}$ is represented by $\overrightarrow{AC}$ and $\mathbf{w}$ by $\overrightarrow{CB}$. Since $\mathbf{u}$ is of length $|x_2 - x_1|$ and is in either the same or the opposite direction as $\mathbf{i}$, depending upon whether $x_2 - x_1$ is positive or negative, it follows that $\mathbf{u} = (x_2 - x_1)\mathbf{i}$. Similarly $\mathbf{w} = (y_2 - y_1)\mathbf{j}$.

Thus

$$\mathbf{v} = (x_2 - x_1)\mathbf{i} + (y_2 - y_1)\mathbf{j}.$$

Example 1   Find the vector $\mathbf{v}$ in the plane represented by the directed line segment from $(3, -2)$ to $(-1, 5)$. Sketch $\mathbf{v}$.

Solution

$$\begin{aligned}\mathbf{v} &= (x_2 - x_1)\mathbf{i} + (y_2 - y_1)\mathbf{j} \\ &= (-1 - 3)\mathbf{i} + (5 + 2)\mathbf{j} \\ &= -4\mathbf{i} + 7\mathbf{j}\end{aligned}$$

The vector $\mathbf{v}$ may now be sketched as a directed line segment from $(3, -2)$ to $(-1, 5)$ or from $(0, 0)$ to $(-4, 7)$ as shown in Figure 2.7. Of course any other parallel displacement of either of these directed line segments is an equally valid representation of $\mathbf{v}$.

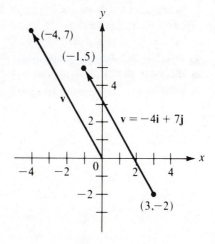

Figure 2.7

Theorem 2.5 (a) $(a_1\mathbf{i} + b_1\mathbf{j}) + (a_2\mathbf{i} + b_2\mathbf{j}) = (a_1 + a_2)\mathbf{i} + (b_1 + b_2)\mathbf{j}$
(b) $(a_1\mathbf{i} + b_1\mathbf{j}) - (a_2\mathbf{i} + b_2\mathbf{j}) = (a_1 - a_2)\mathbf{i} + (b_1 - b_2)\mathbf{j}$
(c) $d(a\mathbf{i} + b\mathbf{j}) = da\mathbf{i} + db\mathbf{j}$
(d) $|a\mathbf{i} + b\mathbf{j}| = \sqrt{a^2 + b^2}$

The proof is left to the student.

Example 2   If $\mathbf{u} = 3\mathbf{i} - 2\mathbf{j}$ and $\mathbf{v} = -\mathbf{i} + 5\mathbf{j}$, find $\mathbf{u} + \mathbf{v}$, $\mathbf{u} - \mathbf{v}$, $|\mathbf{v}|$, and $2\mathbf{v}$.

Solution

$$\begin{aligned}\mathbf{u} + \mathbf{v} &= (3 - 1)\mathbf{i} + (-2 + 5)\mathbf{j} = 2\mathbf{i} + 3\mathbf{j} \\ \mathbf{u} - \mathbf{v} &= (3 - (-1))\mathbf{i} + (-2 - 5)\mathbf{j} = 4\mathbf{i} - 7\mathbf{j} \\ |\mathbf{v}| &= \sqrt{(-1)^2 \times 5^2} = \sqrt{26} \\ 2\mathbf{v} &= 2(-\mathbf{i} + 5\mathbf{j}) = -2\mathbf{i} + 10\mathbf{j}\end{aligned}$$

Example 3   Using the vectors of Example 2, give graphical
representations of **u**, **v**, **u** + **v**, **u** − **v**, and 2**u**.

Solution   **u** = 3**i** − 2**j** is represented by the directed
line segment from the origin to (3, −2)
(see Figure 2.8). Similarly, **v** = −**i** + 5**j**
is represented by the directed line segment from the origin to the point
(−1, 5). Now we complete the parallelogram with **u** and **v** as two sides. The two
diagonals are **u** + **v** and **u** − **v**, **u** + **v**
being directed from the origin to the
opposite vertex and **u** − **v** from the head
of **v** to the head of **u**. Finally, 2**u** is the
directed line segment from the origin, in
the same direction as **u**, but twice as
long as **u**.

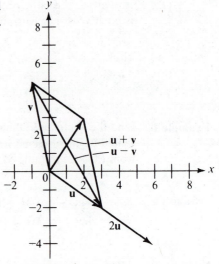

Figure 2.8

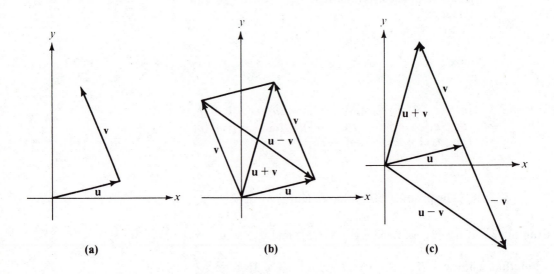

(a)   (b)   (c)

Figure 2.9

Example 4   Given the vectors of Figure 2.9a, sketch $\mathbf{u} + \mathbf{v}$ and $\mathbf{u} - \mathbf{v}$.

Solution   One way to handle this problem is to take a parallel displacement of $\mathbf{v}$ with its tail at the origin, and then use the parallelogram law. This was done in Figure 2.9b. However, let us first note that when the head of $\mathbf{u}$ and the tail of $\mathbf{v}$ correspond, then $\mathbf{u} + \mathbf{v}$ goes from the tail of $\mathbf{u}$ to the head of $\mathbf{v}$. Thus there is no need to draw the entire parallelogram; we can simply draw $\mathbf{u} + \mathbf{v}$ directly. Finally, it is easily seen that $\mathbf{u} - \mathbf{v} = \mathbf{u} + (-\mathbf{v})$. But $-\mathbf{v}$ is a vector whose length is the same as the length of $\mathbf{v}$ and whose direction is opposite that of $\mathbf{v}$. By sketching $-\mathbf{v}$ with the same tail as $\mathbf{v}$, we can then use $\mathbf{u} - \mathbf{v} = \mathbf{u} + (-\mathbf{v})$ to get the result we want as shown in Figure 2.9c.

It might be noted that the two different methods gave $\mathbf{u} - \mathbf{v}$ in two different positions. Nevertheless, they have the same length and direction; they are both $\mathbf{u} - \mathbf{v}$. It is sometimes convenient to have the tail of $\mathbf{u} - \mathbf{v}$ at the origin, and the second method above puts it in that position directly.

Example 5   Find the endpoint A of $\mathbf{v}$, represented by $\overrightarrow{AB}$, if $\mathbf{v} = 4\mathbf{i} - 2\mathbf{j}$ and $B = (-2, 1)$.

Solution   Since $\mathbf{v} = (x_2 - x_1)\mathbf{i} + (y_2 - y_1)\mathbf{j}$, where $A = (x_1, y_1)$ and $B = (x_2, y_2)$, it follows that

$$4 = -2 - x_1 \qquad -2 = 1 - y_1$$
$$x_1 = -6 \qquad\qquad y_1 = 3.$$

Thus $A = (-6, 3)$. See Figure 2.10.

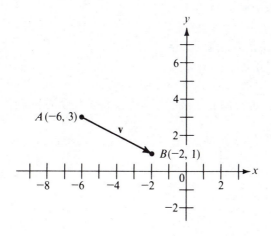

Figure 2.10

PROBLEMS

A   *In Problems* 1–8, *give in component form the vector* **v** *that is represented by* $\overrightarrow{AB}$.

   1. $A = (4, 3), B = (-2, 1)$ **2.** $A = (2, 5), B = (0, 1)$

   3. $A = (-3, 2), B = (4, 3)$ **4.** $A = (-2, 4), B = (0, 4)$

   5. $A = (1, -2), B = (0, 3)$ **6.** $A = (0, 2), B = (1, 4)$

   7. $A = (-3, 2), B = (1, -1)$ **8.** $A = (4, -3), B = (0, 5)$

In Problems 9–16, *give the unit vector in the direction of* **v**.

   9. $\mathbf{v} = 3\mathbf{i} - \mathbf{j}$ **10.** $\mathbf{v} = 2\mathbf{i} + 4\mathbf{j}$

  11. $\mathbf{v} = -\mathbf{i} + 2\mathbf{j}$ **12.** $\mathbf{v} = 3\mathbf{j}$

  13. $\mathbf{v} = \mathbf{i} + 2\mathbf{j}$ **14.** $\mathbf{v} = 3\mathbf{i} - \mathbf{j}$

  15. $\mathbf{v} = -4\mathbf{i} + 3\mathbf{j}$ **16.** $\mathbf{v} = 4\mathbf{j}$

In Problems 17–24, *find the endpoints of the representative* $\overrightarrow{AB}$ *of* **v** *from the given information.*

  17. $\mathbf{v} = 3\mathbf{i} - \mathbf{j}, A = (1, 4)$

  18. $\mathbf{v} = 2\mathbf{i} + 3\mathbf{j}, A = (-1, 3)$

  19. $\mathbf{v} = -\mathbf{i} + 2\mathbf{j}, B = (4, 2)$

  20. $\mathbf{v} = 2\mathbf{i} - 4\mathbf{j}, B = (0, 3)$

  21. $\mathbf{v} = \mathbf{i} - \mathbf{j}, A = (5, 1)$

  22. $\mathbf{v} = 2\mathbf{i} + \mathbf{j}, A = (-2, 0)$

  23. $\mathbf{v} = 3\mathbf{i} - \mathbf{j}, B = (4, 2)$

  24. $\mathbf{v} = 2\mathbf{i} + 2\mathbf{j}, B = (1, 1)$

In Problems 25–32, *find* **u** + **v** *and* **u** − **v**. *Draw a diagram showing* **u**, **v**, **u** + **v**, *and* **u** − **v**.

  25. $\mathbf{u} = 3\mathbf{i} - \mathbf{j}, \mathbf{v} = \mathbf{i} + 2\mathbf{j}$

  26. $\mathbf{u} = 2\mathbf{i} + 3\mathbf{j}, \mathbf{v} = 2\mathbf{i} - \mathbf{j}$

  27. $\mathbf{u} = \mathbf{i} - \mathbf{j}, \mathbf{v} = 2\mathbf{i} + 2\mathbf{j}$

  28. $\mathbf{u} = 3\mathbf{i} + \mathbf{j}, \mathbf{v} = 2\mathbf{i}$

  29. $\mathbf{u} = \mathbf{i} - 3\mathbf{j}, \mathbf{v} = 2\mathbf{i} + 4\mathbf{j}$

  30. $\mathbf{u} = 4\mathbf{i} + \mathbf{j}, \mathbf{v} = \mathbf{i} + 2\mathbf{j}$

  31. $\mathbf{u} = 3\mathbf{i} - \mathbf{j}, \mathbf{v} = 2\mathbf{i} - 2\mathbf{j}$

  32. $\mathbf{u} = \mathbf{i} - \mathbf{j}, \mathbf{v} = 2\mathbf{i} + \mathbf{j}$

33. Given the vectors of Figure 2.11, sketch **u** + **v**.

34. Given the vectors of Figure 2.11, sketch **u** − **v**.

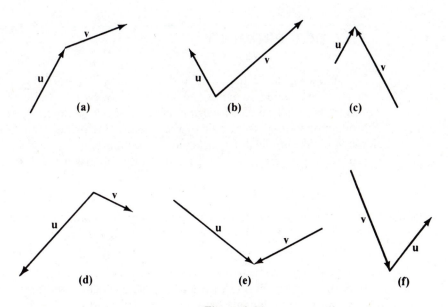

Figure 2.11

B   *In Problems 35–38, find the end points of* $\mathbf{v} = \overrightarrow{AB}$ *from the given information.*

35. $\mathbf{v} = 3\mathbf{i} + 5\mathbf{j}, (4, 1)$ is the midpoint of AB

36. $\mathbf{v} = 4\mathbf{i} - 6\mathbf{j}, (2, 5)$ is the midpoint of AB

37. $\mathbf{v} = \mathbf{i} + \mathbf{j}, (2, 0)$ is the midpoint of AB

38. $\mathbf{v} = 2\mathbf{i} - \mathbf{j}, (0, -2)$ is the midpoint of AB

In Problems 39–42, find the vector indicated and sketch **u**, **v**, *and the new vector.*

39. $\mathbf{u} = 5\mathbf{i} + \mathbf{j}, \mathbf{v} = -2\mathbf{i} + 3\mathbf{j}$; find $2\mathbf{u} + \mathbf{v}$.

40. $\mathbf{u} = -2\mathbf{i} + 5\mathbf{j}, \mathbf{v} = 3\mathbf{i} - \mathbf{j}$; find $\mathbf{u} - 3\mathbf{v}$.

41. $\mathbf{u} = 3\mathbf{i} + \mathbf{j}, \mathbf{v} = \mathbf{i} - \mathbf{j}$; find $-2\mathbf{u} + \mathbf{v}$.

42. $\mathbf{u} = -\mathbf{i} + 2\mathbf{j}, \mathbf{v} = 3\mathbf{i} - 4\mathbf{j}$; find $3\mathbf{u} + 2\mathbf{v}$.

C   **43.** Prove Theorem 2.1.

      44. Prove Theorem 2.2.

45. Prove Theorem 2.5.

46. Show that the sum of two vectors is well defined (see the paragraph following the definition of the sum).

47. Show that $a\mathbf{i} + b\mathbf{j} = c\mathbf{i} + d\mathbf{j}$ if and only if $a = c$ and $b = d$.

2.2

THE DOT PRODUCT

We have considered the sums and differences of pairs of vectors, but the only product considered so far is the scalar multiple—the product of a scalar and a vector. We shall now consider the product of two vectors. There are two different product operations for a pair of vectors, the dot product and the cross product. We shall take up the dot product in this section but defer a discussion of the cross product to Chapter 10 (since it requires three dimensions). First let us consider the angle between two vectors.

Definition *The **angle** between two nonzero vectors $\mathbf{u}$ and $\mathbf{v}$ is the smaller angle between the representatives of $\mathbf{u}$ and $\mathbf{v}$ having their tails at the origin.*

Note that the angle between two vectors is nondirected. That is, we do not consider the angle from one vector to another, which would imply a preferred direction, but rather the angle between two vectors. If θ is the angle between two vectors, then

$$0° \leq \theta \leq 180°$$

Theorem 2.6 *If $\mathbf{u} = a_1\mathbf{i} + b_1\mathbf{j}$ and $\mathbf{v} = a_2\mathbf{i} + b_2\mathbf{j}$ ($\mathbf{u} \neq \mathbf{0}$ and $\mathbf{v} \neq \mathbf{0}$) and if θ is the angle between them, then*

$$\cos\theta = \frac{a_1a_2 + b_1b_2}{|\mathbf{u}|\,|\mathbf{v}|}.$$

Proof By the law of cosines (see Figure 2.12),

$$|\mathbf{v} - \mathbf{u}|^2 = |\mathbf{u}|^2 + |\mathbf{v}|^2 - 2|\mathbf{u}|\,|\mathbf{v}|\cos\theta.$$

Since $\mathbf{v} - \mathbf{u} = (a_2 - a_1)\mathbf{i} + (b_2 - b_1)\mathbf{j}$, we have

$$(a_2 - a_1)^2 + (b_2 - b_1)^2 = a_1^2 + b_1^2 + a_2^2 + b_2^2 - 2|\mathbf{u}|\,|\mathbf{v}|\cos\theta,$$

$$|\mathbf{u}|\,|\mathbf{v}|\cos\theta = a_1a_2 + b_1b_2,$$

$$\cos\theta = \frac{a_1a_2 + b_1b_2}{|\mathbf{u}|\,|\mathbf{v}|}.$$

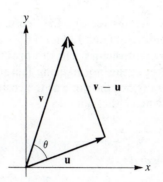

Figure 2.12

Example 1 Find the cosine of the angle between
$\mathbf{u} = 3\mathbf{i} - 4\mathbf{j}$ and $\mathbf{v} = 5\mathbf{i} + 12\mathbf{j}$.

Solution
$$\cos \theta = \frac{a_1 a_2 + b_1 b_2}{|\mathbf{u}| \, |\mathbf{v}|}$$

$$= \frac{(3)(5) + (-4)(12)}{\sqrt{9 + 16} \sqrt{25 + 144}}$$

$$= -\frac{33}{65}$$

$$= -0.5077$$
$$\theta = 120.5°$$

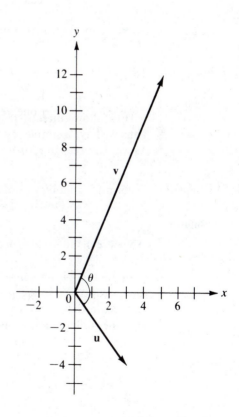

Figure 2.13

If **u** and **v** are unit vectors, the denominator of the expression for cos θ is one— it may be omitted. Thus $1/|\mathbf{u}||\mathbf{v}|$ is a normalizing factor, which is not needed if **u** and **v** are unit vectors. The dot product of two vectors is simply the expression for cos θ without this normalizing factor. Thus, while the dot product of two vectors is generally not the same as the cosine of the angle between them, it does equal this cosine when either $|\mathbf{u}||\mathbf{v}| = 1$ or (more significantly) when cos $\theta = 0$.

Definition   *If* $\mathbf{u} = a_1\mathbf{i} + b_1\mathbf{j}$ *and* $\mathbf{v} = a_2\mathbf{i} + b_2\mathbf{j}$, *then the* **dot product** (*scalar product, inner product*) *of* **u** *and* **v** *is*

$$\mathbf{u} \cdot \mathbf{v} = a_1 a_2 + b_1 b_2.$$

Note that the dot product of two vectors is *not* another vector; it is a scalar.

Example 2   Find the dot product of $\mathbf{u} = 3\mathbf{i} - 2\mathbf{j}$ and $\mathbf{v} = \mathbf{i} + \mathbf{j}$.

Solution
$$\mathbf{u} \cdot \mathbf{v} = (3)(1) + (-2)(1) = 1$$

Let us now consider some applications of the dot product.

Theorem 2.7   *The vectors* **u** *and* **v** (*not both* **0**) *are* **orthogonal** (**perpendicular**) *if and only if* $\mathbf{u} \cdot \mathbf{v} = 0$ (*the zero vector is taken to be orthogonal to every other vector*).

This follows directly from Theorem 2.6 and the definition of the dot product. Thus we have a simple test for the orthogonality (perpendicularity) of two vectors. As we shall see later, orthogonality of vectors is an important concept.

Example 3   Determine whether or not $\mathbf{u} = 2\mathbf{i} - \mathbf{j}$ and $\mathbf{v} = \mathbf{i} + 2\mathbf{j}$ are orthogonal.

Solution
$$\mathbf{u} \cdot \mathbf{v} = (2)(1) + (-1)(2) = 0$$
They are orthogonal, since $\mathbf{u} \cdot \mathbf{v} = 0$.

Again there is the question of whether or not the dot product of vectors has the same properties as the product of numbers. The definition itself shows one difference, in that the dot product of two vectors is not itself a vector. While there are other differences, let us first note the similarities.

Theorem 2.8   *If* **u**, **v**, *and* **w** *are vectors, then*

$$\mathbf{u} \cdot \mathbf{v} = \mathbf{v} \cdot \mathbf{u}$$

and

$$(\mathbf{u} + \mathbf{v}) \cdot \mathbf{w} = \mathbf{u} \cdot \mathbf{w} + \mathbf{v} \cdot \mathbf{w}.$$

This is easily proved from the definitions. The proof is left to the student. It might be noted that the dot product of three vectors $\mathbf{u} \cdot \mathbf{v} \cdot \mathbf{w}$ is meaningless, since the dot product of any pair of them is a scalar.

Theorem 2.9 *If $\mathbf{u}$ and $\mathbf{v}$ are vectors and θ is the angle between them, then*

$$\mathbf{u} \cdot \mathbf{v} = |\mathbf{u}|\,|\mathbf{v}|\,\cos\theta$$

and

$$\mathbf{v} \cdot \mathbf{v} = |\mathbf{v}|^2.$$

The proof is left to the student. It might be noted that $\mathbf{u} \cdot \mathbf{v} = |\mathbf{u}|\,|\mathbf{v}|\,\cos\theta$ is often used by physicists as the definition of the dot product.

Example 4 Find $\mathbf{u} \cdot \mathbf{v}$ for the vectors of Figure 2.14.

Solution $\mathbf{u} \cdot \mathbf{v} = |\mathbf{u}|\,|\mathbf{v}|\,\cos\theta$
$= 6 \cdot 10 \cos 60°$
$= 6 \cdot 10 \cdot \dfrac{1}{2}$
$= 30$

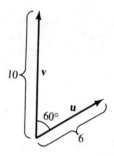

Figure 2.14

We might note some special cases of Theorem 2.9. First we need a definition.

Definition *The nonzero vectors $\mathbf{u}$ and $\mathbf{v}$ are parallel if $\mathbf{u} = k\mathbf{v}$ for some scalar k.*

If $\mathbf{u}$ and $\mathbf{v}$ are orthogonal, $\theta = 90°$ and $\mathbf{u} \cdot \mathbf{v} = 0$, as we have seen before. If $\mathbf{u}$ and $\mathbf{v}$ are parallel, $\theta = 0°$ or $\theta = 180°$ and $\mathbf{u} \cdot \mathbf{v} = \pm |\mathbf{u}|\,|\mathbf{v}|$.

The projection of one vector upon another is determined by the angle between them or the dot product.

Definition *The projection of $\mathbf{u}$ on $\mathbf{v}(\mathbf{v} \neq \mathbf{0})$ is a vector $\mathbf{w}$ such that if $\overrightarrow{AB}$ is a representative of $\mathbf{u}$ and $\overrightarrow{AC}$ is a representative of $\mathbf{v}$, then a representative of $\mathbf{w}$ is a directed line segment $\overrightarrow{AD}$ lying on the line determined by AC with BD perpendicular to that line (see Figure 2.15).*

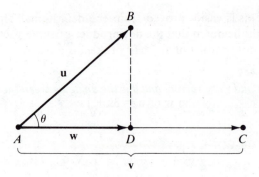

Figure 2.15

The projection of **u** on **v** is defined in terms of representatives of these vectors. Thus we again have the question of whether or not the projection is well defined. Theorem 2.1 and the congruence of triangles show that the projection of **u** on **v** is independent of the representatives considered.

Theorem 2.10   *If* **w** *is the projection of* **u** *on* **v** *and* θ *is the angle between* **u** *and* **v**, *then*

$$|\mathbf{w}| = \frac{|\mathbf{u} \cdot \mathbf{v}|}{|\mathbf{v}|} \quad and \quad \mathbf{w} = \left(\frac{\mathbf{u} \cdot \mathbf{v}}{|\mathbf{v}|}\right)\frac{\mathbf{v}}{|\mathbf{v}|} = \frac{\mathbf{u} \cdot \mathbf{v}}{|\mathbf{v}|^2}\mathbf{v}.$$

Proof   We can see from Figure 2.15, that

$$|\mathbf{w}| = |\mathbf{u}| \, |\cos \theta|$$
$$= \frac{|\mathbf{u}| \, |\mathbf{v}| \, |\cos \theta|}{|\mathbf{v}|}$$
$$= \frac{|\mathbf{u} \cdot \mathbf{v}|}{|\mathbf{v}|} \quad \text{(by Theorem 2.9).}$$

Now we can easily find **w**. Its length is $|\mathbf{u} \cdot \mathbf{v}| \, / \, |\mathbf{v}|$, and its direction is determined by **v**, since the projection is upon **v**. Thus

$$\mathbf{w} = \pm \left(\frac{|\mathbf{u} \cdot \mathbf{v}|}{|\mathbf{v}|}\right)\frac{\mathbf{v}}{|\mathbf{v}|}.$$

Now if the angle θ between **u** and **v** is less than $\pi/2$, **u** $\cdot$ **v** > 0. But **w** and **v** have the same direction in this case and

$$\mathbf{w} = + \left(\frac{|\mathbf{u} \cdot \mathbf{v}|}{|\mathbf{v}|}\right)\frac{\mathbf{v}}{|\mathbf{v}|} = \left(\frac{\mathbf{u} \cdot \mathbf{v}}{|\mathbf{v}|}\right)\frac{\mathbf{v}}{|\mathbf{v}|}.$$

If $\theta > \pi/2$, $\mathbf{u} \cdot \mathbf{v} < 0$ and $\mathbf{w}$ and $\mathbf{v}$ have opposite directions. This leads to

$$\mathbf{w} = -\left(\frac{|\mathbf{u} \cdot \mathbf{v}|}{|\mathbf{v}|}\right)\frac{\mathbf{v}}{|\mathbf{v}|} = \left(\frac{\mathbf{u} \cdot \mathbf{v}}{|\mathbf{v}|}\right)\frac{\mathbf{v}}{|\mathbf{v}|}.$$

Example 4   Find the projection $\mathbf{w}$ of $\mathbf{u} = 3\mathbf{i} + \mathbf{j}$ on $\mathbf{v} = 3\mathbf{i} + 4\mathbf{j}$.

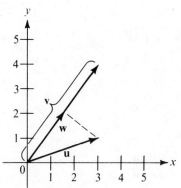

$\mathbf{u} \cdot \mathbf{v} = (3)(3) + (1)(4) = 13$,
$|\mathbf{v}| = \sqrt{3^2 + 4^2} = 5$.
Thus

$$\mathbf{w} = \left(\frac{\mathbf{u} \cdot \mathbf{v}}{|\mathbf{v}|}\right)\frac{\mathbf{v}}{|\mathbf{v}|}$$

$$= \frac{13}{5}\frac{3\mathbf{i} + 4\mathbf{j}}{5} = \frac{39}{25}\mathbf{i} + \frac{52}{25}\mathbf{j}.$$

Figure 2.16 gives a graphical representation of $\mathbf{u}$, $\mathbf{v}$, and $\mathbf{w}$.

Figure 2.16

PROBLEMS

A   *In Problems 1–8, find the angle θ between the given vectors.*

1. $\mathbf{u} = 3\mathbf{i} - \mathbf{j}, \mathbf{v} = \mathbf{i} + 2\mathbf{j}$ 2. $\mathbf{u} = 4\mathbf{i} + \mathbf{j}, \mathbf{v} = \mathbf{i} + 2\mathbf{j}$

3. $\mathbf{u} = -\mathbf{i} + 2\mathbf{j}, \mathbf{v} = 2\mathbf{i} + \mathbf{j}$ 4. $\mathbf{u} = \mathbf{i} + \mathbf{j}, \mathbf{v} = 2\mathbf{i} - \mathbf{j}$

5. $\mathbf{u} = 2\mathbf{i} - \mathbf{j}, \mathbf{v} = \mathbf{i} + 2\mathbf{j}$ 6. $\mathbf{u} = 3\mathbf{i} + 2\mathbf{j}, \mathbf{v} = \mathbf{i} - \mathbf{j}$

7. $\mathbf{u} = 2\mathbf{i} - 2\mathbf{j}, \mathbf{v} = 4\mathbf{i} + \mathbf{j}$ 8. $\mathbf{u} = \mathbf{i} + \mathbf{j}, \mathbf{v} = 2\mathbf{i} + 4\mathbf{j}$

In Problems 9–16, find $\mathbf{u} \cdot \mathbf{v}$ and indicate whether or not $\mathbf{u}$ and $\mathbf{v}$ are orthogonal.

9. $\mathbf{u} = \mathbf{i} - \mathbf{j}, \mathbf{v} = 2\mathbf{i} + \mathbf{j}$ 10. $\mathbf{u} = 2\mathbf{i} + \mathbf{j}, \mathbf{v} = \mathbf{i} - 2\mathbf{j}$

11. $\mathbf{u} = 3\mathbf{i} + 2\mathbf{j}, \mathbf{v} = 2\mathbf{i} - \mathbf{j}$ 12. $\mathbf{u} = 2\mathbf{i} - 4\mathbf{j}, \mathbf{v} = 2\mathbf{i} + \mathbf{j}$

13. $\mathbf{u} = \mathbf{i} - \mathbf{j}, \mathbf{v} = 3\mathbf{i} + 4\mathbf{j}$ 14. $\mathbf{u} = \mathbf{i} + \mathbf{j}, \mathbf{v} = 2\mathbf{i} - 3\mathbf{j}$

15. $\mathbf{u} = 2\mathbf{i} - 3\mathbf{j}, \mathbf{v} = 3\mathbf{i} + \mathbf{j}$ 16. $\mathbf{u} = 4\mathbf{i}, \mathbf{v} = \mathbf{i} + \mathbf{j}$

B   *In Problems 17–24, find the projection of $\mathbf{u}$ on $\mathbf{v}$.*

17. $\mathbf{u} = 2\mathbf{i} - \mathbf{j}, \mathbf{v} = \mathbf{i} + \mathbf{j}$ 18. $\mathbf{u} = \mathbf{i} - 3\mathbf{j}, \mathbf{v} = 2\mathbf{i} + \mathbf{j}$

19. $\mathbf{u} = 2\mathbf{i} + 4\mathbf{j}, \mathbf{v} = \mathbf{i} - 2\mathbf{j}$ **20.** $\mathbf{u} = 4\mathbf{i} + \mathbf{j}, \mathbf{v} = 2\mathbf{i} + \mathbf{j}$

21. $\mathbf{u} = \mathbf{i} - \mathbf{j}, \mathbf{v} = 2\mathbf{i} + \mathbf{j}$ **22.** $\mathbf{u} = 2\mathbf{i} - 3\mathbf{j}, \mathbf{v} = 3\mathbf{i} + 2\mathbf{j}$

23. $\mathbf{u} = 2\mathbf{i} + \mathbf{j}, \mathbf{v} = 4\mathbf{i} - 2\mathbf{j}$ **24.** $\mathbf{u} = 3\mathbf{i} - \mathbf{j}, \mathbf{v} = 2\mathbf{i} + 2\mathbf{j}$

25. Find $\mathbf{u} \cdot \mathbf{v}$ for the vectors of Figure 2.17.

26. Find $\mathbf{u} \cdot \mathbf{v}$ for the vectors of Figure 2.18.

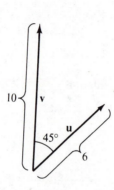

Figure 2.17

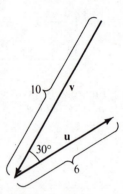

Figure 2.18

In Problems 27–38, determine the value(s) of a so that the given conditions are satisfied.

27. $\mathbf{u} = 3\mathbf{i} - \mathbf{j}, \mathbf{v} = \mathbf{i} + a\mathbf{j}, \mathbf{u}$ and $\mathbf{v}$ are perpendicular.

28. $\mathbf{u} = \mathbf{i} + \mathbf{j}, \mathbf{v} = 3\mathbf{i} - a\mathbf{j}, \mathbf{u}$ and $\mathbf{v}$ are perpendicular.

29. $\mathbf{u} = 4\mathbf{i} + \mathbf{j}, \mathbf{v} = 2\mathbf{i} + a\mathbf{j}, \mathbf{u}$ and $\mathbf{v}$ are perpendicular.

30. $\mathbf{u} = 2\mathbf{i} - \mathbf{j}, \mathbf{v} = a\mathbf{i} + \mathbf{j}, \mathbf{u}$ and $\mathbf{v}$ are perpendicular.

31. $\mathbf{u} = \mathbf{i} - 2\mathbf{j}, \mathbf{v} = a\mathbf{i} + \mathbf{j}, \mathbf{u}$ and $\mathbf{v}$ are parallel.

32. $\mathbf{u} = a\mathbf{i} - \mathbf{j}, \mathbf{v} = 2\mathbf{i} + a\mathbf{j}, \mathbf{u}$ and $\mathbf{v}$ are parallel.

33. $\mathbf{u} = \mathbf{i} + \mathbf{j}, \mathbf{v} = a\mathbf{i} - \mathbf{j}, \mathbf{u}$ and $\mathbf{v}$ are parallel.

34. $\mathbf{u} = a\mathbf{i} + 3\mathbf{j}, \mathbf{v} = 2\mathbf{i} + \mathbf{j}, \mathbf{u}$ and $\mathbf{v}$ are parallel.

35. $\mathbf{u} = a\mathbf{i} + 2\mathbf{j}, \mathbf{v} = \mathbf{i} - \mathbf{j}$, the angle between $\mathbf{u}$ and $\mathbf{v}$ is $\pi/3$.

36. $\mathbf{u} = 3\mathbf{i} - a\mathbf{j}, \mathbf{v} = 2\mathbf{i} + \mathbf{j}$, the angle between $\mathbf{u}$ and $\mathbf{v} = \pi/4$.

37. $\mathbf{u} = 2\mathbf{i} + \mathbf{j}, \mathbf{v} = a\mathbf{i} - \mathbf{j}$, the angle between $\mathbf{u}$ and $\mathbf{v}$ is $2\pi/3$.

38. $\mathbf{u} = \mathbf{i} - \mathbf{j}, \mathbf{v} = 4\mathbf{i} + a\mathbf{j}$, the angle between $\mathbf{u}$ and $\mathbf{v}$ is $\pi/6$.

In Problems 39–42, let **u** *be represented by* $\overrightarrow{AB}$, **v** *by* $\overrightarrow{AC}$, *and* **w** *by* $\overrightarrow{BC}$. *Find the projections of* **v** *and* **w** *on* **u**.

39. $A = (0, 0), B = (1, 4), C = (2, -1)$ **40.** $A = (2, 3), B = (-3, -1), C = (4, 2)$

41. $A = (4, 1), B = (3, -1), C = (0, 2)$ **42.** $A = (2, 0), B = (5, 5), C = (3, 5)$

C   **43.** Prove Theorem 2.7.   **44.** Prove Theorem 2.8.

45. Prove Theorem 2.9.   **46.** Show that $|\mathbf{u} \cdot \mathbf{v}| \le |\mathbf{u}|\,|\mathbf{v}|$.

2.3

APPLICATIONS OF VECTORS

Let us now consider some applications of vectors. Vector methods can be used in some of the proofs of elementary geometry. Sometimes the resulting proof is much shorter than either an analytic or a synthetic argument.

Example 1   Using vector methods, prove that the line joining the midpoints of two sides of a triangle is parallel to and one-half the length of the third side.

Solution   Suppose *D* and *E* are the midpoints of *AB* and *BC*, respectively. Let us give directions to the line segments involved and consider them to be representatives of vectors as shown in Figure 2.19. Then

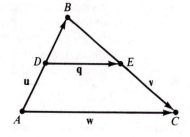

Figure 2.19

$$\mathbf{u} + \mathbf{v} = \mathbf{w},$$

$$\frac{1}{2}\mathbf{u} + \mathbf{q} + \frac{1}{2}\mathbf{v} = \mathbf{w},$$

and

$$\mathbf{q} = \mathbf{w} - \frac{1}{2}(\mathbf{u} + \mathbf{v}) = \mathbf{w} - \frac{1}{2}\mathbf{w} = \frac{1}{2}\mathbf{w}.$$

Since vectors have both magnitude and direction, we have proved that **q** has the same direction as **w** and that it is half the length of **w**.

One reason for the brevity of the foregoing solution is that we were interested in both the direction and magnitude of $\overrightarrow{DE}$. Thus a representation by vectors was very efficient. When perpendicularity is involved, we consider the dot product.

Since this product is defined in terms of the components, it is convenient to give component representations of the vectors.

Example 2   Prove by vector methods that the diagonals of a rhombus are perpendicular.

Solution   Since a rhombus is a parallelogram, the opposite sides are parallel and equal. Thus they have the same vector representation, as shown in Figure 2.20. In addition, since all sides are equal in length, $|\mathbf{u}| = |\mathbf{v}|$, or

$$\sqrt{a^2 + b^2} = \sqrt{c^2 + d^2}.$$

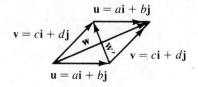

Figure 2.20

Then

$$\mathbf{w} = \mathbf{u} + \mathbf{v} = (a + c)\mathbf{i} + (b + d)\mathbf{j},$$
$$\mathbf{w}' = \mathbf{v} - \mathbf{u} = (c - a)\mathbf{i} + (d - b)\mathbf{j},$$
$$\mathbf{w} \cdot \mathbf{w}' = (a + c)(c - a) + (b + d)(d - b)$$
$$= c^2 - a^2 + d^2 - b^2,$$

But since $a^2 + b^2 = c^2 + d^2$, $\mathbf{w} \cdot \mathbf{w}' = 0$; thus the diagonals are perpendicular.

One of the commonest uses of vectors is in analyzing the forces on an object. If there are several forces acting on a body in different directions, each force can be represented by a vector. A single equivalent force (resultant force) is one that has the same effect on the body as the given forces all acting together; it is simply the vector sum of all of the given forces.

Sometimes this problem must be worked in the other direction; that is, given a single force, find a pair of forces satisfying certain conditions that are equivalent to the single given force. This is often done to find the horizontal and vertical components of a given force.

Example 3   A force of 20 pounds is directed 60° from the horizontal. What is the vector representation for this force?

Solution   In effect we are to find the vectors $\mathbf{v}_1 = a\mathbf{i}$ and $\mathbf{v}_2 = b\mathbf{j}$ such that $\mathbf{v} = \mathbf{v}_1 + \mathbf{v}_2 = a\mathbf{i} + b\mathbf{j}$. It is easily seen from Figure 2.21 that $a = |\mathbf{v}| \cos 60°$ and $b = |\mathbf{v}| \sin 60°$. Thus

$$\mathbf{v} = 20 \cos 60° \, \mathbf{i} + 20 \sin 60° \, \mathbf{j}$$
$$= 10\mathbf{i} + 10\sqrt{3}\mathbf{j}.$$

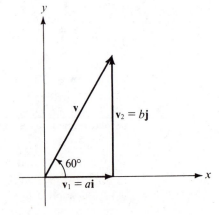

Figure 2.21

Now let us go back to the problem of finding a single force equivalent to several forces acting on a body. As we can see in the following example, the earlier problem is sometimes needed within this one.

Example 4 The following forces are exerted on an object: 5 pounds to the right, 10 pounds upward, 2 pounds upward and to the right, inclined to the horizontal at an angle of 30°. What single force is equivalent to them?

Solution The three given forces are first represented by vectors; the equivalent force is their sum.

$$\mathbf{v}_1 = 5\mathbf{i} \quad \text{and} \quad \mathbf{v}_2 = 10\mathbf{j}$$

For the third vector, we use the method of Example 3 to see that

$$\mathbf{v}_3 = |\mathbf{v}_3| \cos 30° \, \mathbf{i} + |\mathbf{v}_3| \sin 30° \, \mathbf{j}$$
$$= 2 \cdot \frac{\sqrt{3}}{2} \mathbf{i} + 2 \cdot \frac{1}{2} \mathbf{j}$$
$$= \sqrt{3} \, \mathbf{i} + \mathbf{j}$$

This gives $\mathbf{v}_3 = \sqrt{3} \, \mathbf{i} + \mathbf{j}$.

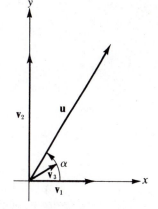

Figure 2.22

$$\mathbf{u} = \mathbf{v}_1 + \mathbf{v}_2 + \mathbf{v}_3 = (5 + \sqrt{3})\mathbf{i} + 11\mathbf{j}$$
$$|\mathbf{u}| = \sqrt{(5 + \sqrt{3})^2 + 11^2}$$
$$= \sqrt{149 + 10\sqrt{3}} = 12.9$$
$$\cos \alpha = \frac{5 + \sqrt{3}}{\sqrt{149 + 10\sqrt{3}}} = 0.5220$$
$$\alpha = 58.53°$$

Thus the three given forces are equivalent to a single force of 12.9 pounds directed upward to the right at an angle of 58.53° with the x axis. See Figure 2.22.

Example 5 Find the resultant force if there are two forces: one with magnitude 6 and directed 120° from the horizontal, and the other with magnitude 10 and directed −135° from the horizontal (see Figure 2.23).

Solution First we express each of the two forces in vector form as in Example 4.

$$v_1 = 6 \cos 120° \, i + 6 \sin 120° \, j$$
$$= -3i + 3\sqrt{3}j$$
$$v_2 = 10 \cos (-135°) \, i + 10 \sin (-135°) \, j$$
$$= -5\sqrt{2}i - 5\sqrt{2}j$$

Now the resultant force is found by adding.

$$v = v_1 + v_2$$
$$= (-3i + 3\sqrt{3}j) + (-5\sqrt{2}i - 5\sqrt{2}j)$$
$$= (-3 - 5\sqrt{2}) \, i + (3\sqrt{3} - 5\sqrt{2}) \, j$$

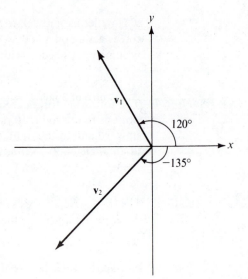

Figure 2.23

There are many other important applications of vectors that are beyond the scope of this book. For example, we may use vectors to analyze the forces on a wire suspended from two points, or on the cables of a suspension bridge. Similarly, vectors may be used to study the motion of planets, finding equations of their paths from a knowledge of the gravitational and centrifugal forces on them.

PROBLEMS

A *In Problems 1–6, the given forces are acting on a body. What single force is equivalent to them?*

1. $f_1 = 4i + 3j, f_2 = i - 2j, f_3 = i + j$

2. $f_1 = 2i - j, f_2 = 3i + 4j, f_3 = -i + 2j$

3. 5 pounds to the right; 10 pounds upward

4. 4 pounds downward; 6 pounds to the right; 2 pounds upward

5. 3 pounds downward; 4 pounds to the right and inclined upward at an angle of 45° with the horizontal

6. 4 pounds to the left; 5 pounds to the right and inclined upward at an angle of 60° with the horizontal

In Problems 7–12, give a vector representation of the given force.

7. A force of 6 pounds directed 30° from the horizontal

8. A force of 8 pounds directed 150° from the horizontal

9. A force of 3 pounds directed −45° from the horizontal

10. A force of 10 pounds directed −120° from the horizontal

11. A force of 2 pounds directed 35° from the horizontal

12. A force of 5 pounds directed −75° from the horizontal

B *In Problems 13–18, the given forces are acting on a body. What additional force will result in equilibrium? (A set of forces is in equilibrium if the sum of all of them is the zero vector.)*

13. $f_1 = 2i + j, f_2 = i − 3j, f_3 = −3i + j$

14. $f_1 = 4i − j, f_2 = i + 3j, f_3 = 2i − 5j$

15. 2 pounds to the left; 5 pounds upward

16. 4 pounds to the right; 6 pounds upward; 6 pounds to the left

17. 3 pounds to the right; 5 pounds to the right and upward inclined at an angle of 45° with the horizontal

18. 3 pounds upward; 2 pounds to the right and inclined upward at an angle of 30° with the horizontal; 4 pounds to the left and inclined downward at an angle of 60° with the horizontal

In Problems 19–24, find the resultant of the forces given.

19. A force of 8 pounds directed 45° from the horizontal; a force of 9 pounds directed 270° from the horizontal

20. A force of 3 pounds directed 150° from the horizontal; a force of 5 pounds directed −60° from the horizontal

21. A force of 6 pounds directed 45° from the horizontal; a force of 8 pounds directed 180° from the horizontal

22. A force of 10 pounds directed 120° from the horizontal; a force of 10 pounds directed −60° from the horizontal

23. A force of 10 pounds directed 120° from the horizontal; a force of 8 pounds directed −30° from the horizontal; a force of 20 pounds directed −90° from the horizontal

24. A force of 8 pounds directed 135° from the horizontal; a force of 10 pounds directed 30° from the horizontal; a force of 2 pounds directed −120° from the horizontal

In Problems 25–33, use vector methods to prove the given theorem.

25. The segment joining the midpoints of the nonparallel sides of a trapezoid is parallel to and one-half the sum of the lengths of the parallel sides.

26. The lines joining consecutive midpoints of a quadrilateral form a parallelogram.

27. If the diagonals of a parallelogram are perpendicular, then it is a rhombus.

28. If the sum of the squares of two sides of a triangle equals the square of the third side, then the triangle is a right triangle.

29. The sum of the squares of the four sides of a parallelogram is equal to the sum of the squares of the two diagonals.

30. The diagonals of a rectangle are equal.

31. The base angles of an isosceles triangle are equal.

32. If one of the parallel sides of a trapezoid is twice the length of the other, then the diagonals intersect at a point of trisection of both of them.

33. The medians of a triangle are concurrent at a point two-thirds of the way from each vertex to the midpoint of the opposite side. [*Hint:* Let $\overrightarrow{BP} = r\overrightarrow{BE}$ and $\overrightarrow{AP} = s\overrightarrow{AD}$ (see Figure 2.24); use the fact that $a\mathbf{u} + b\mathbf{v} = \mathbf{0}$, where neither $\mathbf{u}$ nor $\mathbf{v}$ is a scalar multiple of the other, implies that $a = b = 0$.]

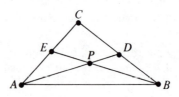

Figure 2.24

REVIEW PROBLEMS

A **1.** Find the unit vector in the direction of $\mathbf{v} = (2, -3)$.

  2. Suppose $\overrightarrow{AB}$ is a representative of $\mathbf{v} = 2\mathbf{i} - 5\mathbf{j}$. Find B when $A = (1, 3)$.

  3. Suppose $\overrightarrow{AB}$ is a representative of $\mathbf{v} = 5\mathbf{i} - 3\mathbf{j}$, and $(-2, 1)$ is the midpoint of AB. Find A and B.

  4. If $\mathbf{u} = (3, -2)$ and $\mathbf{v} = (1, 6)$, find $\mathbf{u} + \mathbf{v}$, $\mathbf{u} - \mathbf{v}$, $3\mathbf{u}$, and $2\mathbf{u} + \mathbf{v}$. Sketch all the vectors.

  5. If $\mathbf{u} = 2\mathbf{i} - \mathbf{j}$ and $\mathbf{v} = 3\mathbf{i} - 2\mathbf{j}$, find $\mathbf{u} \cdot \mathbf{v}$.

  6. Find the angle θ between $\mathbf{u} = 2\mathbf{i} - 3\mathbf{j}$ and $\mathbf{v} = \mathbf{i} + 5\mathbf{j}$.

B **7.** Find the projection $\mathbf{w}$ of $\mathbf{u} = \mathbf{i} + 4\mathbf{j}$ upon $\mathbf{v} = 2\mathbf{i} - 9\mathbf{j}$.

  8. Find the projection $\mathbf{w}$ of $\mathbf{u} = 4\mathbf{i} + \mathbf{j}$ upon $\mathbf{v} = 2\mathbf{i} - 2\mathbf{j}$.

  9. Determine a so that the angle between $\mathbf{u} = 5\mathbf{i} + \mathbf{j}$ and $\mathbf{v} = a\mathbf{i} - \mathbf{j}$ is $\pi/4$.

10. Determine a so that $\mathbf{u} = \mathbf{i} - 4\mathbf{j}$ and $\mathbf{v} = 3\mathbf{i} + a\mathbf{j}$ are orthogonal.

11. Find the resultant force if a force of 6 pounds is directed 90° from the horizontal, 8 pounds is directed 30° from the horizontal, and 10 pounds is directed $-135°$ from the horizontal.

12. Use vectors to prove that the diagonals of an isosceles trapezoid are equal.

13. Use vectors to prove that if the lengths of the parallel sides of a trapezoid are in the ratio $1:n$, then the point of intersection of the diagonals divides both of them in the ratio $1:(n + 1)$.

14. The following forces are exerted on an object: 30 lb to the left, 15 lb downward, $10\sqrt{2}$ lb upward and to the right inclined to the horizontal at an angle of 45°. What single force is equivalent to them?

15. The handle of a lawnmower is inclined to the horizontal at an angle of 30°. If a man pushes forward and down in the direction of the handle with a force of 20 lb, with what force is the mower being pushed forward? With what force is it being pushed into the ground?

C 16. Given $\mathbf{u} = 3\mathbf{i} + 4\mathbf{j}$ and $\mathbf{v} = 12\mathbf{i} - 5\mathbf{j}$, find the vectors $\mathbf{v}_1$ and $\mathbf{v}_2$ such that $\mathbf{v} = \mathbf{v}_1 + \mathbf{v}_2$, $\mathbf{v}_1 = k\mathbf{u}$, and $\mathbf{v}_2 \cdot \mathbf{u} = 0$.

3

THE LINE

3.1

POINT-SLOPE AND TWO-POINT FORMS

The last section of Chapter 1 dealt with the problem of finding an equation of a curve from a description of it. In this chapter, as well as the next two, we shall consider this problem in more detail. Let us begin with a consideration of the line. The two simplest ways of determining a line are by a pair of points or by one point and the slope. Thus, if a line is described in either of these ways, we should be able to give an equation for it. We begin with a line described by its slope and a point on it.

Theorem 3.1 (*Point-slope form of a line.*) *A line that has slope m and contains the point* (x_1, y_1) *has equation*

$$y - y_1 = m(x - x_1).$$

Proof Let (x, y) be any point different from (x_1, y_1) on the given line (see Figure 3.1). Since the line has slope, it is not vertical. Thus $x \neq x_1$, which gives

$$m = \frac{y - y_1}{x - x_1}$$

and

$$y - y_1 = m(x - x_1).$$

Although the formula was derived only for points on the line different from the given point (x_1, y_1), it is easily seen that (x_1, y_1) also satisfies the equation. Thus, every point on the line satisfies the equation. Suppose now that the point (x_2, y_2) satisfies the equation—that is,

$$y_2 - y_1 = m(x_2 - x_1).$$

62

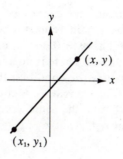

Figure 3.1

If $x_2 = x_1$, then $y_2 - y_1 = 0$, or $y_2 = y_1$. In this case, $(x_2, y_2) = (x_1, y_1)$, which is on the line. If $x_2 \neq x_1$, then

$$\frac{y_2 - y_1}{x_2 - x_1} = m.$$

Thus, the slope of the line joining (x_1, y_1) and (x_2, y_2) is m, and this line has the point (x_1, y_1) in common with the given line. Thus, (x_2, y_2) is on the given line since there can be only one line with slope m containing (x_1, y_1).

Example 1 Find an equation of the line through $(-2, -3)$ with slope 1/2 (see Figure 3.2).

Solution

$$y - y_1 = m(x - x_1)$$
$$y - (-3) = \frac{1}{2}[x - (-2)]$$
$$2y + 6 = x + 2$$
$$x - 2y - 4 = 0$$

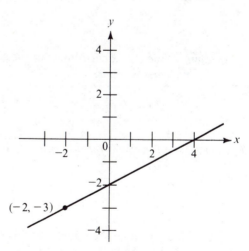

Figure 3.2

Of course vertical lines cannot be represented by the point-slope form, since they have no slope. Again, remember that "no slope" does not mean "zero slope." A horizontal line has $m = 0$, and it can be represented by the point-slope form, which gives $y - y_1 = 0$. There is no x in the resulting equation! But the points on a horizontal line satisfy the condition that they all have the same y coordinate, no matter what the x coordinate is. Similarly, the points on a vertical line satisfy the condition that all have the same x coordinate. Thus, if (x_1, y_1) is one point on a vertical line, then $x = x_1$, or $x - x_1 = 0$ for every point (x, y) on the line.

Example 2   Find an equation of the vertical line through $(3, -2)$ (see Figure 3.3).

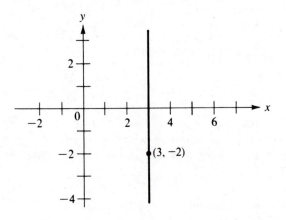

Figure 3.3

Solution   Since the x coordinate of the given point is 3, all points on the line have x coordinates 3. Thus,

$$x = 3 \quad \text{or} \quad x - 3 = 0$$

Theorem 3.2   (*Two-point form of a line.*) *A line through* (x_1, y_1) *and* (x_2, y_2), $x_1 \neq x_2$, *has equation*

$$y - y_1 = \frac{y_2 - y_1}{x_2 - x_1}(x - x_1).$$

It might be noted that this result is often stated in the form

$$\frac{y - y_1}{x - x_1} = \frac{y_2 - y_1}{x_2 - x_1}.$$

While the symmetry of this form is appealing, the form has one serious defect— the point (x_1, y_1) is on the desired line, but it does not satisfy this equation. It does satisfy the equation of Theorem 3.2.

The proof of Theorem 3.2 follows directly from Theorem 3.1 and the fact that $m = (y_2 - y_1)/(x_2 - x_1)$, provided $x_1 \neq x_2$. Actually, this follows so easily from Theorem 3.1 that you may prefer to use the earlier theorem after finding the slope from the two given points. Of course, the designation of the two points as "point 1" and "point 2" is quite arbitrary.

Example 3   Find an equation of the line through $(4, 1)$ and $(-2, 3)$.

Solution

$$y - y_1 = \frac{y_2 - y_1}{x_2 - x_1}(x - x_1)$$

$$y - 1 = \frac{3 - 1}{-2 - 4}(x - 4)$$

$$x + 3y - 7 = 0$$

Example 4   Find the perpendicular bisector of the segment joining $(5, -3)$ and $(1, 7)$.

Solution   First let us find the midpoint.

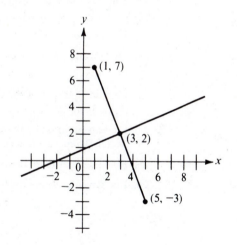

$$x = \frac{5 + 1}{2} = 3 \quad \text{and}$$

$$y = \frac{-3 + 7}{2} = 2$$

The midpoint is $(3, 2)$.

The slope of the line joining $(5, -3)$ and $(1, 7)$ is

$$m = \frac{-3 - 7}{5 - 1} = -\frac{5}{2};$$

the slope of the perpendicular line is $m = 2/5$.

Now we merely need to use the point-slope formula, using $(3, 2)$ and $m = 2/5$.

Figure 3.4

$$y - 2 = \frac{2}{5}(x - 3)$$

$$2x - 5y + 4 = 0$$

The result is shown in Figure 3.4.

PROBLEMS

A   *In Problems 1–16, find an equation of the line indicated and sketch the graph.*

1. Through $(2, -4)$; $m = -2$ **2.** Through $(5, 3)$; $m = 4$

3. Through $(2, 2)$; $m = 1$ **4.** Through $(-4, 6)$; $m = 5$

5. Through $(9, 0)$; $m = 1$ **6.** Through $(0, 3)$; $m = 2$

7. Through $(4, -2)$; $m = 0$ **8.** Through $(2, 5)$; no slope

9. Through $(1, 4)$ and $(3, 5)$ **10.** Through $(2, -1)$ and $(4, 4)$

11. Through $(3, 3)$ and $(1, 1)$ **12.** Through $(2, 1)$ and $(-3, 3)$

13. Through $(0, 0)$ and $(1, 5)$ **14.** Through $(0, 1)$ and $(-2, 0)$

15. Through $(2, 3)$ and $(5, 3)$ **16.** Through $(5, 1)$ and $(5, 3)$

B   **17.** Find equations of the three sides of the triangle with vertices $(1, 4)$, $(3, 0)$, and $(-1, -2)$.

18. Find equations of the medians of the triangle of Problem 17.

19. Find equations of the altitudes of the triangle of Problem 17.

20. Find the vertices of the triangle with sides $x - 5y + 8 = 0$, $4x - y - 6 = 0$, and $3x + 4y + 5 = 0$.

21. Find equations of the medians of the triangle of Problem 20.

22. Find equations of the altitudes of the triangle of Problem 20.

23. Find an equation of the chord of the circle $x^2 + y^2 = 25$ which joins $(-3, 4)$ and $(5, 0)$. Sketch the circle and its chord.

24. Find an equation of the chord of the parabola $y = x^2$ which joins $(-1, 1)$ and $(2, 4)$. Sketch the curve and its chord.

25. Find an equation of the perpendicular bisector of the segment joining $(4, 2)$ and $(-2, 6)$.

26. Find an equation of the line through the points of intersection of the circles

$$x^2 + y^2 + 2x - 19 = 0 \quad \text{and} \quad x^2 + y^2 - 10x - 12y + 41 = 0.$$

Look over your work. Is there any easier way?

27. Repeat Problem 26 for the circles

$$x^2 + y^2 + 4x + 2y + 3 = 0 \quad \text{and} \quad x^2 + y^2 - 6x - 8y + 21 = 0.$$

What is wrong?

28. Find an equation of the line through the centers of the two circles of Problem 26.

29. What condition must the coordinates of a point satisfy in order that it be equidistant from $(2, 5)$ and $(4, -1)$?

30. Find the center and radius of the circle through the points $(1, 3)$, $(4, -6)$, and $(-3, 1)$.

31. Consider the triangle with vertices $A = (3, 1)$, $B = (0, 5)$, and $C = (7, 4)$. Find equations of the altitude and the median from A. What do your results tell us about the triangle?

32. The pressure within a partially evacuated container is being measured by means of an open-end manometer. This gives the difference between the pressure in the container and atmospheric pressure. It is known that a difference of 0 mm of mercury corresponds to a pressure of 1 atmosphere and that if the pressure in the container were reduced to 0 atmospheres, a difference of 760 mm of mercury would be observed. Assuming that the difference D in mm of mercury and the pressure P in atmospheres are related by a linear relation, determine what such a relation is.

33. Knowing that water freezes at 0°C, or 32°F, that it boils at 100°C, or 212°F, and that the relation between the temperature in degrees centigrade C and in degrees Fahrenheit F is linear, find that relation.

34. The amount of a given commodity that consumers are willing to buy at a given price is called the demand for that commodity corresponding to the given price; the relationship between the price and the demand is called a demand equation. Similarly the amount that manufacturers are willing to offer for sale at a given price is called the supply corresponding to the given price, and the relationship between the price and the supply is called a supply equation. Market equilibrium exists when the supply and demand are equal. The demand and supply equations for a given commodity are

$$2p + x - 100 = 0 \quad \text{and} \quad p - x + 10 = 0,$$

respectively, where p is the price of the commodity and x is its supply or demand. At what price will there be market equilibrium? Graph both equations with p on the vertical axis. What happens to the demand as the price increases? What happens to the supply as the price increases?

C **35.** Show that a line through points (x_1, y_1) and (x_2, y_2) can be represented by

$$\begin{vmatrix} x & y & 1 \\ x_1 & y_1 & 1 \\ x_2 & y_2 & 1 \end{vmatrix} = 0.$$

The expression on the left-hand side of this equation is a determinant. Some authors use the notation

$$\det \begin{bmatrix} x & y & 1 \\ x_1 & y_1 & 1 \\ x_2 & y_2 & 1 \end{bmatrix}$$

for this determinant.

36. Show that the points $(x_1, y_1), (x_2, y_2), (x_3, y_3)$ are collinear if and only if

$$\begin{vmatrix} x_1 & y_1 & 1 \\ x_2 & y_2 & 1 \\ x_3 & y_3 & 1 \end{vmatrix} = 0.$$

37. Show that if no pair of the equations

$$A_1 x + B_1 y + C_1 = 0$$
$$A_2 x + B_2 y + C_2 = 0$$
$$A_3 x + B_3 y + C_3 = 0$$

represent parallel lines, then the lines are concurrent if and only if

$$\begin{vmatrix} A_1 & B_1 & C_1 \\ A_2 & B_2 & C_2 \\ A_3 & B_3 & C_3 \end{vmatrix} = 0.$$

3.2

SLOPE-INTERCEPT AND INTERCEPT FORMS

The x and y intercepts of a line are the points at which the line crosses the x and y axes, respectively. These points are of the form $(a, 0)$ and $(0, b)$ (see Figure 3.5), but they are usually represented simply by a and b, since the 0's are understood by their position on the axes. We shall continue using the convention that the x and y intercepts of a line are represented by the symbols a and b, respectively. It might be noted that lines parallel to the x axis have no x intercept and those parallel to the y axis have no y intercept. While a line on the x axis has infinitely many points in common with the x axis, we shall adopt the convention that it has no x intercept. Similarly, a line on the y axis has no y intercept. Thus no horizontal line has an x intercept and no vertical line has a y intercept. One other special case is that of a line through the origin which is neither horizontal nor vertical; it has a single point (the origin) which is both its x and y intercept. In this case $a = b = 0$. With these special points defined, we now introduce two more forms of a line.

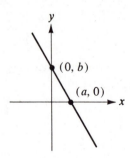

Figure 3.5

Theorem 3.3   (*Slope-intercept form of a line.*) *A line with slope m and y intercept b has equation*
$$y = mx + b.$$

Proof   Since the y intercept is really the point $(0, b)$, the use of the point-slope form gives
$$y - b = m(x - 0) \qquad \text{or} \qquad y = mx + b.$$

Theorem 3.4 (*Intercept form of a line.*) *A line with nonzero intercepts a and b has equation*

$$\frac{x}{a} + \frac{y}{b} = 1.$$

Proof Since the intercepts are the points $(a, 0)$ and $(0, b)$, the line has slope

$$m = -\frac{b}{a}.$$

Using the slope-intercept form, we have

$$y = -\frac{b}{a}x + b.$$

Dividing each term by b gives

$$\frac{y}{b} = \frac{-x}{a} + 1, \qquad \text{or} \qquad \frac{x}{a} + \frac{y}{b} = 1.$$

It might be noted that these two forms are merely special cases of the point-slope and two-point forms; thus, the earlier forms may be used in place of these at any time. However, these forms, especially the slope-intercept form, are so convenient to use that it is well to remember them. We shall see an example of their use shortly.

Example 1 Find an equation of the line with slope 2 and y intercept 5 (see Figure 3.6).

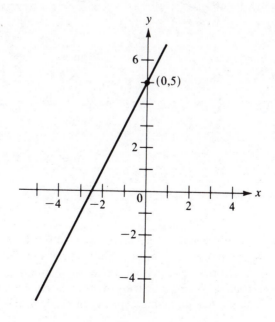

Figure 3.6

Solution
$$y = mx + b$$
$$y = 2x + 5$$
$$2x - y + 5 = 0$$

There is no commonly used special form for a line with a given slope and x intercept. Although one can easily be derived, it has not proved to be as convenient as the slope-intercept form. If you know the slope and the x intercept, simply use the point-slope form with the point $(a, 0)$.

Example 2   Find an equation of the line with x and y intercepts 5 and -2, respectively (see Figure 3.7).

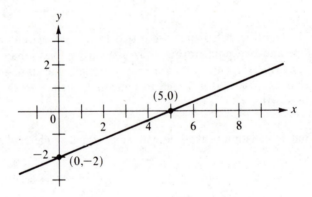

Figure 3.7

Solution
$$\frac{x}{a} + \frac{y}{b} = 1$$

$$\frac{x}{5} + \frac{y}{-2} = 1$$

$$-2x + 5y = -10$$

$$2x - 5y - 10 = 0$$

Just as it was true that vertical lines could not be represented by the point-slope form, we see that vertical lines cannot be represented by the slope-intercept form, since vertical lines have neither slope nor y intercept. The intercept form is even more restrictive, accommodating neither horizontal nor vertical lines, because a horizontal line has no x intercept and a vertical line has no y intercept. Furthermore, no line through the origin can be put into the intercept form, since $a = b = 0$ gives zeros in the denominators.

In all of the examples we have considered so far, we used the special forms only as a starting point; the final form was always $Ax + By + C = 0$. The question arises, whether every equation representing a line can be put into such a form and if every equation in such a form represents a line.

Theorem 3.5   *(General form of a line.) Every line can be represented by an equation of the form*

$$Ax + By + C = 0,$$

where A and B are not both zero, and any such equation represents a line.

Proof   Any line we consider is either vertical or can be put into slope-intercept form. Thus any line can be represented by either

$$x = k \qquad \text{or} \qquad y = mx + b.$$

Thus any line is in the form

$$x - k = 0 \qquad \text{or} \qquad mx - y + b = 0.$$

Both are special cases of $Ax + By + C = 0$.

Suppose we have an equation of the form $Ax + By + C = 0$, where A and B are not both 0. Let us consider two cases.

Case I: $B = 0$. Then

$$Ax + C = 0 \qquad \text{and} \qquad x = -\frac{C}{A}$$

(since $B = 0$ and A and B are not both 0, we know that $A \neq 0$ and we may divide by A). This represents an equation of a vertical line.

Case II: $B \neq 0$. Solving $Ax + By + C = 0$ for y, we have

$$y = -\frac{A}{B}x - \frac{C}{B}$$

(since $B \neq 0$, we may divide by B). This represents an equation of a line with slope $-A/B$ and y intercept $-C/B$.

Theorem 3.5 has the following implication for graphing: any equation of the form $Ax + By + C = 0$ represents a line, and its graph can be determined by two of its points. Since the intercepts are so easily found, finding the line through these two points (if there are two) is the quickest way of sketching a line. Of course, vertical or horizontal lines do not have two intercepts, but these are easily sketched. The only problem comes from lines through the origin. The origin is both the x and y intercept; so just find a second point in any convenient way.

Example 3 Sketch the line $2x - 3y - 6 = 0$.

Solution When $y = 0$, $x = 3$, and when $x = 0$, $y = -2$. We
did not put the equation into intercept form in order
to determine the intercepts, although we might have
done so; however, we can find the intercepts by inspection by setting y and x equal to zero in turn and
solving for the other. Actually, this represents a convenient way of putting the line into intercept form.
Since $a = 3$ and $b = -2$, the intercept form of
$2x - 3y - 6 = 0$ is

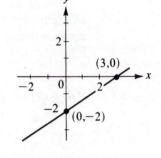

Figure 3.8

$$\frac{x}{3} + \frac{y}{-2} = 1.$$

The graph of this equation is given in Figure 3.8.

The proof of Theorem 3.5 also shows us that the slope of the line $Ax + By + C = 0$
is $-A/B$ provided $B \neq 0$. This implies that the slope of a line is determined entirely
by the coefficients of x and y; the constant term has nothing to do with the slope. Thus

$$Ax + By + C_1 = 0 \qquad \text{and} \qquad Ax + By + C_2 = 0$$

both have the same slope, namely $-A/B$. Furthermore,

$$Ax + By + C_1 = 0 \qquad \text{and} \qquad Bx - Ay + C_2 = 0$$

are perpendicular since they have slopes $-A/B$ and $-B/(-A) = B/A$, respectively.
This gives us an easy way to write an equation of a line parallel (or perpendicular) to
a given line.

Example 4 Find an equation of the line that is (a) parallel and (b) perpendicular to $3x + 2y -
5 = 0$ and contains the point $(3, 1)$.

Solution Any line parallel to the given line has the form $3x + 2y + C_1 = 0$ since such a line has
the same slope as the given line. Now all we need to do is to determine C_1. Since the
line contains the point $(3, 1)$, its equation is satisfied by $(3, 1)$. Thus we have

$$3 \cdot 3 + 2 \cdot 1 + C_1 = 0$$
$$C_1 = -11$$

and the desired line is

$$3x + 2y - 11 = 0.$$

Any line perpendicular to the given line has the form $2x - 3y + C_2 = 0$. Again the
point $(3, 1)$ satisfies the equation. Thus

$$2 \cdot 3 - 3 \cdot 1 + C_2 = 0$$
$$C_2 = -3$$

and the desired line is

$$2x - 3y - 3 = 0.$$

It might be noted that, with very little practice, most of the above can be done in one's head and the answer written directly.

PROBLEMS

A *In Problems 1–20, find an equation of the line described and express it in general form with integer coefficients. Sketch the line.*

1. $m = 4, b = 2$
2. $m = -1, b = 3$
3. $m = 5, b = 1/2$
4. $m = 2/3, b = -1/3$
5. $m = 3/4, b = 2/3$
6. $m = -1/6, b = -5/4$
7. $m = 5, a = -2$
8. $m = 6, a = 3$
9. $m = 0, b = -3$
10. No m, $a = 2$
11. $a = 4, b = 2$
12. $a = -1, b = 3$
13. $a = 2, b = 1/2$
14. $a = 1/2, b = 1/2$
15. $a = 2/3, b = -2/5$
16. $a = -3/4, b = 2/3$
17. $a = b = 0$, through $(2, 5)$
18. $a = b = 0$, through $(-2, -3)$
19. $a = 4$, no b
20. No a, $b = -3$

21. Find an equation of the line parallel to $2x - 5y + 1 = 0$ and containing the point $(2, 3)$.

22. Find an equation of the line perpendicular to $x + 2y - 5 = 0$ and containing the point $(4, 1)$.

B *In Problems 23–26, find an equation of the line described and express it in the general form with integer coefficients. Sketch the line.*

23. $a = b \neq 0$, through $(2, 5)$
24. $a = 3b \neq 0$, through $(5, -4)$
25. $a + b = 8$, through $(3, 1)$
26. $ab = 6$, through $(-3, 4)$

27. Find an equation of the line parallel to $4x + y + 2 = 0$ with y intercept 3.

28. Find an equation of the line perpendicular to $4x - y - 3 = 0$ with x intercept 4.

29. Find the center of the circle circumscribed about the triangle with vertices $(1, 3)$, $(4, -2)$, and $(-2, 1)$.

30. Find the center of the circle circumscribed about the triangle with sides $x + y = 2$, $x - y = 0$, and $2x - y = 4$.

31. Find the orthocenter (points of concurrency of the altitudes) of the triangle with vertices $(-10, 11)$, $(8, 2)$, and $(2, -1)$.

32. Prove analytically that the altitudes of a triangle are concurrent.

33. For what value(s) of m does the line $y = mx - 5$ have x intercept 2?

34. For what value(s) of m does the line $y = mx + 2$ contain the point $(4, 5)$?

35. For what value(s) of a does the line $(x/a) - (y/2) = 1$ have slope 2?

36. For what value(s) of b does the line $(x/3) + (y/b) = 1$ have slope -4?

37. Plot the graph of $x^2 - y^2 = 0$.

38. Plot the graph of $xy = 0$.

39. Plot the graph of $x^2 - 5x + 6 = 0$.

40. Plot the graph of $(x + y - 1)(3x - y + 2) = 0$.

41. Show that $v = Ai + Bj$ is perpendicular to $Ax + By + C = 0$.

42. Show that $v = Bi - Aj$ is parallel to $Ax + By + C = 0$.

C **43.** Work Problem 3.5 of the previous section without expanding the determinant. [*Hint:* Use Theorem 3.5.]

44. One vertex of a rectangle is $(6, 1)$; the diagonals intersect at $(2, 4)$; and one side has slope -2. Find the other three vertices.

45. One vertex of a parallelogram is $(1, 4)$; the diagonals intersect at $(2, 1)$; and the sides have slopes 1 and $-1/7$. Find the other three vertices.

3.3

DISTANCE FROM A POINT TO A LINE

Before considering the distance from a point to a line, let us note the result of Problem 39 of the previous section: the vector $v = Ai + Bj$ is perpendicular to $Ax + By + C = 0$. This perpendicularity allows us to find the distance from any point to a given line.

Theorem 3.6   The distance from the point (x_1, y_1) to the line $Ax + By + C = 0$ is

$$d = \frac{|Ax_1 + By_1 + C|}{\sqrt{A^2 + B^2}}.$$

Proof   The distance we are considering here is the shortest, or perpendicular, distance. As noted above, the vector $v = Ai + Bj$ is perpendicular to $Ax + By + C = 0$. Let (x, y) be a point on $Ax + By + C = 0$ and u be the vector represented by the segment from (x, y) to (x_1, y_1) (see Figure 3.9). Thus

$$\mathbf{u} = (x_1 - x)\mathbf{i} + (y_1 - y)\mathbf{j}.$$

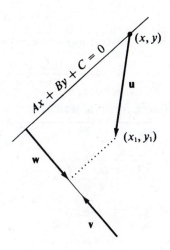

Figure 3.9

The length we seek is the length of the projection $\mathbf{w}$ of $\mathbf{u}$ upon $\mathbf{v}$. By Theorem 2.10,

$$d = |\mathbf{w}| = \frac{|\mathbf{v} \cdot \mathbf{u}|}{|\mathbf{v}|} = \frac{|A(x_1 - x) + B(y_1 - y)|}{\sqrt{A^2 + B^2}}$$

$$= \frac{|Ax_1 + By_1 - (Ax + By)|}{\sqrt{A^2 + B^2}}$$

$$= \frac{|Ax_1 + By_1 + C|}{\sqrt{A^2 + B^2}}.$$

The following is an alternate proof that does not use vectors.

Proof Given the line

$$Ax + By + C = 0$$

and the point (x_1, y_1) then

$$Ax + By - (Ax_1 + By_1) = 0$$

is parallel to the given line and contains (x_1, y_1) (see Figure 3.10). Moreover, $Bx - Ay = 0$ is perpendicular to both of them. The distance we seek is the distance between the points P and

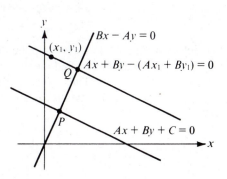

Figure 3.10

Q of Figure 3.10. The point of intersection of $Bx - Ay = 0$ and $Ax + By + C = 0$ is

$$P = \left(\frac{-AC}{A^2 + B^2}, \frac{-BC}{A^2 + B^2} \right),$$

while the point of intersection of $Bx - Ay = 0$ and $Ax + By - (Ax_1 + By_1) = 0$ is

$$Q = \left(\frac{A(Ax_1 + By_1)}{A^2 + B^2}, \frac{B(Ax_1 + By_1)}{A^2 + B^2} \right).$$

Using the distance formula, we have

$$
\begin{aligned}
d &= \sqrt{ \left(\frac{A(Ax_1 + By_1)}{A^2 + B^2} + \frac{AC}{A^2 + B^2} \right)^2 + \left(\frac{B(Ax_1 + By_1)}{A^2 + B^2} + \frac{BC}{A^2 + B^2} \right)^2 } \\
&= \sqrt{ \frac{(Ax_1 + By_1 + C)^2}{A^2 + B^2} } \\
&= \frac{|Ax_1 + By_1 + C|}{\sqrt{A^2 + B^2}}.
\end{aligned}
$$

Example 1   Find the distance from the point $(1, 4)$ to the line $3x - 5y + 2 = 0$.

Solution
$$
\begin{aligned}
d &= \frac{|Ax_1 + By_1 + C|}{\sqrt{A^2 + B^2}} \\
&= \frac{|3 \cdot 1 - 5 \cdot 4 + 2|}{\sqrt{3^2 + (-5)^2}} \\
&= \frac{15}{\sqrt{34}}
\end{aligned}
$$

Example 2   Find the distance between the parallel lines
$$2x - 5y - 10 = 0 \quad \text{and} \quad 2x - 5y + 4 = 0.$$

Solution   First let us select a point on one of the two lines. We may select *any* point in whatever way that we find most convenient. For example, if we take $x = 0$ on the first line, then $y = -2$. Thus we have $(0, -2)$ on the first line (see Figure 3.11). Now all we have to do is find the distance from $(0, -2)$ to the other line, $2x - 5y + 4 = 0$. The result is

$$
\begin{aligned}
d &= \frac{|Ax_1 + By_1 + C|}{\sqrt{A^2 + B^2}} \\
&= \frac{|2 \cdot 0 - 5(-2) + 4|}{\sqrt{2^2 + (-5)^2}} = \frac{14}{\sqrt{29}}.
\end{aligned}
$$

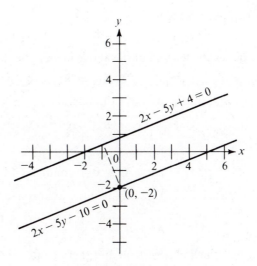

Figure 3.11

The absolute value in the distance formula is sometimes very inconvenient in practice. We could get rid of it if we knew whether $Ax_1 + By_1 + C$ were positive or negative. The following theorem gives us a method of determining this.

Theorem 3.7 *If $P(x_1, y_1)$ is a point not on the line $Ax + By + C = 0\,(B \neq 0)$, then*

(a) B and $Ax_1 + By_1 + C$ agree in sign if P is above the line.
(b) B and $Ax_1 + By_1 + C$ have opposite signs if P is below the line.

Proof *Case I: $B > 0$.* Let Q be the point on the given line with abscissa x_1 (see Figure 3.12). If P is above the line, then $y_1 > y$. Since $B > 0$, $By_1 > By$. Therefore,

$$Ax_1 + By_1 + C > Ax_1 + By + C.$$

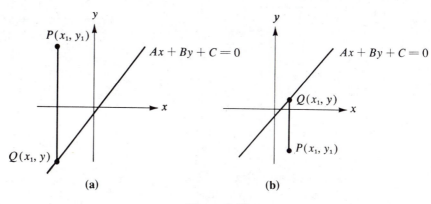

(a)   **(b)**

Figure 3.12

Since (x_1, y) is on the line,

$$Ax_1 + By + C = 0 \quad \text{and} \quad Ax_1 + By_1 + C > 0.$$

If P is below the line, all of the above inequalities are reversed and

$$Ax_1 + By_1 + C < 0.$$

Case II: $B < 0$. If P is above the line, then $y_1 > y$. Since $B < 0$, $By_1 < By$. Thus,

$$Ax_1 + By_1 + C < Ax_1 + By + C.$$

Again

$$Ax_1 + By + C = 0 \quad \text{and} \quad Ax_1 + By_1 + C < 0.$$

As with Case I, all of these inequalities are reversed if P is below the line, and

$$Ax_1 + By_1 + C > 0.$$

If $B = 0$, the line is vertical and there is no "above" nor "below." Theorem 3.7 does not apply to this case, but the distance from a point to a vertical line is easily found without using Theorem 3.6. Other methods of determining the sign of $Ax_1 + By_1 + C$ are given in Problems 32 and 33.

Example 3 Find an equation of the line bisecting the angle from $3x - 4y - 3 = 0$ to $5x + 12y + 1 = 0$.

Solution If (x, y) is any point on the desired line (see Figure 3.13), then it is equidistant from the two given lines. By Theorem 3.6,

$$\frac{|5x + 12y + 1|}{\sqrt{5^2 + 12^2}}$$
$$= \frac{|3x - 4y - 3|}{\sqrt{3^2 + (-4)^2}}$$
$$5|5x + 12y + 1|$$
$$= 13|3x - 4y - 3|.$$

Now let us apply Theorem 3.7. Since P is above $5x + 12y + 1 = 0$ and the coefficient of y is positive, $5x + 12y + 1$ is also positive. Similarly, since P is above $3x - 4y - 3 = 0$ and B is negative,

$$3x - 4y - 3 < 0.$$

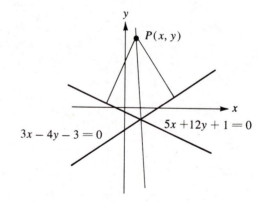

Figure 3.13

Thus

$$5(5x + 12y + 1) = -13(3x - 4y - 3) \qquad \text{or} \qquad 32x + 4y - 17 = 0.$$

Perhaps you object to the designation of P above both lines. Not every point on the bisector is above them. While this is true, the points on the bisector that are not above both are below both. Thus, we still have one expression positive and the other negative, and the result is the same.

It might be noted that we can avoid the use of Theorem 3.7 by considering both cases; that is, $5x + 12y + 1$ and $3x - 4y - 3$ either agree in signs or have opposite signs. In the one case we have $7x - 56y - 22 = 0$; the other gives $32x + 4y - 17 = 0$. Now we need to select one of them. We do so by a comparison of slopes. The first has slope $1/8$ and the second, -8. Since Figure 3.13 shows that the line we want is nearly vertical, it must have a slope that is numerically large. Thus the answer we want is $32x + 4y - 17 = 0$ with slope -8. It might be noted that the two lines above with slopes $1/8$ and -8 are perpendicular. The line $32x + 4y - 17 = 0$ bisects the angle from $3x - 4y - 3 = 0$ to $5x + 12y + 1 = 0$, while $7x - 56y - 22 = 0$ bisects the angle from the second line to the first.

PROBLEMS

A   *In Problems* 1–10, *find the distance from the given point to the given line.*

1. $x + y - 5 = 0, (2, 5)$
2. $2x - 4y + 2 = 0, (1, 3)$
3. $4x + 5y - 3 = 0, (-2, 4)$
4. $x - 3y + 5 = 0, (1, 2)$
5. $3x + 4y - 5 = 0, (1, 1)$
6. $5x + 12y + 13 = 0, (0, 2)$
7. $2x - 5y = 3, (3, -3)$
8. $2x + y = 5, (4, -1)$
9. $3x + 4 = 0, (2, 4)$
10. $y = 3, (1, 5)$

In Problems 11–16, *find the distance between the given parallel lines.*

11. $2x - 5y + 3 = 0, 2x - 5y + 7 = 0$
12. $x + 2y - 2 = 0, x + 2y + 5 = 0$
13. $2x + y + 2 = 0, 4x + 2y - 3 = 0$
14. $4x - y + 2 = 0, 12x - 3y + 1 = 0$
15. $2x - y + 1 = 0, 2x - y - 7 = 0$
16. $3x + 2y = 0, 6x + 4y - 5 = 0$

B **17.** Find the altitudes of the triangle with vertices $(1, 2)$, $(5, 5)$, and $(-1, 7)$.

18. Find the altitudes of the triangle with sides $x + y - 3 = 0$, $x - 2y + 4 = 0$, and $2x + 3y = 5$.

19. Find the area of the triangle of Problem 17.

20. Find the area of the triangle of Problem 18.

In Problems 21–26, find an equation of the line bisecting the angle from the first line to the second.

21. $3x - 4y - 2 = 0, 4x - 3y + 4 = 0$

22. $8x + 15y - 5 = 0, 5x - 12y + 1 = 0$

23. $24x - 7y + 1 = 0, 3x + 4y - 5 = 0$

24. $12x + 35y - 4 = 0, 15y - 8x + 3 = 0$

25. $x + y - 2 = 0, 2x - 3 = 0$

26. $2x + y + 3 = 0, y + 5 = 0$

27. For what value(s) of m is the line $y = mx + 1$ at a distance 3 from $(4, 1)$?

28. For what value(s) of m is the line $y = mx + 5$ at a distance 4 from the origin?

29. For what value(s) of b is the line $(x/3) + (y/b) = 1$ at a distance 1 from the origin?

30. For what value(s) of a is the line $(x/a) + (y/2) = 1$ at a distance 2 from the point $(5, 4)$?

31. The center of the circle inscribed in a triangle is the incenter of the triangle. The center of a circle which is tangent to one side and the extensions of the other two sides is an excenter of the triangle. Find the incenter and the three excenters of the triangle with vertices $(0, 0)$, $(4, 0)$, and $(0, 3)$.

C **32.** Prove that if $P = (x_1, y_1)$ is a point not on the line $Ax + By + C = 0$ $(A \neq 0)$, then

  a. A and $Ax_1 + By_1 + C$ agree in sign if P is to the right of the line.

  b. A and $Ax_1 + By_1 + C$ have opposite signs if P is to the left of the line.

33. Prove that if $P = (x_1, y_1)$ is a point not on the line $Ax + By + C = 0$ $(C \neq 0)$, then

  a. C and $Ax_1 + By_1 + C$ agree in sign if P and the origin are on the same side of the line.

  b. C and $Ax_1 + By_1 + C$ have opposite signs if P and the origin are on opposite sides of the line.

34. Find the center of the circle inscribed in the triangle with vertices $(0, 0)$, $(4, 0)$, and $(0, 3)$.

35. A board leaning against a fence makes an angle of $30°$ with the horizontal. If the board is 4 feet long (see Figure 3.14), what is the diameter of the largest pipe which will fit between the board, the fence, and the ground?

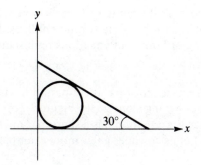

Figure 3.14

36. Suppose that α is the inclination of a line perpendicular (or normal) to the line l and p is the directed distance of l from the origin, p being positive if l is above the origin and negative if l is below (see Figure 3.15). Show that l can be put into the form

$$x \cos \alpha + y \sin \alpha - p = 0.$$

This is called the **normal form** of the line.

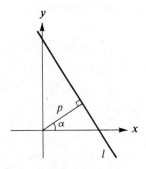

Figure 3.15

3.4

FAMILIES OF LINES

The equation

$$y = 2x + b$$

is in the form $y = mx + b$, with $m = 2$; thus, it represents a line with slope 2 and *y* intercept *b*. But what is *b*? Clearly we could substitute many different values for *b* and get equations of many different lines. It is of interest then to consider the following set, or family, of equations representing lines.

$$M = \{ y = 2x + b \mid b \text{ real} \}$$

M represents a set of parallel lines all having slope 2; in fact, it represents the set of *all* lines having slope 2 (see Figure 3.16). The *b* in $y = 2x + b$ is called a parameter. Since the equation has a single parameter, *M* is called a one-parameter family of lines. Let us consider a few more examples.

Example 1 $\{ y - 2 = m\,(x - 1) \mid m \text{ real} \}$ represents a family of lines through the point (1, 2); however, it does not represent all such lines. The vertical line $x = 1$ (which has no slope) is not a member of this family (see Figure 3.17). The set of *all* lines through the point (1, 2) is $\{ y - 2 = m\,(x - 1) \mid m \text{ real} \} \cup \{ x = 1 \}$.

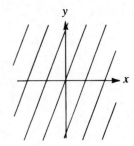

Figure 3.16

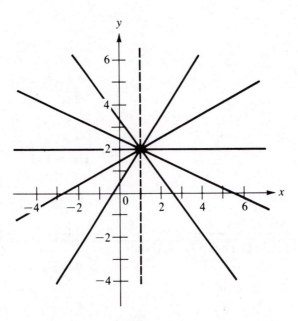

Figure 3.17

Example 2 $\{x/2 + y/b = 1 \mid b$ real, $b \neq 0\}$ represents a family of lines, all having x intercept 2 and some y intercept. It represents all such lines. However, it does not represent all lines having x intercept 2, since the line $x = 2$ is not represented, nor does it represent all lines through $(2, 0)$, since $x = 2$ and $y = 0$ are not included (see Figure 3.18).

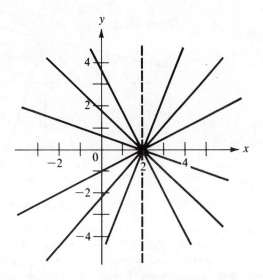

Figure 3.18

Example 3 $\{y = mx + b \mid m, b$ real$\}$ is a two-parameter family of lines representing all non-vertical lines.

Example 4 $\{x = k \mid k$ real$\}$ is the family of all vertical lines.

Example 5 $\{2x + 3y - 6 + k(4x - y + 2) = 0 \mid k$ real$\}$ represents a family of lines (no matter what value we choose for k, the resulting equation is linear) all containing the point of intersection of

$$2x + 3y - 6 = 0 \quad \text{and} \quad 4x - y + 2 = 0$$

(because any point satisfying $2x + 3y - 6 = 0$ and $4x - y + 2 = 0$ must satisfy

$$2x + 3y - 6 + k(4x - y + 2) = 0$$

no matter what value of k we choose). Again, it does not represent *all* such lines; the line $4x - y + 2 = 0$ is not a member of this family (see Figure 3.19).

Example 6 $\{Ax + By + C = 0 \mid A, B, C \text{ real}\}$ is a three-parameter family representing all lines in the plane.

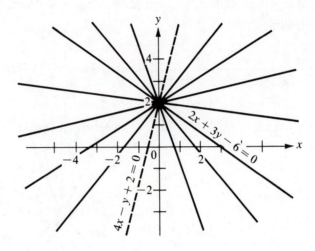

Figure 3.19

Let us now consider the use of families of lines. This concept is most useful in finding an equation of a line which cannot be represented in any of the standard forms that we have seen. Suppose we consider the following example.

Example 7   Find an equation(s) of a line(s) that contains the point $(6, 0)$ and is a distance 5 from the point $(1, 3)$.

Solution $\{y = m(x - 6) \mid m \text{ real}\}$ represents a family of lines all containing the point $(6, 0)$. Note that it does not represent all lines containing the point $(6, 0)$; the only one not represented is the vertical line with equation $x = 6$. Thus, the family of all lines containing $(6, 0)$ is (see Figure 3.20)

$$\{y = m(x - 6) \mid m \text{ real}\} \cup \{x = 6\}.$$

Now we must choose those members of the family that are at a distance 5 from $(1, 3)$. We first consider those lines of the form $y = m(x - 6)$, which can be rewritten in the form

$$mx - y - 6m = 0.$$

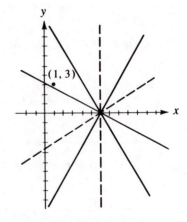

Figure 3.20

The distance from this line to the point $(1, 3)$ is

$$\frac{|m - 3 - 6m|}{\sqrt{m^2 + 1}} = 5.$$

Multiply both sides by $\sqrt{m^2 + 1}$ and square.

$$|-3 - 5m| = 5\sqrt{m^2 + 1}$$
$$9 + 30m + 25m^2 = 25m^2 + 25$$
$$m = \frac{8}{15}$$

Substituting this value back into the original equation, we get

$$y = \frac{8}{15}(x - 6)$$
$$8x - 15y - 48 = 0.$$

Now we must consider the line $x = 6$, which is a distance 5 from the point $(1, 3)$. Thus, the two lines we want are

$$8x - 15y - 48 = 0 \quad \text{and} \quad x - 6 = 0.$$

Example 8   Find an equation(s) of the line(s) parallel to $3x - 5y + 2 = 0$ and containing the point $(3, 8)$.

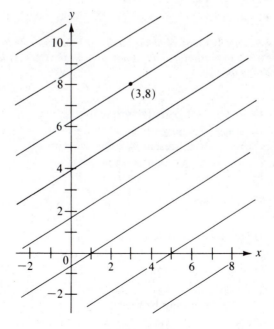

Figure 3.21

Solution   The family of all lines parallel to $3x - 5y + 2 = 0$, (including the given line), is $\{3x - 5y = k | k \text{ real}\}$ (see Figure 3.21). The member of the family which contains $(3, 8)$ satisfies the condition

$$3 \cdot 3 - 5 \cdot 8 = k$$
$$k = -31.$$

The equation desired is $3x - 5y + 31 = 0$. The above procedure is simple enough to do mentally, and a similar procedure can be used for perpendicular lines.

Example 9   Find an equation(s) of the line(s) perpendicular to $3x - 5y = 2 = 0$ and containing the point $(3, 8)$.

Solution   The family of all lines perpendicular to $3x - 5y + 2 = 0$ is $\{5x + 3y = k \mid k \text{ real}\}$. The member that contains $(3, 8)$ satisfies the conditions

$$5 \cdot 3 + 3 \cdot 8 = k$$
$$k = 39.$$

The desired equation is $5x + 3y - 39 = 0$.

PROBLEMS

A   *In Problems 1–14, describe the family of lines given. Indicate whether or not it contains every line of that description, and, if not, give all the lines with that description which are not included in the family.*

1. $\{y - 4 = m(x + 1) | m \text{ real}\}$
2. $\{y = mx - 5 | m \text{ real}\}$
3. $\left\{\dfrac{x}{2} + \dfrac{y}{b} = 1 | b \text{ real}, b \neq 0\right\}$
4. $\{x = ky | k \text{ real}\}$
5. $\{Ax + By = 0 | A, B \text{ real}, A \text{ and } B \text{ not both } 0\}$
6. $\{2x - 3y = k \mid k \text{ real}\}$
7. $\left\{\dfrac{x}{a} + \dfrac{y}{b} = 1 | a, b \text{ real}, a \neq 0, b \neq 0\right\}$
8. $\{y = mx + b | m, b \text{ real}\}$
9. $\{2x + 3y + 1 + k(4x + 2y - 5) = 0 | k \text{ real}\}$

10. $\{x = k \mid k \text{ real}\}$

11. $\left\{\dfrac{x}{a} + \dfrac{y}{2a} = 1 \mid a \text{ real}, a \neq 0\right\}$

12. $\left\{\dfrac{x}{a} + \dfrac{y}{3 - a} = 1 \mid a \text{ real}, a \neq 0, a \neq 3\right\}$

13. $\{y = mx + m \mid m \text{ real}\}$

14. $\{y - a = m(x - a) \mid a, m \text{ real}\}$

In Problems 15–24, *give, in set notation, the family described.*

15. All lines parallel to $3x - 5y - 7 = 0$

16. All lines perpendicular to $3x - 5y - 7 = 0$

17. All lines containing $(2, 5)$

18. All lines with x intercept twice the y intercept

19. All lines containing the point of intersection of $3x - 5y + 1 = 0$ and $2x + 3y - 7 = 0$

20. All horizontal lines

21. All lines containing the origin

22. All lines at a distance 3 from the origin

23. All lines at a distance 5 from $(6, 0)$

24. All lines which form with the coordinate axes a triangle of area 4

In Problems 25–28, *find the lines satisfying the given condition that are (a) parallel and (b) perpendicular, respectively, to the given line.*

25. Containing $(5, 8)$; $3x - 5y + 1 = 0$ **26.** Containing $(3, 2)$; $2x + 3y - 7 = 0$

27. y intercept 5; $4x + 2y - 5 = 0$ **28.** x intercept 2; $3x + y + 2 = 0$

B   **29.** Find an equation(s) of the line(s) with slope 5 at a distance 3 from the origin.

30. Find an equation(s) of the line(s) perpendicular to $3x - 4y + 1 = 0$ and at a distance 4 from $(2, 3)$.

31. Find an equation(s) of the line(s) containing $(5, 4)$ and at a distance 2 from $(-1, -3)$.

32. Find an equation(s) of the line(s) containing $(3, -1)$ and at a distance 4 from $(-1, 3)$.

33. Find an equation(s) of the line(s) containing $(7, 1)$ and at a distance 5 from $(2, -5)$.

34. Find an equation(s) of the line(s) containing $(-4, 3)$ and at a distance 5 from $(-2, 2)$.

35. Find an equation(s) of the line(s) containing the point of intersection of $3x - y - 5 = 0$ and $2x + 2y - 3 = 0$ and having slope 2.

36. Find an equation(s) of the line(s) containing the point of intersection of $4x + 5y - 1 = 0$ and $3x - 2y + 1 = 0$ and the point $(1, 1)$.

37. Find an equation(s) of the line(s) containing $(4, -3)$, such that the sum of the intercepts is 5.

38. Find an equation(s) of the line(s) with slope 3 such that the sum of the intercepts is 12.

39. Find an equation(s) of the line(s) containing $(2, 3)$ and forming a triangle of area 16 with the coordinate axes.

40. Prove analytically that the bisector of an exterior angle determined by the two equal sides of an isosceles triangle is parallel to the third side.

41. An isosceles right triangle is circumscribed about the circle with center $(2, 2)$ and radius 2. The coordinate axes are two of the sides. What is the third?

42. An isosceles right triangle is circumscribed about the circle with center $(4, 2)$ and radius 2. The x axis is the hypotenuse. What are the other two sides?

43. An equilateral triangle is circumscribed about the circle with center $(4, 2)$ and radius 2. The x axis is one side. What are the other two?

3.5

FITTING A LINE TO EMPIRICAL DATA

In experimental work, one is often called upon to fit a line to a given set of empirical data. For example, the electrical resistance of a wire of a given material and diameter is directly proportional to its length. In symbols,

$$R = kL.$$

Now suppose we have found the following data from a laboratory experiment.

L (cm)	1.3	4.2	7.0	10.1	14.2
R (ohm)	11.8	32.7	58.4	81.4	115.2

It is easily seen (see Figure 3.19) that there is no line containing all of the points determined by the above data. In fact, it is unrealistic to expect a line to contain all of the points, since there must be some experimental error involved. Our problem, then, is to find the line which most nearly fits the given data or, equivalently, to find the best value of k from the given data. This is typical of problems that you are likely to encounter in your lab courses in chemistry, physics, engineering, psychology, and so on.

One way of solving this problem is by the method of selected points. This is a graphical method. The points determined by the data are graphed and the line which seems to fit the data best is drawn using a transparent straightedge. Then k is determined by randomly selecting a pair of points on the line (these will not normally coincide with any of the data points) and using them to find the slope of the line (which is k).

Example 1   Determine the value of k for the data given above using the method of selected points.

Solution   The points determined by the data, as well as a line which seems to fit the data best are given in Figure 3.22. Note that the point $(0, 0)$, which was not given in the data, must be on the line because of the form of the equation. Let us choose the points $(0, 0)$ and $(10, 82.0)$ to determine k.

$$k = \frac{82.0 - 0}{10 - 0} = 8.20$$

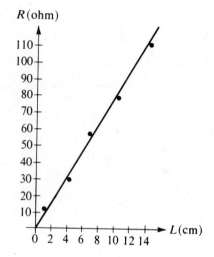

Figure 3.22

The method just used has some serious drawbacks. The selection of the line which best fits the data is strictly a matter of guesswork. As such it is purely subjective—two people might differ on which line fits the data best. Another method, which eliminates the guessing, is the method of averages. Before going into this method, let us define a term used in it.

Definition   The *residual for a given value of x is the observed y coordinate minus the computed y coordinate at that value of x.*

For example, if $y = 4x$ were found from data which included the point $(2, 8.2)$, then the residual at $x = 2$ is $r = 8.2 - 8 = 0.2$, since 8.2 is the observed value at $x = 2$ and 8 is the value computed from $y = 4x$. For Example 1, the residual at $L = 7.0$ is

$$r = 58.4 - (8.2)(7.0) = 1.0.$$

The method of averages simply directs that the value of m in $y = mx$ be chosen in such a way that the sum of the residuals is zero. Suppose we are given the data $(x_1, y_1), (x_2, y_2), \ldots, (x_n, y_n)$. Then

$$r_1 = y_1 - mx_1$$
$$r_2 = y_2 - mx_2$$
$$\vdots$$
$$r_n = y_n - mx_n.$$

Adding, we have

$$r_1 + r_2 + \cdots + r_n = y_1 + y_2 + \cdots + y_n - m(x_1 + x_2 + \cdots + x_n),$$

or, using the shorter notation

$$\sum r_i = r_1 + r_2 + \cdots + r_n$$
$$\sum y_i = y_1 + y_2 + \cdots + y_n$$
$$\sum x_i = x_1 + x_2 + \cdots + x_n,$$

we have

$$\sum r_i = \sum y_i - m \sum x_i.$$

Since $\sum r_i = 0$,

$$m = \frac{\sum y_i}{\sum x_i}.$$

Example 2   Use the method of averages to find the value of k in Example 1.

Solution   Since we have equation $R = kL$ instead of $y = mx$, the formula $m = \sum y_i / \sum x_i$ becomes $k = \sum R_i / \sum L_i$.

$$\sum L_i = 1.3 + 4.2 + 7.0 + 10.1 + 14.2 = 36.8$$
$$\sum R_i = 11.8 + 32.7 + 58.4 + 81.4 + 115.2 = 299.5$$

Thus,

$$k = \frac{\sum R_i}{\sum L_i} = \frac{299.5}{36.8} = 8.14.$$

In the preceding examples, the situation was relatively simple, since we merely wanted to determine m in $y = mx$. Suppose now we consider the problem of determining m and b in $y = mx + b$. The method of selected points is essentially unchanged. Of course, the origin is not necessarily on the line and both the slope and the y intercept are to be found. The method of averages must be altered somewhat when dealing with the equation $y = mx + b$. If we have n points given, then the addition of n equations of the form $r_i = y_i - (mx_i + b)$ leads to

$$\sum r_i = \sum y_i - m \sum x_i - nb = 0.$$

Since there is only one equation in two unknowns (m and b), we cannot solve for either. This is easily remedied by dividing the given data into two sets. Each set leads to an equation of the form just shown, and the two resulting equations can be solved simultaneously for m and b.

Example 3 Find m and b of $y = mx + b$ by both methods of this section from the data given.

x	-1	0	1	2	3	4
y	-3.6	-1.4	1.3	3.1	5.4	8.5

Solution Figure 3.23 shows the data points and a line that is a reasonable fit. Now $b = -1.25$, as can be read from the graph directly; and, using $(0, -12.5)$ and $(4, 8.25)$, we have $m = 2.38$.

For the method of averages, let us divide the data into two sets, using the first three points for the first set and the last three points for the other. The first set gives

$$-3.7 - 3b = 0$$

and the second gives

$$17 - 9m - 3b = 0.$$

Solving simultaneously, we have $b = -1.23$ and $m = 2.30$.

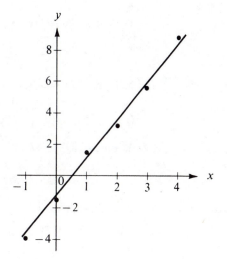

Figure 3.23

In many cases, a nonlinear equation can be handled by the above methods we have used, as the following example illustrates.

Example 4 Find k of $y = kx^3$ from the given data.

x	0	1	2	3	4
y	0	1.35	10.40	36.01	84.52

Solution Since the given equation is nonlinear, our first problem is to convert it to a linear equation. This is easily done by the substitution $z = x^3$. Thus the equation becomes $y = kz$ and the data are as follows.

z	0	1	8	27	64
y	0	1.35	10.40	36.01	84.52

Now, by the method of averages,

$$k = \frac{\sum y_i}{\sum z_i} = \frac{132.28}{100} = 1.323.$$

In addition to the methods given here, there are other, more sophisticated methods which yield somewhat better results.*

*For a discussion of the method of least squares and the method of moments, see Ivan S. Sokolnikoff and Elizabeth S. Sokolnikoff, *Higher Mathematics for Engineers and Physicists*, 2d ed., (New York: McGraw-Hill Book Company, Inc., 1941), pp. 536–545.

PROBLEMS

A   *In Problems 1–10, find the unknown constant(s) from the data given (a) by the method of selected points and (b) by the method of averages.*

1. Find m of $y = mx$.

x	0	1	2	3	4
y	0	4.1	8.5	12.8	16.3

2. Find m of $y = mx$.

x	-3	-2	-1	0	1	2
y	7.2	4.5	2.4	0	-2.3	-4.7

3. Find k of $P = kT$.

T	50	71	102	140	182
P	20.8	29.4	42.0	58.0	74.2

4. Find k of $C = kt$.

t	5	10	20	40	80
C	2.0	4.4	8.3	16.4	33.9

B   **5.** Find m and b of $y = mx + b$.

x	1	3	5	7	10	15
y	6.4	10.9	14.5	19.8	26.0	35.2

6. Find m and b of $y = mx + b$.

x	1	2	5	7	10	15
y	-3.3	1.0	14.1	21.8	34.7	57.0

7. Find p and q of $y = px + q$.

x	2	10	14	22	31	45
y	8.6	29.0	37.2	56.8	78.0	112.9

8. Find k and c of $\gamma = kt + c$.

t	0	20	40	60	80	100
γ	75.6	72.8	69.6	66.2	62.6	58.9

9. Find k of $P = k(1/V)$.

V	9.0	4.8	2.5	1.3	0.7
P	1.2	2.3	4.2	8.8	16.2

10. Find k of $y = kx^2$.

x	1	2	3	4	5
y	5.3	20.5	47.0	82.9	131.0

C 11. The relationship between the vapor pressure P of a liquid and its absolute temperature, T, is given by the Clausius-Clapeyron equation,

$$2.303 \log_{10} P = \frac{-\Delta H}{R} \cdot \frac{1}{T} + C,$$

where ΔH is the molar heat of vaporization of the liquid and R is the ideal gas constant, 1.987 calories degree^{-1} mole^{-1}. The following data were found.

$1/T$	0.00364	0.00357	0.00341	0.00328	0.00319
$\log_{10} P$	0.0000218	0.0000230	0.0000250	0.0000272	0.0000287

What is the molar heat of vaporization of the liquid?

12. The Freundlich equation for adsorption is

$$y = kC^{1/n},$$

where y represents the weight in grams of substance adsorbed, C the concentration in moles/liter of the solute. In logarithmic form, the equation is

$$\log_{10} y = \log_{10} k + \frac{1}{n} \log_{10} C.$$

Experimentation with the adsorption of acetic acid from water solutions by charcoal gave the following results.

C	0.079	0.036	0.019	0.0097	0.0045
y	0.054	0.038	0.029	0.022	0.016

What are k and n?

REVIEW PROBLEMS

A 1. Write an equation (in general form with integer coefficients) for each of the following lines.
   (a) The line through $(1, 5)$ and $(-2, 3)$
   (b) The line with slope 2 and x intercept 3
   (c) The line with inclination $135°$ and y intercept $1/3$
   (d) The line through $(2, 3)$ and $(2, 8)$

2. Write an equation (in general form with integer coefficients) for each of the following lines.
   (a) The line through $(4, 2)$ and parallel to $3x - y + 4 = 0$
   (b) The line with x intercept $1/2$ and y intercept $-5/4$
   (c) The horizontal line through $(3, -2)$

3. Find the slope and intercepts of each of the following lines.
   (a) $x - 4y + 1 = 0$ (b) $2x + 3y + 5 = 0$
   (c) $5x + 2y = 0$ (d) $3x + 1 = 0$

4. Find the distance from $(2, -5)$ to $12x + 5y + 7 = 0$.

5. Find the distance between $3x - y + 5 = 0$ and $6x - 2y - 7 = 0$.

6. Describe the family of lines given. Indicate whether or not it contains every line of that description, and, if not, give all lines with that description that are not included in the family.
   (a) $\{y - 1 = m(x + 3) \mid m \text{ real}\}$
   (b) $\{y = 3x + b \mid b \text{ real}\}$
   (c) $\{x/a - y/3 = 1 \mid a \text{ real}, a \neq 0\}$

7. Give, in set notation, the family described.
   (a) All lines containing $(5, -1)$
   (b) All lines perpendicular to $3x + 2y - 6 = 0$
   (c) All lines at a distance 3 from $(2, 5)$

8. Find k of $u = kv$ from the given data (a) by the method of selected points and (b) by the method of averages.

v	-1	2	4	7	8
u	-1.45	2.90	5.71	9.89	11.32

B 9. Find an equation of the perpendicular bisector of the segment joining $(4, 1)$ and $(0, -3)$.

10. A triangle has vertices $(1, 5)$, $(-2, 3)$, and $(4, -1)$. Find equations for the three altitudes.

11. Find the medians of the triangle of Problem 10.

12. Find equations for the three medians of the triangle of Problem 10.

13. If the line l has slope 3 and contains the point $(-1, 1)$, at what points does it cross the coordinate axes?

14. Find an equation of the line bisecting the angle from $x + y - 5 = 0$ to $x - 7y + 3 = 0$.

15. Find an equation(s) of the line(s) containing $(5, 1)$ and at a distance 1 from the origin.

16. Find an equation(s) of the line(s) with slope 3 and containing the point of intersection of $2x + 3y - 5 = 0$ and $3x - 7y + 5 = 0$.

17. Sketch $x^2 - xy + 3x - 3y = 0$.

18. Find m and b of $y = mx + b$ from the given data (a) by the method of selected points and (b) by the method of averages.

x	1	3	7	10	12	15
y	7.5	13.3	26.5	36.0	42.8	51.6

4

THE CIRCLE

4.1

THE STANDARD FORM FOR AN EQUATION OF A CIRCLE

The standard form for an equation of a circle is a direct consequence of the definition and the length formula.

Definition A **circle** *is the set of all points in a plane at a fixed positive distance* (*radius*) *from a fixed point* (*center*).

Theorem 4.1 *A circle with center* (h, k) *and radius* r *has equation*

$$(x - h)^2 + (y - k)^2 = r^2.$$

Proof If (x, y) is any point on the circle, then the distance from the center (h, k) to (x, y) is r (see Figure 4.1).

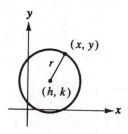

Figure 4.1

$$r = \sqrt{(x - h)^2 + (y - k)^2}$$

Squaring, we have

$$(x - h)^2 + (y - k)^2 = r^2.$$

Since the steps above are reversible, we see that every point satisfying the equation of Theorem 4.1 is on the circle described.

Example 1   Give an equation for the circle with center $(3, -5)$ and radius 2.

Solution   From Theorem 4.1, an equation is

$$(x - 3)^2 + [y - (-5)]^2 = 2^2,$$

or

$$(x - 3)^2 + (y + 5)^2 = 4.$$

Although the above form is a convenient one, in that it shows at a glance the center and radius of the circle, another form is usually used. Called the general form, it is comparable to the general form of a line. Let us first illustrate this form with the result of Example 1. Squaring the two binomials and combining similar terms, we have

$$(x - 3)^2 + (y + 5)^2 = 4$$
$$x^2 - 6x + 9 + y^2 + 10y + 25 = 4$$
$$x^2 + y^2 - 6x + 10y + 30 = 0.$$

Normally an equation of a circle will be given in this form. Let us now repeat the manipulation, starting with the standard form of Theorem 4.1.

$$(x - h)^2 + (y - k)^2 = r^2$$
$$x^2 - 2hx + h^2 + y^2 - 2ky + k^2 = r^2$$
$$x^2 + y^2 - 2hx - 2ky + (h^2 + k^2 - r^2) = 0$$

The last equation is in the form

$$x^2 + y^2 + D'x + E'y + F' = 0.$$

Upon multiplication by a nonzero constant, A, we have

$$Ax^2 + Ay^2 + Dx + Ey + F = 0 \qquad (A \neq 0),$$

as the following theorem states.

Theorem 4.2   *Every circle can be represented in the general form*
$$Ax^2 + Ay^2 + Dx + Ey + F = 0 \qquad (A \neq 0).$$

It is a simple matter to take an equation of a circle in the standard form and reduce it to the general form. We have already seen an example of this. However, it is somewhat more difficult to go from the general form to the standard form. The latter is accomplished by the process of "completing the square." To see how this is accomplished, suppose we consider

$$(x + a)^2 = x^2 + 2ax + a^2.$$

The constant term a^2 and the coefficient of x have a definite relationship; namely, the constant term is the square of one-half the coefficient of x. Thus,

$$a^2 = \left[\frac{1}{2}(2a) \right]^2.$$

Note, however, that this relationship holds only when the coefficient of x^2 is 1.

This relationship suggests the following procedure. If the coefficients of x^2 and y^2 are not one, make them one by division. Group the x terms and the y terms on one side of the equation and take the constant to the other side. Then complete the square on both the x and the y terms. Remember that whatever is added to one side of an equation must be added to the other in order to maintain equality.

Example 2 Express $2x^2 + 2y^2 - 2x + 6y - 3 = 0$ in the standard form. Sketch the graph of the equation.

Solution

$$2x^2 + 2y^2 - 2x + 6y - 3 = 0$$

$$x^2 + y^2 - x + 3y - \frac{3}{2} = 0$$

$$(x^2 - x \quad\;) + (y^2 + 3y \quad\;) = \frac{3}{2}$$

$$\left(x^2 - x + \frac{1}{4} \right) + \left(y^2 + 3y + \frac{9}{4} \right) = \frac{3}{2} + \frac{1}{4} + \frac{9}{4}$$

$$\left(x - \frac{1}{2} \right)^2 + \left(y + \frac{3}{2} \right)^2 = 4$$

Thus, the original equation represents a circle with center $(1/2, -3/2)$ and radius 2. The graph is given in Figure 4.2.

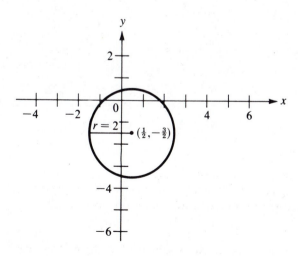

Figure 4.2

The next two examples show that the converse of Theorem 4.2 is not true: that is, an equation of the form

$$Ax^2 + Ay^2 + Dx + Ey + F = 0$$

does not necessarily represent a circle.

Example 3 Express $x^2 + y^2 + 4x - 6y + 13 = 0$ in standard form. Sketch the graph of the equation.

Solution

$$x^2 + y^2 + 4x - 6y + 13 = 0$$

$$(x^2 + 4x \quad) + (y^2 - 6y \quad) = -13$$

$$(x^2 + 4x + 4) + (y^2 - 6y + 9)$$
$$= -13 + 4 + 9$$
$$(x + 2)^2 + (y - 3)^2 = 0$$

Since neither of the two expressions on the left-hand side of the last equation can be negative, their sum can be zero only if both expressions are zero. This is possible only when $x = -2$ and $y = 3$. Thus, the point $(-2, 3)$ is the only point in the plane that satisfies the original equation. The graph is given in Figure 4.3.

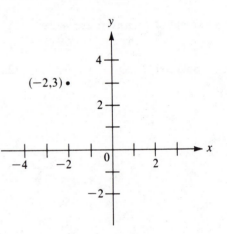

Figure 4.3

Example 4 Express $x^2 + y^2 + 2x + 8y + 19 = 0$ in standard form.

Solution

$$x^2 + y^2 + 2x + 8y + 19 = 0$$
$$(x^2 + 2x \quad) + (y^2 + 8y \quad) = -19$$
$$(x^2 + 2x + 1) + (y^2 + 8y + 16) = -19 + 1 + 16$$
$$(x + 1)^2 + (y + 4)^2 = -2$$

Again, since neither expression on the left-hand side of the last equation can be negative, their sum cannot possibly be negative. There is no point in the plane satisfying this equation. It has no graph.

The results illustrated by the last three examples are stated in the next theorem.

Theorem 4.3 *The graph of every equation of the form*

$$Ax^2 + Ay^2 + Dx + Ey + F = 0 \quad (A \neq 0)$$

is either a circle or a point, or contains no points. (The last two cases are called the ***degenerate cases*** *of a circle.)*

PROBLEMS

A *In Problems 1–16, write an equation of the circle described in both the standard form and the general form. Sketch the graph of each equation.*

1. Center $(1, 3)$; radius 5   **2.** Center $(0, 0)$; radius 1

3. Center $(5, -2)$; radius 2   **4.** Center $(0, 3)$; radius $1/2$

5. Center $(1/2, -3/2)$; radius 2   **6.** Center $(-2/3, -1/2)$; radius $3/2$

B   **7.** Center $(4, -2)$; $(3, 3)$ on the circle   **8.** Center $(-1, 0)$; $(4, -3)$ on the circle

9. $(2, -3)$ and $(-2, 0)$ are the end points of a diameter.

10. $(-3, 5)$ and $(2, 4)$ are the end points of a diameter.

11. Radius 3; in the first quadrant and tangent to both axes

12. Radius 5; in the fourth quadrant and tangent to both axes

13. Radius 2; tangent to $x = 2$ and $y = -1$ and above and to the right of these lines

14. Radius 3; tangent to $x = -3$ and $y = 4$ and below and to the left of these lines

15. Tangent to both axes at $(4, 0)$ and $(0, -4)$

16. Tangent to $x = -2$ and $y = 2$ at $(-2, 0)$ and $(-4, 2)$

In Problems 17–28, express the equation in standard form. Sketch if the graph is nonempty.

17. $x^2 + y^2 - 2x - 4y + 1 = 0$ **18.** $x^2 + y^2 + 4x - 6y - 3 = 0$

19. $x^2 + y^2 + 6x - 16 = 0$ **20.** $x^2 + y^2 - 10x + 4y + 29 = 0$

21. $4x^2 + 4y^2 - 4x - 12y + 1 = 0$ **22.** $9x^2 + 9y^2 - 12x - 24y - 13 = 0$

23. $5x^2 + 5y^2 - 8x - 4y - 121 = 0$ **24.** $9x^2 + 9y^2 - 18x - 12y - 23 = 0$

25. $9x^2 + 9y^2 - 6x + 18y + 11 = 0$ **26.** $36x^2 + 36y^2 - 36x + 24y - 23 = 0$

27. $36x^2 + 36y^2 - 48x - 36y + 25 = 0$ **28.** $8x^2 + 8y^2 + 24x - 4y + 19 = 0$

29. Find the point(s) of intersection of

$$x^2 + y^2 - x - 3y - 6 = 0 \quad \text{and} \quad 4x - y - 9 = 0.$$

30. Find the point(s) of intersection of

$$x^2 + y^2 + 4x - 12y + 6 = 0 \quad \text{and} \quad 3x - 5y + 2 = 0.$$

31. Find the point(s) of intersection of

$$x^2 + y^2 + 5x + y - 26 = 0 \quad \text{and} \quad x^2 + y^2 + 2x - y - 15 = 0.$$

32. Find the point(s) of intersection of

$$x^2 + y^2 + x + 12y + 8 = 0 \quad \text{and} \quad 2x^2 + 2y^2 - 4x + 9y + 4 = 0.$$

33. What happens when we try to solve

$$x^2 + y^2 - 2x + 4y + 1 = 0 \quad \text{and} \quad x - 2y + 2 = 0$$

simultaneously? Interpret geometrically.

34. What happens when we try to solve

$$x^2 + y^2 - 4x - 2y + 1 = 0 \quad \text{and} \quad x^2 + y^2 + 6x - 6y + 14 = 0$$

simultaneously? Interpret geometrically.

35. Find the line through the points of intersection of

$$x^2 + y^2 - x + 3y - 10 = 0 \quad \text{and} \quad x^2 + y^2 - 2x + 2y - 11 = 0.$$

36. For what value(s) of k is the line $x + 2y + k = 0$ tangent to the circle

$$x^2 + y^2 - 2x + 4y + 1 = 0?$$

37. Prove analytically that if P_1 and P_2 are the ends of a diameter of a circle and Q is any point on the circle, then $\angle P_1 Q P_2$ is a right angle.

38. A set of points in the plane has the property that every point in it is twice as far from $(1, 1)$ as it is from $(5, 3)$. What equation must be satisfied by every point (x, y) in the set?

C 39. Find the relation between A, D, E, and F of Theorem 4.2 in order that the equation represent

a. a circle   **b.** a point   **c.** no graph.

If the equation represents a circle, find h, k, and r in terms of A, D, E, and F.

40. In general, squaring both sides of an equation is not reversible (if $x = 2$, then $x^2 = 4$; but if $x^2 = 4$, then $x = \pm 2$). Yet, in the proof of Theorem 4.1, the argument was declared to be reversible even though both sides of an equation were squared. Why?

4.2

CONDITIONS TO DETERMINE A CIRCLE

We have seen two forms for equations of a circle: the standard form,

$$(x - h)^2 + (y - k)^2 = r^2,$$

with the three parameters h, k, and r, and the general form,

$$Ax^2 + Ay^2 + Dx + Ey + F = 0 \qquad (A \neq 0),$$

with the parameters, A, D, E, and F. However, since $A \neq 0$, we can divide through by A to obtain

$$x^2 + y^2 + D'x + E'y + F' = 0,$$

which, like the standard form, has only three parameters. Thus we need three equations in h, k, and r or in D', E', and F' in order to determine these parameters and give the equation desired. Since each condition on a circle determines one such equation, three conditions are required to determine a circle.

Example 1 Find an equation of the circle through points $(1, 5)$, $(-2, 3)$, and $(2, -1)$.

Solution The desired equation is

$$x^2 + y^2 + D'x + E'y + F' = 0$$

for suitable choices of D', E', and F'. Since the three given points are on the circle, they satisfy this equation. Thus

$$1 + 25 + D' + 5E' + F' = 0$$
$$4 + 9 - 2D' + 3E' + F' = 0$$
$$4 + 1 + 2D' - E' + F' = 0$$

or

$$D' + 5E' + F' = -26,$$
$$-2D' + 3E' + F' = -13,$$
$$2D' - E' + F' = -5.$$

Solving simultaneously, we have $D' = -9/5$, $E' = -19/5$, and $F' = -26/5$. Thus the circle is

$$x^2 + y^2 - \frac{9}{5}x - \frac{19}{5}y - \frac{26}{5} = 0$$

or

$$5x^2 + 5y^2 - 9x - 19y - 26 = 0.$$

The example above illustrates the use of the general form to find the desired equation. The general form is rarely used because the constants D', E', and F' have no easily discernible geometric significance. The problem of finding an equation of a circle through three given points is the only one using this form. Even this problem can be solved using the standard form if we recall that the perpendicular bisector of a chord of a circle contains the center. Let us use this on the preceding problem.

Since the points $(1, 5)$ and $(-2, 3)$ are on the circle, the segment from one to the other is a chord of the desired circle (see Figure 4.4). For the chord from $(1, 5)$ to $(-2, 3)$:

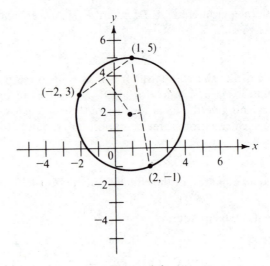

Figure 4.4

$$\text{midpoint} = \left(-\frac{1}{2}, 4 \right)$$

$$m = \frac{5 - 3}{1 + 2} = \frac{2}{3}$$

and the perpendicular bisector is

$$y - 4 = -\frac{3}{2} \left(x + \frac{1}{2} \right)$$

$$4y - 16 = -6x - 3$$

$$6x + 4y = 13.$$

Repeating for the chord from $(1, 5)$ to $(2, -1)$:

$$\text{midpoint} = \left(\frac{3}{2}, 2 \right)$$

$$m = \frac{5 + 1}{1 - 2} = -6$$

and the perpendicular bisector is

$$y - 2 = \frac{1}{6}\left(x - \frac{3}{2}\right)$$
$$12y - 24 = 2x - 3$$
$$2x - 12y = -21.$$

Thus the center is on $6x + 4y = 13$ and $2x - 12y = -21$. Solving simultaneously, we see that the center is $(9/10, 19/10)$. The radius is the distance from the center to any of the given points, say $(1, 5)$.

$$r = \sqrt{\frac{1}{100} + \frac{961}{100}} = \sqrt{\frac{962}{100}}$$

Thus the desired equation is

$$\left(x - \frac{9}{10}\right)^2 + \left(y - \frac{19}{10}\right)^2 = \frac{962}{100}$$

or

$$5x^2 + 5y^2 - 9x - 19y - 26 = 0.$$

Example 2   Find an equation(s) of the circle(s) of radius 4 with center on the line $4x + 3y + 7 = 0$ and tangent to $3x + 4y + 34 = 0$.

Solution   The three conditions lead to the following three relations involving h, k, and r (see Figure 4.5).

(1)
$$r = 4$$

(2)
$$4h + 3k + 7 = 0$$

(3)
$$\frac{|3h + 4k + 34|}{\sqrt{3^2 + 4^2}} = r$$

The first and third give

$$|3h + 4k + 34| = 20.$$

Solving the second for k, we have

$$k = -\frac{4h + 7}{3},$$

and substituting into

$$|3h + 4k + 34| = 20,$$

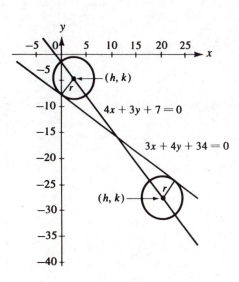

Figure 4.5

we have

$$\left| 3h - \frac{16h + 28}{3} + 34 \right| = 20$$

$$|74 - 7h| = 60$$

$$74 - 7h = \pm 60$$

$$h = 2 \quad \text{or} \quad h = \frac{134}{7};$$

and $k = -5$ or $k = -195/7$, respectively. Thus the two solutions are

$$(x - 2)^2 + (y + 5)^2 = 16 \quad \text{and} \quad \left(x - \frac{134}{7} \right)^2 + \left(y + \frac{195}{7} \right)^2 = 16,$$

or

$$x^2 + y^2 - 4x + 10y + 13 = 0$$

and

$$49x^2 + 49y^2 - 1876x + 2730y + 55{,}197 = 0.$$

This problem can also be solved in the following way. Since the desired circle has radius 4 and is tangent to $3x + 4y + 34 = 0$, its center is on a line parallel to $3x + 4y + 34 = 0$ and at a distance 4 from it. There are two such lines (see Figure 4.6) given by

$$\frac{|3x + 4y + 34|}{5} = 4$$

$$3x + 4y + 34 = \pm 20$$

$$3x + 4y + 14 = 0 \quad \text{or} \quad 3x + 4y + 54 = 0.$$

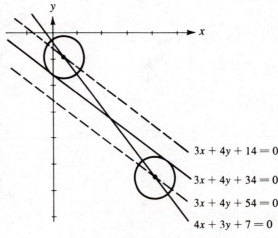

$$3x + 4y + 14 = 0$$
$$3x + 4y + 34 = 0$$
$$3x + 4y + 54 = 0$$
$$4x + 3y + 7 = 0$$

Figure 4.6

Since the center is also on $4x + 3y + 7 = 0$, we can find its coordinates by solving this equation simultaneously with each of the two equations above. From

$$3x + 4y + 14 = 0 \quad \text{and} \quad 4x + 3y + 7 = 0,$$

we get center $(2, -5)$; from

$$3x + 4y + 54 = 0 \quad \text{and} \quad 4x + 3y + 7 = 0,$$

we get center $(134/7, -195/7)$. Using these centers with the given radius, 4, we have the desired circles.

Example 3 Find an equation(s) of the circle(s) tangent to both axes and containing the point $(-8, -1)$.

Solution The three conditions give

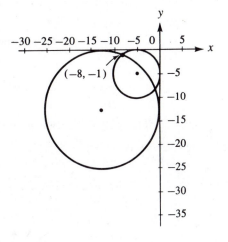

(1) $\qquad\qquad\qquad h = -r$

(2) $\qquad\qquad\qquad k = -r$

(3) $(-8 - h)^2 + (-1 - k)^2 = r^2$

(see Figure 4.7). Substituting (1) and (2) into (3), we have the following.

$$(-8 + r)^2 + (-1 + r)^2 = r^2$$
$$r^2 - 18r + 65 = 0$$
$$(r - 5)(r - 13) = 0$$
$$r = 5 \quad \text{or} \quad r = 13$$

Figure 4.7

Thus we have the circle with radius 5 and center $(-5, -5)$ with equation

$$(x + 5)^2 + (y + 5)^2 = 25$$

or

$$x^2 + y^2 + 10x + 10y + 25 = 0;$$

or we have the circle with radius 13 and center $(-13, -13)$ with equation

$$(x + 13)^2 + (y + 13)^2 = 169$$

or

$$x^2 + y^2 + 26x + 26y + 169 = 0.$$

Example 4 Find an equation(s) of the circle(s) tangent to $3x - 4y - 4 = 0$ at $(0, -1)$ and containing the point $(-1, -8)$.

Solution The center of the desired circle is on the line perpendicular to the tangent line at $(0, -1)$ (see Figure 4.8). An equation of this perpendicular is

$$4x + 3y = 4 \cdot 0 + 3(-1) \qquad \text{or} \qquad 4x + 3y + 3 = 0.$$

Thus, for center (h, k) we have

$$(1) \qquad\qquad\qquad 4h + 3k + 3 = 0.$$

The center is also on the perpendicular bisector of the line joining $(0, -1)$ and $(-1, -8)$ (see Figure 4.8). The slope of the line joining $(0, -1)$ and $(-1, -8)$ is 7; thus the slope of a perpendicular line is $-1/7$. The midpoint of the segment from $(0, -1)$ to $(-1, -8)$ is $(-1/2, -9/2)$. By the point-slope formula, the perpendicular bisector is

$$y + \frac{9}{2} = -\frac{1}{7}\left(x + \frac{1}{2}\right)$$

or

$$x + 7y + 32 = 0.$$

Thus,

$$(2) \qquad\qquad\qquad h + 7k + 32 = 0.$$

Solving (1) and (2) simultaneously, we have

$$h = 3 \qquad \text{and} \qquad k = -5.$$

Using this point with $(0, -1)$, we find the radius

$$r = \sqrt{(3 - 0)^2 + (-5 + 1)^2} = 5.$$

Thus, the desired equation is

$$(x - 3)^2 + (y + 5)^2 = 25$$

or

$$x^2 + y^2 - 6x + 10y + 9 = 0.$$

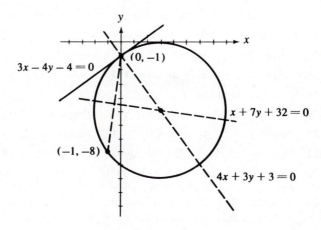

$3x - 4y - 4 = 0$

$(0, -1)$

$x + 7y + 32 = 0$

$(-1, -8)$

$4x + 3y + 3 = 0$

Figure 4.8

PROBLEMS

B   *In Problems 1–23, find an equation(s) of the circle(s) described.*

1. Through $(-1, 2)$, $(3, 4)$, and $(2, -1)$
2. Through $(-2, -1)$, $(0, 3)$, and $(2, 0)$
3. Circumscribed about the triangle with vertices $(2, 3)$, $(0, 5)$, and $(1, -1)$
4. Circumscribed about the triangle with vertices $(1, 1)$, $(-2, 1)$, and $(1, 4)$
5. Circumscribed about the triangle with sides $x - y = 0$, $x + 2y = 0$, and $4x + y = 35$
6. Through $(2, 1)$, $(-4, 4)$, and $(6, -1)$; [*Watch out!*]
7. Tangent to the x axis; center on $2x + y - 1 = 0$; radius 5
8. Tangent to $2x + 3y + 13 = 0$ and $2x - 3y - 1 = 0$; contains $(0, 4)$
9. Tangent to $3x + 4y - 15 = 0$ at $(5, 0)$; contains $(-2, -1)$
10. Tangent to $5x - 12y + 89 = 0$ at $(-1, 7)$; contains $(16, 0)$
11. Tangent to $x + y = 0$ and $x - y - 6 = 0$; center on $3x - y + 3 = 0$
12. Tangent to $x - 3y - 7 = 0$ and $3x + y - 21 = 0$; center on $x - 3y + 3 = 0$
13. Tangent to $x - 3y = 0$ at $(0, 0)$; center on $2x + y + 1 = 0$
14. Tangent to $x - y = 0$ at $(2, 2)$; center on $2x + 3y - 7 = 0$

15. Contains $(-1, 4)$ and $(3, 2)$; center on $3x - y + 3 = 0$

16. Contains $(5, 2)$ and $(-1, 6)$; center on $x = y$

17. Tangent to $2x + 3y - 5 = 0$ at $(1, 1)$; tangent to $2x + 3y + 10 = 0$

18. Tangent to $y = 0$ at $(4, 0)$; tangent to $3x - 4y - 17 = 0$

19. Tangent to both axes; radius 3

20. Tangent to $x = 0$; center on $x + y = 10$; contains $(2, 9)$

21. Tangent to $3x - 4y + 3 = 0$ at $(-1, 0)$; radius 7

22. Tangent to $x^2 + y^2 - 22x + 20y + 77 = 0$ at $(91/17, 10/17)$; containing $(0, 1)$

23. Tangent to $x^2 + y^2 - 8x - 22y + 112 = 0$ and $3x + 4y + 19 = 0$; radius 5

C   **24.** Show that if (x_1, y_1), (x_2, y_2), and (x_3, y_3) are three noncollinear points, then the circle containing these three points has equation

$$\begin{vmatrix} x^2 + y^2 & x & y & 1 \\ x_1^2 + y_1^2 & x_1 & y_1 & 1 \\ x_2^2 + y_2^2 & x_2 & y_2 & 1 \\ x_3^2 + y_3^2 & x_3 & y_3 & 1 \end{vmatrix} = 0$$

25. Show that if (x_1, y_1), (x_2, y_2), and (x_3, y_3) are three collinear points, then the determinant of Problem 24 is linear.

26. Find and equation(s) of the line(s) tangent to $x^2 + y^2 + 4x - 10y + 4 = 0$ from the point $(3, 2)$.

27. Find an equation(s) of the line(s) tangent to $x^2 + y^2 - 8x + 2y - 152 = 0$ and having slope $1/3$.

4.3

FAMILIES OF CIRCLES

We can, of course, consider families of circles just as we did families of lines. In particular, let us suppose that

$$Ax^2 + Ay^2 + Dx + Ey + F = 0$$

and

$$A'x^2 + A'y^2 + D'x + E'y + F' = 0$$

represent two circles which intersect at the points P_1 and P_2. Now let us consider the family

$$M = \{Ax^2 + Ay^2 + Dx + Ey + F$$
$$+ k(A'x^2 + A'y^2 + D'x + E'y + F') = 0 \mid k \text{ real}\}.$$

Since the coordinates of P_1 satisfy the equations of both of the given circles, they must satisfy the equation of the family M no matter what value k might have. Thus the point P_1 belongs to every member of the family. Similarly, P_2 belongs to every member. Since the coefficients of x^2 and y^2 are the same, the members of M consist of circles containing P_1 and P_2, together with the line containing these two points (when $k = -A/A'$). Note, however, that the family does not include *all* circles through P_1 and P_2; no value of k gives

$$A'x^2 + A'y^2 + D'x + E'y + F' = 0.$$

Example 1 Find an equation of the circle containing $(2, -1)$ and the points of intersection of

$$x^2 + y^2 - 2x - 4y + 1 = 0 \quad \text{and} \quad x^2 + y^2 - 6x - 2y + 9 = 0.$$

Solution Since neither circle contains $(2, -1)$, the desired circle must be one of the family

$$\{x^2 + y^2 - 2x - 4y + 1 + k(x^2 + y^2 - 6x - 2y + 9) = 0 \mid k \text{ real}\}$$

(see Figure 4.9). If $(2, -1)$ is on the desired circle, then, for the proper choice of k, $(2, -1)$ satisfies the family equation.

$$4 + 1 - 4 + 4 + 1 + k(4 + 1 - 12 + 2 + 9) = 0$$
$$6 + 4k = 0$$
$$k = -\frac{3}{2}$$

Thus the circle is

$$x^2 + y^2 - 2x - 4y + 1 - \frac{3}{2}(x^2 + y^2 - 6x - 2y + 9) = 0$$

or

$$x^2 + y^2 - 14x + 2y + 25 = 0.$$

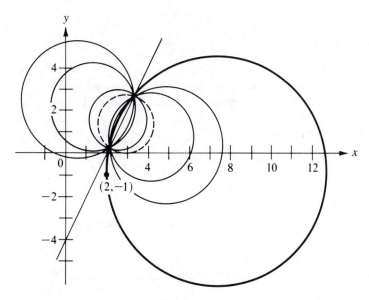

Figure 4.9

We might note the procedure necessary to solve this problem without using families of circles. First we would have to find the points of intersection of the given circles. These are $(3, 2)$ and $(11/5, 2/5)$. Then, using these two points together with the point $(2, -1)$, we would have to find the circle containing them all (see Example 1, page 101). This would be considerably more difficult than the method above. Furthermore, if the points of intersection had been irrational, this method would be even more tedious. Thus the use of families of circles is quite convenient here.

Example 2   Find the line through the points of intersection of the circles

$$x^2 + y^2 + 4x - 2y - 4 = 0 \quad \text{and} \quad x^2 + y^2 - 4x = 0.$$

Solution   The line desired is the one line of the family

$$\{x^2 + y^2 + 4x - 2y - 4 + k(x^2 + y^2 - 4x) = 0 \mid k \text{ real}\}$$

(see Figure 4.10). We get the line by choosing $k = -1$ so that the second-degree terms cancel. Thus we have

$$x^2 + y^2 + 4x - 2y - 4 - (x^2 + y^2 - 4x) = 0$$
$$8x - 2y - 4 = 0$$
$$4x - y - 2 = 0$$

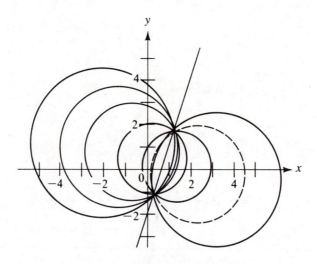

Figure 4.10

We must be careful when using the above family to be sure that the given circles actually intersect at two points. If they are tangent or have no point in common, the set discussed above is still a family of circles (including degenerate circles) together with one line. If the given circles are tangent, the family consists of circles and a line tangent to both of the given circles. If the given circles have no point in common and are not concentric, the family consists of circles with their centers on the line of centers of the given circles, together with one line perpendicular to the line of centers. If the given circles are concentric, the family consists of circles concentric with each other and the given circles (there is no line in the family in this case).

Whenever the given circles are not concentric, the family contains one line, called the **radical axis** of the two circles. This line has some interesting properties. From the discussion above, it follows that the radical axis must be perpendicular to the line of centers of the given circles. Furthermore, if P is any point on the radical axis of two circles C_1 and C_2 and P is outside of both C_1 and C_2, and if T_1 and T_2 are points of C_1 and C_2, respectively, such that PT_1 and PT_2 are tangent to C_1 and C_2, respectively, then $\overline{PT_1} = \overline{PT_2}$ (see Figure 4.11). The proof of this statement is left to the student (see Problems 23, 28, and 29).

Of course, other families of circles may also be considered. Some of these are given in the problems that follow (see Problems 15–20).

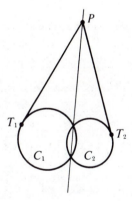

Figure 4.11

PROBLEMS

A   **1.** Find an equation(s) of the circle(s) containing $(1, -4)$ and the points of intersection of

$$x^2 + y^2 + 2x - 4y + 1 = 0 \quad \text{and} \quad x^2 + y^2 + 4x + 6y - 3 = 0.$$

2. Find an equation(s) of the circle(s) containing $(2, 0)$ and the points of intersection of

$$x^2 + y^2 - 2x + 6y + 6 = 0 \quad \text{and} \quad x^2 + y^2 + 2x - 2y - 7 = 0.$$

3. Find an equation(s) of the circle(s) containing $(-1, -2)$ and the points of intersection of

$$x^2 + y^2 + 3x - 4y - 7 = 0 \quad \text{and} \quad x^2 + y^2 - 2x - 5y - 5 = 0.$$

4. Find an equation(s) of the circle(s) containing $(2, 1)$ and the points of intersection of

$$2x^2 + 2y^2 + 3x - y - 8 = 0 \quad \text{and} \quad x^2 + y^2 - 2x + y - 5 = 0.$$

B **5.** Find an equation(s) of the circle(s) with center on $x + y - 2 = 0$ and containing the points of intersection of

$$x^2 + y^2 + 4x + 6y - 3 = 0 \quad \text{and} \quad x^2 + y^2 + 2x + 2y - 2 = 0.$$

6. Find an equation(s) of the circle(s) with center on $x - 2y + 5 = 0$ and containing the points of intersection of

$$x^2 + y^2 - 4x + 2y + 2 = 0 \quad \text{and} \quad x^2 + y^2 - 10x + 4 = 0.$$

7. Find an equation(s) of the circle(s) with center $(3, -1)$ and containing the points of intersection of

$$x^2 + y^2 - 4x - 6y + 9 = 0 \quad \text{and} \quad x^2 + y^2 - 2x - 14y + 15 = 0.$$

8. Find an equation(s) of the circle(s) with center $(4, 0)$ and containing the points of intersection of

$$x^2 + y^2 - 2x - 2y - 2 = 0 \quad \text{and} \quad x^2 + y^2 - 5x - y + 4 = 0.$$

9. Find an equation(s) of the circle(s) with radius 2 and containing the points of intersection of

$$x^2 + y^2 + 2x - 2y - 3 = 0 \quad \text{and} \quad x^2 + y^2 - x - 1 = 0.$$

10. Find an equation(s) of the circle(s) with radius 3 and containing the points of intersection of

$$x^2 + y^2 - 4x - 2y - 1 = 0 \quad \text{and} \quad 2x^2 + 2y^2 - 8y + 3 = 0.$$

11. Find an equation(s) of the line(s) containing the points of intersection of

$$x^2 + y^2 - 2x - 8y + 8 = 0 \quad \text{and} \quad x^2 + y^2 + 2x - 3 = 0.$$

12. Find an equation(s) of the line(s) containing the points of intersection of

$$x^2 + y^2 - 2x + 4y = 0 \quad \text{and} \quad x^2 + y^2 - 6x - 2y + 6 = 0.$$

13. Suppose you are asked to use the method of this section to find an equation(s) of the circle(s) containing $(3, 0)$ and the points of intersection of

$$x^2 + y^2 + 2x + 2y - 7 = 0 \quad \text{and} \quad x^2 + y^2 - 4x - 6y + 9 = 0.$$

What is the result? Does this represent all possible circles satisfying the given conditions? If not, why not? Sketch the given circles and the result.

14. Suppose you are asked to use the method of this section to find an equation(s) of the circle(s) containing $(-2, -1)$ and the points of intersection of

$$x^2 + y^2 + 4x - 8y + 16 = 0 \quad \text{and} \quad x^2 + y^2 - 4x - 2y + 1 = 0.$$

What is the result? Does this represent all possible circles satisfying the given conditions? If not, why not? Sketch the given circles and the result.

In Problems 15–20, describe the family of circles given. Indicate whether or not it contains every circle of that description, and, if not, give all circles with that description which are not included in the family.

15. $\{(x - h)^2 + (y - 1)^2 = 1 \mid h \text{ real}\}$

16. $\{(x - h)^2 + (y - k)^2 = k^2 \mid h, k \text{ real}\}$

17. $\{(x - h)^2 + (y - h)^2 = h^2 \mid h \text{ real}\}$

18. $\{x^2 + y^2 + Dx + Ey = 0 \mid D, E \text{ real}\}$

19. $\{(x - h)^2 + (y - k)^2 = 1 \mid h, k \text{ real}, \ h^2 + k^2 = 1\}$

20. $\{(x - h)^2 + (y - k)^2 = 1 \mid h, k \text{ real}, \ |h| = 2, |k| = 2\}$

In Problems 21–24, use the result of Problem 27 to find the length of the tangent (PT) from the given point to the given circle.

21. $(x - 1)^2 + (y - 2)^2 = 4; \quad (5, 3)$

22. $(x + 3)^2 + (y - 1)^2 = 9; \quad (4, 3)$

23. $x^2 + y^2 - 4x + 2y + 1 = 0; \quad (3, 5)$

24. $x^2 + y^2 + 8x - 4y + 11 = 0; \quad (1, -3)$

C

25. Given two points P_1 and P_2 such that three different circles all contain P_1 and P_2, show that the centers of the circles are collinear.

26. Show that the radical axis of a pair of circles is perpendicular to their line of centers.

27. Show that if $P = (x_1, y_1)$ is outside the circle $(x - h)^2 + (y - k)^2 = r^2$ and T is a point of the circle such that PT is tangent to the circle, then the length of PT is

$$\sqrt{(x_1 - h)^2 + (y_1 - k)^2 - r^2}$$

(see Figure 4.12).

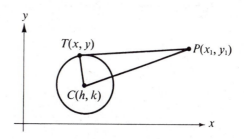

Figure 4.12

28. Use the result of Problem 27 to show that, if P has equal tangents to the circles

$$x^2 + y^2 + Dx + Ey + F = 0 \quad \text{and} \quad x^2 + y^2 + D'x + E'y + F' = 0,$$

then P is on the radical axis of the two circles.

29. Show that the argument of Problem 28 is reversible, so that if P is on the radical axis and there exist tangents from P to the two circles, then the tangents are equal.

REVIEW PROBLEMS

A *In Problems 1–4, put the equation into standard form and identify it as a circle, a point, or no graph. If it is a circle, give its center and radius. If it is a point, give its coordinates.*

1. $x^2 + y^2 - 10x + 4y + 13 = 0$

2. $x^2 + y^2 + 6x - 2y + 10 = 0$

3. $36x^2 + 36y^2 - 24x + 108y + 85 = 0$

4. $4x^2 + 4y^2 - 4x + 12y - 15 = 0$

5. Find an equation of the line through the points of intersection of $x^2 + y^2 - 4x + 2y + 1 = 0$ and $x^2 + y^2 - 8x - 2y + 8 = 0$.

6. Find an equation of the circle containing the origin and the points of intersection of the circles of Problem 5.

B **7.** Find an equation of the circle with center $(4, 1)$ and tangent to $3x + 4y - 2 = 0$.

8. Find an equation of the circle of radius 4 that is tangent to the x axis and has its center on $x - 2y = 2$.

9. Find an equation of the circle with center on $x + 2y + 1 = 0$ and containing the points of intersection of $x^2 + y^2 + 2x - 4y - 4 = 0$ and $x^2 + y^2 - 6x - 2y + 6 = 0$.

10. Find an equation of the circle inscribed in the triangle with vertices $(5, 4)$, $(-15, -1)$, and $(23/3, -20/3)$.

11. Prove analytically that the perpendicular bisector of a chord of a circle contains the center.

12. Find an equation of the line tangent to $x^2 + y^2 = 25$ at $(4, -3)$.

5

CONIC SECTIONS

5.1

INTRODUCTION

In Chapter 3 we saw that an equation of the first degree always represents a line, and every line can always be represented by an equation of the first degree. Now let us consider second degree equations and their geometric representation. The general equation of the second degree has the form

$$Ax^2 + Bxy + Cy^2 + Dx + Ey + F = 0,$$

where A, B, and C are not all zero. We shall see that equations of the second degree represent (with two trivial exceptions) conic sections: that is, curves formed by the intersection of a plane with a right circular cone. Conversely, all conic sections are represented by second degree equations.

The conic sections are shown in Figure 5.1. Note that a cone has two portions, or *nappes,* separated from each other by the vertex. Note that a cone has no base or end; it extends infinitely far in both directions. Thus some of the conic sections are unbounded. The traditional conic sections are the parabola, ellipse, and hyperbola; a circle is a special case of the ellipse. The remaining situations are called *degenerate* conics. In addition, there are two other situations represented by second degree equations: a pair of parallel lines, and no graph at all. No matter what the cone or the plane, there must be some intersection; and it cannot be a pair of parallel lines.

The definitions that we use for the various conic sections give no hint of their connections with cones; this is done to simplify the presentation. A proof that parabolas, ellipses, and hyperbolas really are conic sections is given in Section 5.6. Let us now consider the individual conic sections.

5.2

THE PARABOLA

Definition *A **parabola** is the set of all points in a plane equidistant from a fixed point (**focus**) and a fixed line (**directrix**) not containing the focus.*

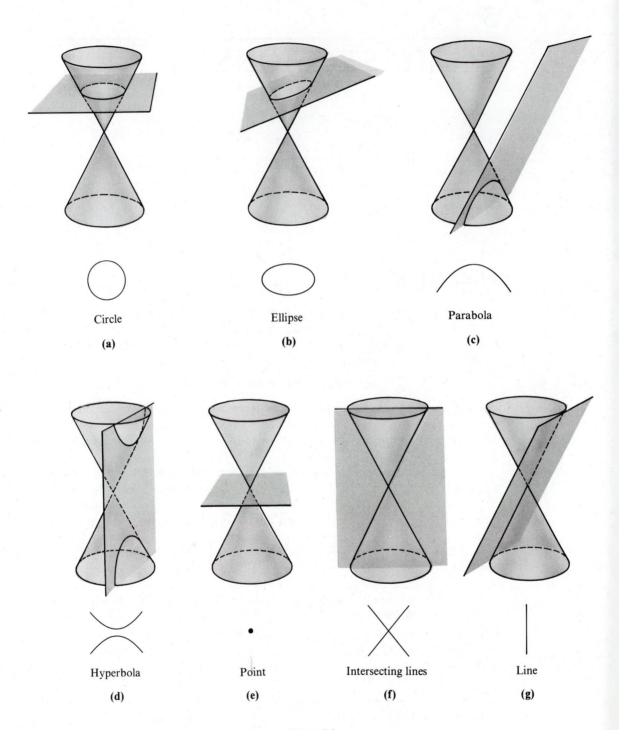

Figure 5.1

Suppose we choose the focus to be the point $(c, 0)$ and we choose the directrix to be $x = -c$, $c \neq 0$ (see Figure 5.2). Let us choose a point (x, y) on the parabola and see what condition must be satisfied by x and y. From the definition, we have

$$\overline{PF} = \overline{PD}$$
$$\sqrt{(x - c)^2 + y^2} = |x + c| \qquad \text{(See Note 1)}$$
$$(x - c)^2 + y^2 = (x + c)^2 \qquad \text{(See Note 2)}$$
$$x^2 - 2cx + c^2 + y^2 = x^2 + 2cx + c^2$$
$$y^2 = 4cx$$

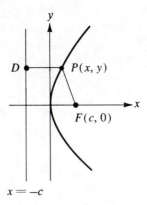

Figure 5.2

Note 1: Since $\overline{PD}$ is a horizontal distance,

$$\overline{PD} = |x - (-c)| = |x + c|.$$

You might feel that we should drop the absolute value signs, since it is clear from Figure 5.2 that $x + c$ must be positive. However, we did not insist that c be positive (although Figure 5.2 is given for a positive value of c). If c is negative, x is also negative and $x + c$ is negative.

Note 2: When we square both sides of an equation, there is a possibility of introducing extraneous roots. For instance, $(0, 1)$ is not a root of $x + y = x - y$, but it is a root of $(x + y)^2 = (x - y)^2$. The reason here is that $x + y = 1$, while $x - y = -1$ for $(0, 1)$, and $1^2 = (-1)^2 = 1$. This situation cannot occur when $(x + y)^2 = (x - y)^2$ and $x + y$ and $x - y$ are either both positive, both negative, or both zero. Since $\sqrt{(x - c)^2 + y^2}$ and $|x + c|$ must both be positive in any case, we have introduced no extraneous roots; that is, any point satisfying

$$(x - c)^2 + y^2 = (x + c)^2$$

must also satisfy

$$\sqrt{(x - c)^2 + y^2} = |x + c|.$$

We see that if a point is on the parabola with focus $(c, 0)$ and directrix $x = -c$, it must satisfy the equation $y^2 = 4cx$. Furthermore, since Note 2 indicates that all steps in the above argument are reversible, any point satisfying the equation $y^2 = 4cx$ is on the given parabola.

Theorem 5.1   *A point (x, y) is on the parabola with focus $(c, 0)$ and directrix $x = -c$ if and only if it satisfies the equation*

$$y^2 = 4cx.$$

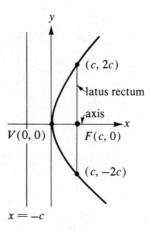

Figure 5.3

Let us observe some properties of this parabola before considering others. First of all, the x axis is a line of symmetry: that is, the portion below the x axis is the mirror image of the portion above. This line is called the **axis** of the parabola. It is perpendicular to the directrix and contains the focus (see Figure 5.3). The point of intersection of the axis and the parabola is the **vertex**. The vertex of the parabola $y^2 = 4cx$ is the origin. Finally, the line segment through the focus, perpendicular to the axis and having both ends on the parabola is the **latus rectum** (literally, straight side). Since the latus rectum of $y^2 = 4cx$ must be vertical and since it contains $(c, 0)$, the x coordinate of both ends is c. Substituting $x = c$ into $y^2 = 4cx$, we have

$$y^2 = 4c^2$$
$$y = \pm 2c.$$

Thus one end of the latus rectum is $(c, 2c)$ and the other $(c, -2c)$; its length is $4|c|$.

Finally the role of the x and y may be reversed throughout, as the next theorem states.

Theorem 5.2 *A point (x, y) is on the parabola with focus $(0, c)$ and directrix $y = -c$ if and only if it satisfies the equation*

$$x^2 = 4cy.$$

Example 1 Sketch and discuss $y^2 = 8x$.

Solution The equation is of the form

$$y^2 = 4cx,$$

with $c = 2$. Thus, it represents a parabola with vertex at the origin and axis on the x axis. The focus is at $(2, 0)$, and the directrix is $x = -2$. Finally, the length of the latus rectum is 8. This length may be used to determine the ends, $(2, \pm 4)$, of the latus rectum, which helps in sketching the curve (see Figure 5.4).

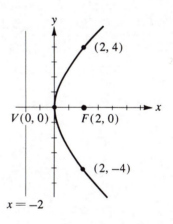

Figure 5.4

Example 2 Sketch and discuss $x^2 = -12y$.

Solution This equation is in the form $x^2 = 4cy$, with $c = -3$. Thus, it is a parabola with vertex at the origin and axis on the y axis. The focus is $(0, -3)$, the length of the latus rectum is 12, and the equation of the directrix is $y = 3$ (see Figure 5.5).

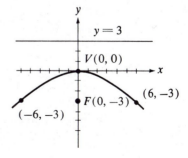

Figure 5.5

It might be observed from these two examples that the sign of c gives the direction in which the parabola opens. If c is positive, then the parabola opens in the positive direction (to the right or upward); if c is negative, then the parabola opens in the negative direction (to the left or downward).

Example 3   Find an equation(s) of the parabola(s) with vertex at the origin and focus $(-4, 0)$.

Solution   Since the focus and vertex are on the x axis, the x axis is the axis of the parabola (see Figure 5.6). Thus the equation is in the form $y^2 = 4cx$. Since the focus is $(-4, 0)$, $c = -4$ and the equation is $y^2 = -16x$.

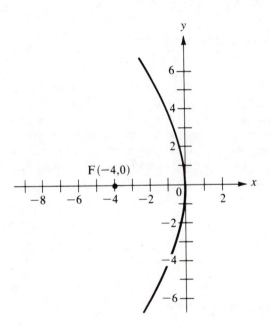

Figure 5.6

A frequently encountered problem is that of finding the tangent to a curve at a given point. This problem is easily solved by the use of calculus. We can solve this problem here without calculus by using a property of conic sections: namely, a tangent to a nondegenerate conic section has only one point in common with it. Let us use this property to solve the following problem.

Example 4   Find an equation of the line tangent to $y^2 = -8x$ at the point $(-2, 4)$ (see Figure 5.7).

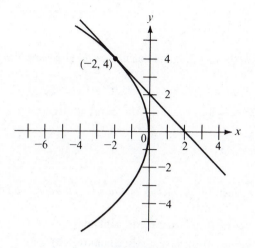

Figure 5.7

Solution The line with slope m through the point $(-2, 4)$ is

$$y - 4 = m(x + 2).$$

This is the desired line. Now all we need to do is determine m. Since the tangent line and the parabola have only one point in common, let us solve simultaneously. Solving the equation of the line for y and substituting into the equation of the parabola, we have

$$(4 + mx + 2m)^2 = -8x$$
$$16 + m^2x^2 + 4m^2 + 8mx + 16m + 4m^2x = -8x$$
$$m^2x^2 + (4m^2 + 8m + 8)x + (4m^2 + 16m + 16) = 0$$

This quadratic equation has only one solution provided $B^2 - 4AC = 0$, where A, B, and C are the coefficients of x^2 and x, and the constant term, respectively, of the equation.

$$(4m^2 + 8m + 8)^2 - 4m^2(4m^2 + 16m + 16) = 0$$
$$64m^2 + 128m + 64 = 0$$
$$64(m + 1)^2 = 0$$
$$m = -1$$

Thus the desired line is

$$y - 4 = -(x + 2)$$
$$x + y - 2 = 0.$$

It might be noted that there is one line through the point $(-2, 4)$ that has only $(-2, 4)$ in common with the parabola, and that line is the horizontal line $y = 4$. Thus it is not true that *every* line through $(-2, 4)$ having only one point in common with the parabola is a tangent line. See Problem 31 for more on this line.

PROBLEMS

A   *In Problems 1–12, sketch and discuss the given parabola.*

1. $y^2 = 16x$ **2.** $y^2 = -12x$ **3.** $x^2 = 4y$

4. $x^2 = -8y$ **5.** $y^2 = 10x$ **6.** $x^2 = -7y$

7. $x^2 = 5y$ **8.** $y^2 = -9x$ **9.** $x^2 = -2y$

10. $y^2 = 3x$ **11.** $x^2 = 6y$ **12.** $y^2 = -5x$

B   *In Problems 13–20, find an equation(s) of the parabola(s) described.*

13. Vertex: $(0,0)$; axis: x axis; contains $(1,5)$

14. Vertex: $(0,0)$; axis: y axis; contains $(1,5)$

15. Vertex: $(0,0)$; axis: x axis; length of latus rectum: 5

16. Vertex: $(0,0)$; focus: $(0,5)$

17. Focus: $(-3,0)$; directrix: $x = 3$

18. Focus: $(0,8)$; directrix: $y = -8$

19. Vertex: $(0,0)$; contains $(2,3)$ and $(-2,3)$

20. Vertex: $(0,0)$; contains $(-3,-4)$ and $(-3,4)$

In Problems 21–24, the required parabola is not in the standard position; so that Theorems 5.1 and 5.2 cannot be used. Instead, go back to the definition of a parabola.

21. Find an equation of the parabola with focus $(4,0)$ and directrix $x = 0$.

22. Find an equation of the parabola with focus $(2,4)$ and directrix $y = -2$.

23. Find an equation of the parabola with focus $(0,0)$ and directrix $x + y = 4$.

24. Find an equation of the parabola with focus $(1,1)$ and directrix $x + y = 0$.

25. Find an equation of the line tangent to $y = x^2$ at $(1,1)$.

26. Find an equation of the line tangent to $x^2 = -5y$ at $(5,-5)$.

27. Find an equation of the line tangent to $y^2 = -16x$ and parallel to $x + y = 1$.

28. Find equations of the lines tangent to $y^2 = 4x$ and containing $(-2,1)$.

C   **29.** Prove Theorem 5.2.

30. Prove that the ordinate of any point P of the parabola $y^2 = 4cx$ is the mean proportional between the length of the latus rectum and the abscissa of P.

31. Note that in Example 4 the horizontal line $y = 4$ has only one point in common with the given parabola. Why was it not found in Example 4? [*Hint:* An assumption was made at one point that prevented us from finding it.]

32. Show that the tangent line to $y^2 = 4cx$ at (x_0, y_0) is $yy_0 = 2c(x + x_0)$.

5.3

THE ELLIPSE

Definition   An *ellipse* is the set of all points (x, y) such that the sum of the distances from (x, y) to a pair of distinct fixed points (*foci*) is a fixed constant.

Let us choose the foci to be $(c, 0)$ and $(-c, 0)$ (see Figure 5.8) and let the fixed constant be $2a$. If (x, y) represents a point on the ellipse, we have the following.

$$\sqrt{(x - c)^2 + y^2} + \sqrt{(x + c)^2 + y^2} = 2a$$

$$\sqrt{(x - c)^2 + y^2} = 2a - \sqrt{(x + c)^2 + y^2}$$

$$x^2 - 2cx + c^2 + y^2 = 4a^2 - 4a\sqrt{(x + c)^2 + y^2} + x^2$$
$$+ 2cx + c^2 + y^2$$

$$4a\sqrt{(x + c)^2 + y^2} = 4a^2 + 4cx$$

$$\sqrt{(x + c)^2 + y^2} = a + \frac{cx}{a}$$

$$x^2 + 2cx + c^2 + y^2 = a^2 + 2cx + \frac{c^2 x^2}{a^2}$$

$$\frac{a^2 - c^2}{a^2} x^2 + y^2 = a^2 - c^2$$

$$\frac{x^2}{a^2} + \frac{y^2}{a^2 - c^2} = 1$$

The triangle of Figure 5.8, with vertices $(c, 0)$, $(-c, 0)$, and (x, y), has one side of length $2c$. The sum of the lengths of the other two sides is $2a$. Thus

$$2a > 2c$$

$$a > c$$

$$a^2 > c^2$$

$$a^2 - c^2 > 0$$

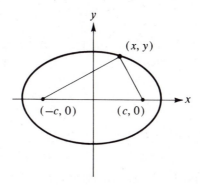

Figure 5.8

Since $a^2 - c^2$ is positive, we may replace it by another positive number, b^2. Thus

$$\frac{x^2}{a^2} + \frac{y^2}{b^2} = 1, \quad \text{where} \quad b^2 = a^2 - c^2.$$

Observe that we squared both sides of the equation at two of the steps. In both cases, both sides of the equation are nonnegative. Thus we have introduced no extraneous roots, and the steps may be reversed (see Note 2 on page 117).

Note that there are two axes of symmetry: the x axis and the y axis. Furthermore, $(\pm a, 0)$ are the x intercepts and $(0, \pm b)$ are the y intercepts, where $a > b$ (since $b^2 = a^2 - c^2$). Thus the x axis is called the **major axis** and the y axis is the **minor axis**. The points $(\pm a, 0)$ on the major axis are called the **vertices**, the points $(0, \pm b)$ on the minor axis are the **covertices**, and the point of intersection of the axes, $(0, 0)$, is called the **center** (see Figure 5.9). The **foci** $(\pm c, 0)$ are on the major axis.

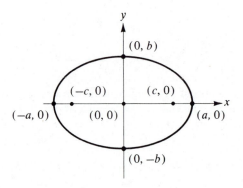

Figure 5.9

Theorem 5.3   *A point (x, y) is on the ellipse with vertices $(\pm a, 0)$ and foci $(\pm c, 0)$ if and only if it satisfies the equation*

$$\frac{x^2}{a^2} + \frac{y^2}{b^2} = 1,$$

where $b^2 = a^2 - c^2$.

An ellipse has two **latera recta** (plural of latus rectum), which are chords of the ellipse perpendicular to the major axis and containing the foci. If $x = \pm c$, then

$$\frac{c^2}{a^2} + \frac{y^2}{b^2} = 1$$

$$\frac{y^2}{b^2} = \frac{a^2 - c^2}{a^2} = \frac{b^2}{a^2}$$

$$y^2 = \frac{b^4}{a^2}$$

$$y = \pm \frac{b^2}{a}.$$

Thus, one latus rectum has endpoints $(c, \pm b^2/a)$, while the other has endpoints $(-c, \pm b^2/a)$. In both cases the length is $2b^2/a$. As with the parabola, this length may be used as an aid in sketching; however, the vertices and covertices allow one to make a reasonable sketch.

Again, the role of the x and y may be reversed.

Theorem 5.4   *A point (x, y) is on the ellipse with vertices $(0, \pm a)$ and foci $(0, \pm c)$ if and only if it satisfies the equation*

$$\frac{y^2}{a^2} + \frac{x^2}{b^2} = 1,$$

where $b^2 = a^2 - c^2$.

One question that immediately arises is how we can tell whether we have

$$\frac{x^2}{a^2} + \frac{y^2}{b^2} = 1 \qquad \text{or} \qquad \frac{y^2}{a^2} + \frac{x^2}{b^2} = 1.$$

The numbers in the denominator are not labeled a and b, so how do we know which is a and which is b? The answer is "size." In both cases, $a > b$. Thus the larger denominator is a^2, and the smaller is b^2.

Example 1   Sketch and discuss $9x^2 + 25y^2 = 225$.

Solution   First, we put the equation into standard form by dividing through by 225:

$$\frac{x^2}{25} + \frac{y^2}{9} = 1.$$

Now

$$a^2 = 25, \qquad b^2 = 9,$$

and

$$c^2 = a^2 - b^2 = 16.$$

This ellipse has center (0.0), vertices $(\pm 5, 0)$, covertices $(0, \pm 3)$, and foci $(\pm 4, 0)$. The latera recta have length $2b^2/a = 2 \cdot 9/5 = 3.6$ (see Figure 5.10).

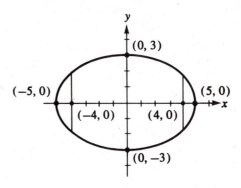

Figure 5.10

Example 2 Sketch and discuss $25x^2 + 16y^2 = 400$.

Solution Putting the equation into standard form gives

$$\frac{x^2}{16} + \frac{y^2}{25} = 1.$$

Now

$$a^2 = 25, \qquad b^2 = 16,$$

and

$$c^2 = a^2 - b^2 = 9.$$

This ellipse has center $(0,0)$, vertices $(0, \pm5)$, covertices $(\pm4, 0)$, and foci $(0, \pm3)$. The latera recta have length $2b^2/a = 2 \cdot 16/5 = 6.4$ (see Figure 5.11).

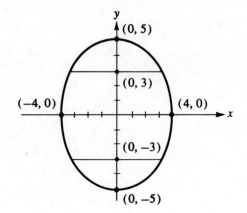

Figure 5.11

Example 3 Find an equation of the ellipse with vertices $(0, \pm8)$ and foci $(0, \pm5)$.

Solution Since the vertices are on the y axis (see Figure 5.12), we have the form

$$\frac{y^2}{a^2} + \frac{x^2}{b^2} = 1.$$

Furthermore, $a = 8$ and $c = 5$; thus $b^2 = a^2 - c^2 = 64 - 25 = 39$. The final result is

$$\frac{y^2}{64} + \frac{x^2}{39} = 1.$$

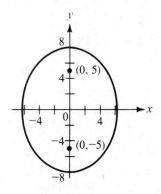

Figure 5.12

In addition to the quantities named, each ellipse is associated with a number, called the **eccentricity.** For any ellipse the eccentricity is

$$e = \frac{c}{a}.$$

The eccentricity of an ellipse satisfies the inequalities $0 < e < 1$. It gives a measure of the shape of the ellipse: the closer the eccentricity is to 0, the more nearly circular is the ellipse. For instance, in Example 1, $e = 4/5$, while in Example 2, $e = 3/5$. The ellipse of Example 2 is more nearly circular than the ellipse of Example 1, as can be easily seen by the sketches.

There is also a **directrix** associated with each focus of an ellipse. Associated with the focus $(c, 0)$ of the ellipse

$$\frac{x^2}{a^2} + \frac{y^2}{b^2} = 1$$

is the directrix

$$x = \frac{a}{e} = \frac{a^2}{c}.$$

If P is any point of the ellipse, the distance from P to the focus divided by the distance from P to the directrix is equal to the eccentricity. This is sometimes used as the definition of an ellipse.

Suppose we start with focus $(c, 0)$, directrix $x = a^2/c$, and eccentricity $e = c/a$. Now let us find the set of all points $P = (x, y)$ such that the distance from P to the focus divided by the distance from P to the directrix equals the eccentricity.

$$\frac{\sqrt{(x - c)^2 + y^2}}{\frac{a^2}{c} - x} = \frac{c}{a}$$

$$\sqrt{(x - c)^2 + y^2} = a - \frac{cx}{a}$$

$$x^2 - 2cx + c^2 + y^2 = a^2 - 2cx + \frac{c^2 x^2}{a^2}$$

$$\frac{a^2 - c^2}{a^2} x^2 + y^2 = a^2 - c^2$$

$$\frac{x^2}{a^2} + \frac{y^2}{a^2 - c^2} = 1$$

With $b^2 = a^2 - c^2$, this becomes

$$\frac{x^2}{a^2} + \frac{y^2}{b^2} = 1.$$

The same result can be obtained using focus $(-c, 0)$, directrix $x = -a^2/c$, and eccentricity $e = c/a$.

For a parabola, the distance from a point on the parabola to the focus divided by the distance of the point from the directrix is always 1. Thus we define $e = 1$ for every parabola.

In Section 5.1 we indicated that the circle is a special case of the ellipse. If $c = 0$, then $b = a$ in Theorems 5.3 and 5.4. Thus the equations in those theorems become $x^2 + y^2 = a^2$, an equation of a circle. With $c = 0$, the eccentricity is also zero, which agrees with our statement that an ellipse with an eccentricity near zero is nearly circular. Note however, that there is no directrix when $c = 0$; the focus-directrix definition cannot be used in this special case.

We can again find tangent lines by using the property that a tangent to an ellipse has only one point in common with the ellipse.

Example 4   Find equations of the lines containing $(5, 1)$ and tangent to $9x^2 + 25y^2 = 225$.

Solution   Note first of all that $(5, 1)$ does not satisfy the given equation; it is not on the ellipse (see Figure 5.13). Thus we can expect not one, but two tangent lines. A line through

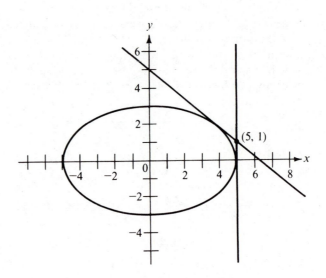

Figure 5.13

$(5, 1)$ with slope m has equation

$$y - 1 = m(x - 5).$$

Let us solve this equation and the given equation simultaneously. Solving this equation for y and substituting into the other, we have

$$9x^2 + 25(1 + mx - 5m)^2 = 225$$
$$9x^2 + 25(1 + m^2x^2 + 25m^2 + 2mx - 10m - 10m^2x) = 225$$
$$(25m^2 + 9)x^2 + (-250m^2 + 50m)x + (625m^2 - 250m - 200) = 0$$

Since a tangent line and the ellipse should have only one point in common, this quadratic equation should have only one solution. Thus

$$B^2 - 4AC = 0$$
$$(-250m^2 + 50m)^2 - 4(25m^2 + 9)(625m^2 - 250m - 200) = 0$$
$$62{,}500m^4 - 25{,}000m^3 + 2500m^2 - 62{,}500m^4$$
$$+ 25{,}000m^3 - 2500m^2 + 9000m + 7200 = 0$$
$$9000m + 7200 = 0$$
$$m = -\frac{4}{5}$$

Substituting this back into

$$y - 1 = m(x - 5),$$

we have

$$y - 1 = -\frac{4}{5}(x - 5)$$
$$4x + 5y - 25 = 0.$$

Recall that we had indicated that there are two answers, but we have found only one. This is because the use of the point-slope form for the tangent line assumes that there is a slope. It is easily seen from Figure 5.13 that the other tangent line is vertical; its equation is

$$x = 5.$$

PROBLEMS

A   *In Problems 1–10, sketch and discuss the given ellipse.*

1. $\dfrac{x^2}{169} + \dfrac{y^2}{25} = 1$

2. $\dfrac{x^2}{144} + \dfrac{y^2}{169} = 1$

3. $\dfrac{x^2}{25} + \dfrac{y^2}{4} = 1$

4. $\dfrac{x^2}{36} + \dfrac{y^2}{16} = 1$

5. $\dfrac{x^2}{25} + \dfrac{y^2}{49} = 1$

6. $x^2 + 4y^2 = 4$

7. $9x^2 + 4y^2 = 36$

8. $9x^2 + y^2 = 9$

9. $16x^2 + 9y^2 = 144$

10. $4x^2 + 25y^2 = 100$

B   *In Problems* 11–18, *find an equation(s) of the ellipse(s) described.*

11. Center: $(0, 0)$; vertex: $(0, 13)$; focus: $(0, -5)$

12. Center: $(0, 0)$; covertex: $(0, 5)$, focus: $(-12, 0)$

13. Center: $(0, 0)$; vertex: $(5, 0)$; contains $(\sqrt{15}, 2)$

14. Center: $(0, 0)$; axes on the coordinate axis; contains $(2, 2)$ and $(-4, 1)$

15. Vertices: $(\pm 6, 0)$; length of latus rectum: 3

16. Covertices: $(\pm 2, 0)$; length of latus rectum: 2

17. Foci: $(\pm 6, 0)$; $e = 3/5$

18. Foci: $(\pm 2, 0)$; directrices: $x = \pm 8$

19. The earth moves in an elliptical orbit about the sun, with the sun at one focus. The least and greatest distances of the earth from the sun are 91,446,000 miles and 94,560,000 miles, respectively. What is the eccentricity of the ellipse?

20. Find an equation of the line tangent to $x^2 + 4y^2 = 20$ at $(2, 2)$.

21. Find an equation of the line tangent to $2x^2 + 3y^2 = 11$ at $(2, 1)$.

22. Find an equation of the line containing $(3, -2)$ and tangent to $4x^2 + y^2 = 8$.

23. Find an equation of the line containing $(2, 4)$ and tangent to $3x^2 + 8y^2 = 84$.

C 24. Suppose, in Figure 5.14, that A and B are fixed pins on the arm ABP and that AP and BP have lengths a and b, respectively. Show that if A is free to slide in channel XX' and B in channel YY', the point P traces an ellipse.

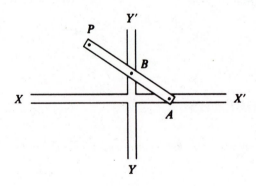

Figure 5.14

25. Given the focus $(-c, 0)$, directrix $x = -a^2/c$, and eccentricity $e = c/a$, show that these define the ellipse

$$\frac{x^2}{a^2} + \frac{y^2}{a^2 - c^2} = 1.$$

26. Prove Theorem 9.7.

27. Show that the line tangent to

$$\frac{x^2}{a^2} + \frac{y^2}{b^2} = 1$$

at (x_0, y_0) is

$$\frac{xx_0}{a^2} + \frac{yy_0}{b^2} = 1.$$

5.4

THE HYPERBOLA

Definition   *A hyperbola is the set of all points (x, y) in a plane such that the positive difference between the distances from (x, y) to a pair of distinct fixed points (foci) is a fixed constant.*

Again, let us choose the foci to be $(c, 0)$ and $(-c, 0)$ (see Figure 5.15) and choose the fixed constant to be $2a$. If (x, y) represents a point on the ellipse, we have the following.

$$\sqrt{(x - c)^2 + y^2} - \sqrt{(x + c)^2 + y^2} = \pm 2a$$
$$\sqrt{(x - c)^2 + y^2} = \sqrt{(x + c)^2 + y^2} \pm 2a$$
$$x^2 - 2cx + c^2 + y^2 = x^2 + 2cx + c^2 + y^2$$
$$\pm 4a \sqrt{(x + c)^2 + y^2} + 4a^2$$
$$\mp 4a \sqrt{(x + c)^2 + y^2} = 4a^2 + 4cx$$
$$\mp \sqrt{(x + c)^2 + y^2} = a + \frac{cx}{a}$$
$$x^2 + 2cx + c^2 + y^2 = a^2 + 2cx + \frac{c^2 x^2}{a^2}$$
$$\frac{c^2 - a^2}{a^2} x^2 - y^2 = c^2 - a^2$$
$$\frac{x^2}{a^2} - \frac{y^2}{c^2 - a^2} = 1$$

In the triangle PCC' of Figure 5.15,

$$\overline{PC'} < \overline{PC} + \overline{CC'}$$
$$\overline{PC'} - \overline{PC} < \overline{CC'}$$
$$2a < 2c$$
$$a < c$$
$$c^2 - a^2 > 0.$$

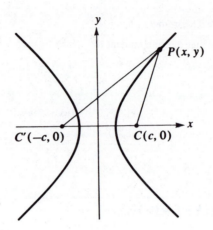

Figure 5.15

Since $c^2 - a^2$ is positive, we may replace it by another positive number, b^2. Thus

$$\frac{x^2}{a^2} - \frac{y^2}{b^2} = 1,$$

where $b^2 = c^2 - a^2$.

Again we squared both sides of the equation at two of the steps. The first time, both sides of the equation were positive; the second time, they were either both positive or both negative. Thus we have introduced no extraneous roots, and the steps may be reversed (see Note 2 on page 117).

Again, both the x axis and the y axis are axes of symmetry and again $(\pm a, 0)$ are the x intercepts. However, there are no y intercepts; when $x = 0$, we have

$$-\frac{y^2}{b^2} = 1,$$

which is not satisfied by any real number y. The x axis (containing two points of the hyperbola) is called the **transverse axis**; the y axis is called the **conjugate axis**. The points $(\pm a, 0)$ on the transverse axis are called the **vertices**, and the point of intersection of the axes, $(0, 0)$, is called the **center** (see Figure 5.16).

Theorem 5.5 *A point (x, y) is on the hyperbola with vertices $(\pm a, 0)$ and foci $(\pm c, 0)$ if and only if it satisfies the equation*

$$\frac{x^2}{a^2} - \frac{y^2}{b^2} = 1,$$

where $b^2 = c^2 - a^2$.

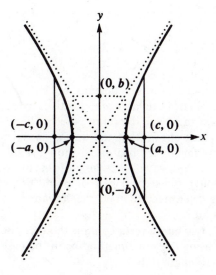

Figure 5.16

For every hyperbola there are two lines that the curve approaches more and more closely at its extremities. These two lines are called **asymptotes** (see Figure 5.16, in which the slanting dotted lines are the asymptotes). It might be noted that parabolas do not have asymptotes. Thus a hyperbola is not—as might appear from inaccurate diagrams—a pair of parabolas. The hyperbola

$$\frac{x^2}{a^2} - \frac{y^2}{b^2} = 1, \quad \text{or} \quad y = \pm \frac{b}{a} \sqrt{x^2 - a^2}$$

has asymptotes

$$y = \pm \frac{b}{a} x.$$

Let us verify this, at least for the portion in the first quadrant. Here we are dealing with

$$y = \frac{b}{a} \sqrt{x^2 - a^2} \quad \text{and} \quad y = \frac{b}{a} x$$

for positive values of x. For a given value of x, let us consider the difference d between the y coordinates of the points on the hyperbola and the line.

$$d = \frac{b}{a} x - \frac{b}{a} \sqrt{x^2 - a^2} = \frac{b}{a} (x - \sqrt{x^2 - a^2})$$

Multiplying numerator and denominator by $x + \sqrt{x^2 + a^2}$, we have

$$d = \frac{b}{a} \frac{x^2 - (x^2 - a^2)}{x + \sqrt{x^2 - a^2}} = \frac{ab}{x + \sqrt{x^2 - a^2}}.$$

Now the numerator is a constant; but, for large positive values of x, both terms of the denominator are large and positive. In fact, the larger the value of x, the larger the denominator and, therefore, the smaller d is. Thus d approaches zero as x gets larger, which shows that the line is an asymptote of the hyperbola. Of course, similar arguments can be used to show the same thing in the other three quadrants.

A convenient way of sketching the asymptotes is to plot both $(\pm a, 0)$ and $(0, \pm b)$ (even though the second pair of points is not on the hyperbola) and sketch the rectangle determined by them (see Figure 5.16). The diagonals of this rectangle are the asymptotes.

Again, two **latera recta** contain the foci and are perpendicular to the transverse axis. By using the same method as in the case of the parabola and ellipse, we can show their length to be

$$\frac{2b^2}{a}.$$

As with the parabola and the ellipse, the roles of x and y can be reversed.

Theorem 5.6   *A point (x, y) is on the hyperbola with vertices $(0, \pm a)$ and foci $(0, \pm c)$ if and only if it satisfies the equation*

$$\frac{y^2}{a^2} - \frac{x^2}{b^2} = 1,$$

where $b^2 = c^2 - a^2$.

It might be noted that a and b are determined by the sign of the term in which they appear; a^2 is always the denominator of the positive term and b^2 the denominator of the negative term. There is no requirement that a be greater than b, as there was for an ellipse.

The asymptotes of the hyperbola

$$\frac{y^2}{a^2} - \frac{x^2}{b^2} = 1$$

are

$$y = \pm \frac{a}{b} x.$$

Since the formulas for the asymptotes for the two cases are rather easy to confuse, a method that always works is to replace the 1 by 0 in the standard form and solve for y.

Example 1   Sketch and discuss $\dfrac{x^2}{9} - \dfrac{y^2}{16} = 1$.

Solution   We see that $a^2 = 9$, $b^2 = 16$, and $c^2 = a^2 + b^2 = 25$. This hyperbola has center $(0,0)$, vertices $(\pm 3, 0)$, and foci $(\pm 5, 0)$. The asymptotes are found by replacing the 1 of the standard form by 0 and solving for y.

$$\frac{x^2}{9} - \frac{y^2}{16} = 0$$

$$y^2 = \frac{16x^2}{9}$$

$$y = \pm \frac{4}{3} x$$

The length of the latera recta is $2b^2/a = 32/3$ (see Figure 5.17).

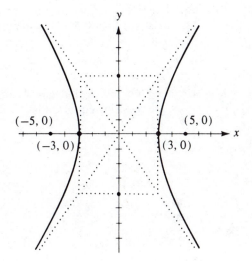

Figure 5.17

Example 2   Sketch and discuss $16x^2 - 9y^2 + 144 = 0$.

Solution   Putting this equation into standard form, we have

$$\frac{y^2}{16} - \frac{x^2}{9} = 1.$$

We see that $a^2 = 16$, $b^2 = 9$, and $c^2 = a^2 + b^2 = 25$. This hyperbola has center $(0,0)$, vertices $(0, \pm 4)$, and foci $(0, \pm 5)$. Its asymptotes are $y = \pm 4x/3$ and the length of the latera recta is $2b^2/a = 9/2$ (see Figure 5.18).

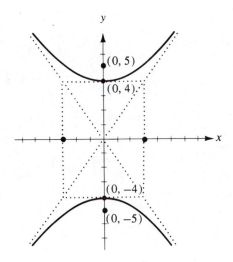

Figure 5.18

Note the relationship between the equations of these two examples when in the standard forms; the left-hand sides are simply opposite in sign. Such hyperbolas are called **conjugate hyperbolas.**

Example 3   Find an equation of the hyperbola with foci $(\pm 4, 0)$ and vertex $(2, 0)$.

Solution   Since the foci are on the transverse axis and we are given that they are on the x axis (see Figure 5.19), we must have the form

$$\frac{x^2}{a^2} - \frac{y^2}{b^2} = 1.$$

The foci tell us that $c = 4$, and the vertex gives $a = 2$; thus $b^2 = c^2 - a^2 = 12$. The resulting equation is

$$\frac{x^2}{4} - \frac{y^2}{12} = 1$$

or

$$3x^2 - y^2 = 12.$$

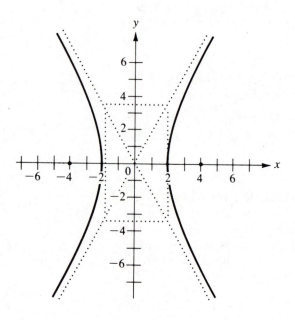

Figure 5.19

Hyperbolas, as well as the other conic sections, can be determined by a single focus, a directrix, and an eccentricity. For the hyperbola

$$\frac{x^2}{a^2} - \frac{y^2}{b^2} = 1,$$

we have **eccentricity**

$$e = \frac{c}{a}$$

and **directrices**

$$x = \pm \frac{a}{e} = \pm \frac{a^2}{c},$$

where $x = a^2/c$ is used in conjunction with the focus $(c, 0)$ and $x = -a^2/c$ with the focus $(-c, 0)$. Since $c > a$, $e = c/a > 1$. Furthermore, a single focus and directrix gives the entire hyperbola—not merely one branch. Either focus with its corresponding directrix generates a hyperbola.

PROBLEMS

A *In Problems 1–14, sketch and discuss each equation.*

1. $\dfrac{x^2}{16} - \dfrac{y^2}{9} = 1$

2. $\dfrac{x^2}{4} - \dfrac{y^2}{1} = 1$

3. $\dfrac{y^2}{9} - \dfrac{x^2}{4} = 1$

4. $\dfrac{y^2}{1} - \dfrac{x^2}{9} = 1$

5. $\dfrac{x^2}{144} - \dfrac{y^2}{25} = 1$

6. $\dfrac{y^2}{25} - \dfrac{x^2}{144} = 1$

7. $\dfrac{y^2}{25} - \dfrac{x^2}{9} = 1$

8. $4x^2 - 9y^2 = 36$

9. $4x^2 - y^2 = 4$

10. $4x^2 - y^2 + 16 = 0$

11. $x^2 - y^2 = 9$

12. $16x^2 - 9y^2 = -36$

13. $36y^2 - 100x^2 = 225$

14. $9x^2 - 4y^2 - 9 = 0$

B *In Problems 15–26, find an equation(s) of the hyperbola(s) described.*

15. Vertices: $(\pm 2, 0)$; focus: $(-4, 0)$

16. Foci: $(0, \pm 5)$; vertex: $(0, 2)$

17. Asymptotes: $y = \pm 2x/3$; vertex: $(6, 0)$

18. Asymptotes: $y = \pm 3x/4$; focus: $(0, -10)$

19. Asymptotes: $y = \pm 4x/3$; contains $(3\sqrt{2}, 4)$

20. Asymptotes: $y = \pm 3x/4$; length of latera recta: $9/2$

21. Vertices: $(\pm 5, 0)$; contains $(9/5, -4)$

22. Foci: $(\pm 2\sqrt{61}, 0)$; contains $(65/6, 5)$

23. Vertices: $(0, \pm 3)$; $e = 5/3$

24. Foci: $(\pm 10, 0)$; $e = 5/2$

25. Directrices: $x = \pm 9/5$; $e = 5/3$

26. Directrices: $y = \pm 25/13$; focus: $(0, -13)$

27. Find an equation of the line tangent to $16x^2 - 9y^2 = 144$ at $(13/4, 5/3)$.

28. Find an equation of the line tangent to $x^2 - y^2 = 16$ at $(-5, 3)$.

29. Find an equation(s) of the line(s) tangent to $x^2 - y^2 = 9$ and containing $(9, 9)$.

30. Find an equation(s) of the line(s) tangent to $4x^2 - 9y^2 = 7$ and containing $(-7, 7)$.

C **31.** Show that there is a number k such that, if P is any point of a hyperbola, the product of the distances of P from the asymptotes of the hyperbola is k.

32. An airplane sends out an impulse that travels at the speed of sound (1100 feet/second). Two receiving stations, whose positions are accurately known, record the times of reception of the impulse (they do not know the time the impulse was sent). How can this information be used to determine the position of the airplane, and to what extent can the position be determined? How many receiving stations are necessary to pinpoint the position of the airplane?

33. A man standing at a point $Q = (x, y)$ hears the crack of a rifle at point $P_1 = (1000, 0)$ and the sound of the bullet hitting the target $P_2 = (-1000, 0)$ at the same time. If the bullet travels at 2000 feet/second and sound travels at 1100 feet/second, find an equation relating x and y.

34. Show that the line tangent to

$$\frac{x^2}{a^2} - \frac{y^2}{b^2} = 1$$

at (x_0, y_0) is

$$\frac{xx_0}{a^2} - \frac{yy_0}{b^2} = 1.$$

5.5

REFLECTION PROPERTIES OF CONICS

The conic sections have several important properties. Among the most frequently encountered of these are the reflection properties. Let us consider them here.

When a ray of light strikes a plane reflecting surface (see Figure 5.20), the ray is reflected in such a way that the **angle of incidence** (the angle between the incident ray and the normal line to the mirror at the point of incidence) equals the **angle of reflection** (the angle between the reflected ray and the normal line). Alternatively, we may say that the incident and reflected rays make equal angles with the reflecting surface. If the reflecting surface is not a plane, the ray of light is reflected in the same way as it would be by a plane tangent to the given surface at the point of incidence. We shall limit our discussion to two dimensions by considering only cross sections, as illustrated in Figure 5.20.

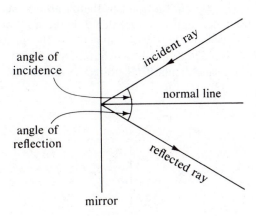

Figure 5.20

Now let us consider reflection by mirrors whose cross sections are conics. For the most part, we shall not use analytic arguments here because of their difficulty. This is due to the fact that we do not have a simple method of determining tangent lines to a curve; this is properly the subject matter of a course in calculus. Nevertheless, parabolic mirrors are considered analytically in Examples 1 and 2.

Let us now consider nonanalytic arguments establishing the reflection properties of conics. We begin with the parabola. Assuming that the parabolic mirror is two-sided, that is, it reflects light striking it from either side, there are four equivalent reflection properties.

Theorem 5.7   (a)   *Light from a source at the focus of the parabola is reflected along a line parallel to the axis.*
(b)   *Light towards the parabola on a line parallel to the axis and on the side of the parabola containing the focus is reflected towards the focus.*
(c)   *Light directed towards the parabola on a line parallel to the axis and on the side of the parabola not containing the focus is reflected away from the focus.*
(d)   *Light directed towards the focus of the parabola and on the side opposite the focus is reflected away from the parabola on a line parallel to the axis.*

Proof   We will justify only Statement (a). Suppose we have the parabola shown in Figure 5.21, with focus F and directrix DD'. If P is any point of the parabola, then $\overline{PF} = \overline{PD}$ (where PD is perpendicular to the directrix). Let QPQ' be the bisector of $\angle DPF$. We first show that this bisector is tangent to the parabola at P by showing that P is the only point that the line and the parabola have in common. Let Q be any point except P on this bisector. Since $\angle QPD = \angle QPF$, $\overline{PD} = \overline{PF}$, and $\overline{QP} = \overline{QP}$, it follows that triangles QPD and QPF are congruent. Hence $\overline{QF} = \overline{QD}$. If QD' is perpendicular to the directrix, then $\overline{QD'} < \overline{QD} = \overline{QF}$. Since $\overline{QD'} \neq \overline{QF}$, Q is not a point of the parabola. Recalling that Q is any point of the bisector other than P, we see that P is the only point that the parabola and QPQ' have in common; QPQ' is tangent to the parabola at P. But since $\angle QPE = \angle DPQ' = \angle FPQ'$, it follows that a ray of light FP is reflected along the line PE, which is perpendicular to the directrix and parallel to the axis of the parabola.

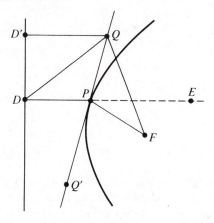

Figure 5.21

This reflection property of the parabola is very important in the reflection of both sound and light. It might be noted, however, that a small portion of a parabola can be closely approximated by a circular arc. Since circles are much easier to construct than parabolas, circular, rather than parabolic, reflectors are often used. Nevertheless, if the reflector is large or if a high degree of accuracy is needed, parabolic reflectors are used. Parabolic reflectors are used in the following situations in which the light source is at the focus: automobile headlights, floodlights, flashlights, and camera flash attachments. Bandshells and some auditoriums have a parabolic shape to reflect sound from the focus. Reflecting telescopes use parabolic mirrors to reflect parallel rays of light to the focus. While circular mirrors are used in inexpensive telescopes, the best reflecting telescopes use parabolic mirrors. The same principle is used in "spy" listening devices that eavesdrop on a conversation some distance away. In this case, the sound waves are reflected to a microphone at the focus. However most such devices use circular, rather than parabolic reflectors. Reflection properties (c) and (d) have no practical use.

The ellipse also has reflection properties that are quite interesting, although they have less practical value.

Theorem 5.8 *Light from one focus of an ellipse is reflected by the ellipse to the other focus. On the other hand, light from a source outside the ellipse and directed towards one focus is reflected away from the other.*

Proof In order to justify this, let us consider Figure 5.22. If P is any point of the ellipse, we extend F_1P to R such that $\overline{PR} = \overline{PF_2}$. Let QPQ' bisect $\measuredangle RPF_2$. It then follows that $\measuredangle Q'PF_1 + \measuredangle RPQ = \measuredangle QPF_2$. Thus the line QPQ' reflects light from

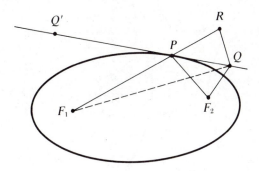

Figure 5.22

one focus to the other. All that we need to show is that QPQ' is tangent to the ellipse at P. We do this by showing that P is the only point that the line and the ellipse have in common. Let Q be any point other than P on this line. Since $\measuredangle RPQ = \measuredangle F_2PQ$, $\overline{PR} = \overline{PF_2}$, and $\overline{PQ} = \overline{PQ}$, it follows that triangles RPQ and F_2PQ are congruent. Thus $\overline{RQ} = \overline{F_2Q}$. Finally,

$$\begin{aligned}
\overline{F_1Q} + \overline{F_2Q} &= \overline{F_1Q} + \overline{RQ} \\
&> \overline{F_1R} \\
&= \overline{F_1P} + \overline{PR} \\
&= \overline{F_1P} + \overline{F_2P}.
\end{aligned}$$

From this, it follows that Q is not on the ellipse. Thus the ellipse and the line have only one point in common, and $Q'PQ$ is tangent to the ellipse.

Buildings having elliptical domes exhibit the reflection properties given above. In such a building, a person can stand at one focus and whisper to another person at the other focus without being overheard by people between them.

Although it is of no practical value, the hyperbola has reflective properties similar to those of the parabola and ellipse.

Theorem 5.9   *A ray of light from one focus is reflected away from the other by a hyperbola; a ray*
of light between the two branches of a hyperbola and directed towards one focus is
reflected towards the other.

Proof   Suppose that a ray of light from focus F_1 strikes the hyperbola at P (see Figure
5.23). Let $Q'PQ$ be the bisector of angle F_1PF_2 with Q any point on it distinct

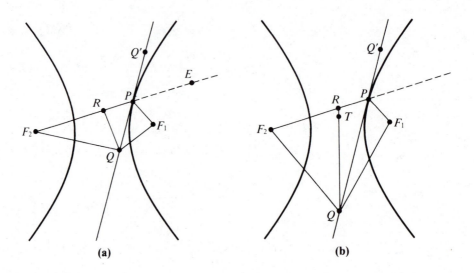

Figure 5.23

from P. Since $\angle Q'PE = \angle F_2PQ = \angle F_1PQ$, a ray of light F_1P is reflected
by $Q'PQ$ away from F_2. Again, all that we need to show is that $Q'PQ$ is tangent
to the hyperbola at P. Clearly $\overline{F_1P} < \overline{F_2P}$. Let R be the point between P and
F_2 such that $\overline{RP} = \overline{F_1P}$. This, together with $\angle RPQ = \angle F_1PQ$ and $\overline{PQ} = \overline{PQ}$,
assures us that triangles RPQ and F_1PQ are congruent. Thus $\overline{RQ} = \overline{F_1Q}$, and
$\overline{F_2P} - \overline{F_1P} = \overline{F_2P} - \overline{RP} = \overline{F_2R}$. Now we divide the argument into two
cases, depending upon the relative distances of Q from F_1 and F_2.

  Case I: $\overline{F_1Q} < \overline{F_2Q}$. Let S (see Figure 5.23a) be the point between F_2 and
Q such that $\overline{SQ} = \overline{F_1Q} = \overline{RQ}$. Then

$$\overline{F_2Q} - \overline{F_1Q} = \overline{F_2Q} - \overline{SQ} = \overline{F_2S}.$$

But

$$\overline{F_2S} + \overline{SQ} < \overline{F_2R} + \overline{RQ} = \overline{F_2R} + \overline{SQ},$$

giving

$$\overline{F_2S} < \overline{F_2R}.$$

Therefore

$$\overline{F_2Q} - \overline{F_1Q} = \overline{F_2S} < \overline{F_2R} = \overline{F_2P} - \overline{F_1P}.$$

It follows that, in this case, Q is not on the hyperbola; $Q'PQ$ is tangent at P.

Case II: $\overline{F_1Q} > \overline{F_2Q}$. $\overline{RQ} = \overline{F_1Q} > \overline{F_2Q}$. Let T (see Figure 5.23b) be the point between R and Q such that $\overline{F_2Q} = \overline{TQ}$. Then

$$\overline{F_1Q} - \overline{F_2Q} = \overline{RQ} - \overline{TQ} = \overline{RT}.$$

But

$$\overline{RT} + \overline{TQ} < \overline{RF_2} + \overline{F_2Q} = \overline{RF_2} + \overline{TQ}.$$

giving

$$\overline{RT} < \overline{TQ}.$$

Therefore

$$\overline{F_1Q} - \overline{F_2Q} = \overline{RT} < \overline{RF_2} = \overline{F_2P} - \overline{F_1P}.$$

Again it follows that Q is not on the hyperbola, and $Q'PQ$ is tangent at P.

The second case given in Theorem 5.9 results in repeated reflections as shown in Figure 5.24.

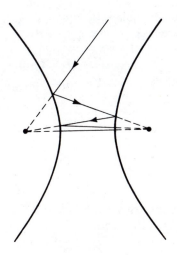

Figure 5.24

We have indicated that analytic arguments were not used because of the difficulty of determining the tangent to a conic at a given point. Although there is some difficulty, it is possible using the same criterion that we used in the above arguments: namely, a tangent to a conic has one and only one point in common with the conic.

Example 1   Show that the tangent to the parabola $y^2 = 4cx$ at the point (x_0, y_0) has slope $m = 2c/y_0$.

Solution   First of all, we note that since the point (x_0, y_0) is on the parabola, $y_0^2 = 4cx_0$ or $x_0 = y_0^2/4c$. The equation of a line through $(y_0^2/4c, y_0)$ is

$$y - y_0 = m\left(x - \frac{y_0^2}{4c}\right).$$

For the proper choice of m, this is the tangent line. If we solve this equation with the equation of the parabola, we get a single solution for this choice of m. Let us then solve these two equations simultaneously to determine what value of m gives a single solution. Solving the equation of the parabola for x and substituting into the equation of the line, we have the following.

$$y - y_0 = m\left(\frac{y^2}{4c} - \frac{y_0^2}{4c}\right)$$
$$4cy - 4cy_0 = my^2 - my_0^2$$
$$my^2 - 4cy + (4cy_0 - my_0^2) = 0$$

Now we have a quadratic equation in y. Since the equation $Ay^2 + By + C = 0$ has exactly one solution if $B^2 - 4AC = 0$, we have the following.

$$(-4c)^2 - 4m(4cy_0 - my_0^2) = 0$$
$$4c^2 - 4cmy_0 + m^2y_0^2 = 0$$
$$(2c - my_0)^2 = 0$$
$$my_0 = 2c$$
$$y_0 = \frac{2c}{m}$$

Once we have this slope, it is relatively easy to verify the reflection properties stated earlier in this section.

Example 2 Show that if $P = (x_0, y_0)$ is a point of the parabola $y^2 = 4cx$, then the line joining P to the focus and the horizontal line through P make equal angles with the tangent to the parabola at P.

Solution We have already seen in Example 1 that the slope of the tangent at $P(x_0, y_0)$ is

$$m_{tan} = \frac{2c}{y_0}.$$

Now we merely need to show that $\theta_1 = \theta_2$ in Figure 5.25. Equivalently, we may show that $\tan \theta_1 = \tan \theta_2$. But since θ_1 is the inclination of the tangent line,

$$\tan \theta_1 = m_{tan} = \frac{2c}{y_0}.$$

The slope of the line PF is $m_{PF} = y_0/(x_0 - c)$. Using the formula for the tangent of an angle from one line to another (see Section 1.6), we have

$$\tan \theta_2 = \frac{m_2 - m_1}{1 + m_1 m_2}$$

$$= \frac{\dfrac{y_0}{x_0 - c} - \dfrac{2c}{y_0}}{1 + \dfrac{y_0}{x_0 - c}\dfrac{2c}{y_0}}$$

$$= \frac{y_0^2 - 2cx_0 + 2c^2}{x_0 y_0 - cy_0 + 2cy_0}$$

$$= \frac{4cx_0 - 2cx_0 + 2c^2}{x_0 y_0 + cy_0}$$

$$= \frac{2c(x_0 + c)}{y_0(x_0 + c)}$$

$$= \frac{2c}{y_0}$$

This gives the desired result.

The conic sections also appear in other interesting situations. For example, a projectile thrown upward follows a nearly parabolic path, the deviation from parabolic caused mostly by air resistance. Planets follow elliptical paths around the sun, as do satellites (real or artificial) about their planets. These ellipses

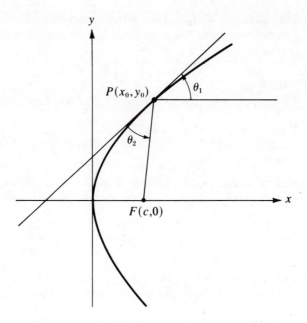

Figure 5.25

have very small eccentricities, making them nearly circular. On the other hand, comets follow orbits that are either parabolas or ellipses with eccentricities near one. All of these things can be shown by an analysis of the forces involved. Such analyses require the use of calculus and are beyond the scope of this book.

PROBLEMS

B   **1.** Show that the slope of the tangent to

$$\frac{x^2}{a^2} + \frac{y^2}{b^2} = 1$$

at (x_0, y_0) is

$$m = -\frac{b^2 x_0}{a^2 y_0}.$$

2. Show that the slope of the tangent to

$$\frac{x^2}{a^2} - \frac{y^2}{b^2} = 1$$

at (x_0, y_0) is

$$m = \frac{b^2 x_0}{a^2 y_0}.$$

3. Show that the line tangent to

$$\frac{x^2}{a^2} + \frac{y^2}{b^2} = 1$$

at (x_0, y_0) is

$$\frac{x x_0}{a^2} + \frac{y y_0}{b^2} = 1.$$

4. Show that the line tangent to

$$\frac{x^2}{a^2} - \frac{y^2}{b^2} = 1$$

at (x_0, y_0) is

$$\frac{x x_0}{a^2} - \frac{y y_0}{b^2} = 1.$$

C **5.** Use the result of Problem 1 to show that the lines from the point $P = (x_0, y_0)$ of an ellipse to its two foci make equal angles with the tangent at P.

6. Use the result of Problem 2 to show that the lines from the point $P = (x_0, y_0)$ of a hyperbola to its two foci make equal angles with the tangent at P.

5.6

CONICS AND A RIGHT CIRCULAR CONE

As we indicated in Section 5.1, the conic sections are curves formed by the intersection of a plane with a right circular cone. We shall prove here that this is really the case. Let us first recall that a cone consists of two portions, or nappes, separated from each other by the vertex.

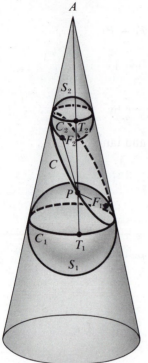

Figure 5.26   **Figure 5.27**

Suppose a plane intersects a right circular cone to give the curve C of Figure 5.26. Let S_1 and S_2 be spheres which are inscribed in the cone and tangent to the plane at the points F_1 and F_2, respectively. (We are considering the case in which there are two distinct tangent spheres which lie in the same nappe of the cone and are tangent at two distinct points. See Problems 1–3 for other cases.) The spheres S_1 and S_2 have the circles C_1 and C_2, respectively, in common with the cone. Of course C_1 and C_2 lie in planes which are perpendicular to the axis of the cone and, therefore, parallel to each other. Let P be any point of the curve C. Consider the line determined by P and the apex A of the cone. This line intersects the circles C_1 and C_2 at points T_1 and T_2, respectively. Since PT_1 and PF_1 are two tangents to the sphere S_1 from P,

$$\overline{PT_1} = \overline{PF_1}.$$

Similarly,

$$\overline{PT_2} = \overline{PF_2}.$$

Thus

$$\overline{PF_1} + \overline{PF_2} = \overline{PT_1} + \overline{PT_2} = \overline{T_1T_2}.$$

Since the circles C_1 and C_2 lie in parallel planes, $\overline{T_1T_2}$ is independent of the choice of the point P. Thus $\overline{PF_1} + \overline{PF_2}$ is constant, proving that C is an ellipse with foci F_1 and F_2.

The conic can also be related to the cone by the focus-directrix property. Suppose a plane x intersects a right circular cone to give the curve C of Figure 5.27. Let S be a sphere inscribed in the cone and tangent to the given plane at F. The sphere S has a circle C_1 in common with the cone, C_1 lying in a plane y, perpendicular to the axis of the cone. Planes x and y intersect at a line d and form the angle α. Let P be any point of C. The line determined by P and the apex A of the cone intersects the circle C_1 at T. Now we drop a perpendicular from P to the plane y, intersecting it at R. Finally RQ is taken perpendicular to d, the line of intersection of planes x and y. Let us consider the triangles PQR and PRT. Angle PQR is α and we shall designate the angle PTR by β. Thus

$$\sin \alpha = \frac{\overline{PR}}{\overline{PQ}}$$

and

$$\sin \beta = \frac{\overline{PR}}{\overline{PT}},$$

giving

$$\frac{\sin \alpha}{\sin \beta} = \frac{\overline{PT}}{\overline{PQ}}.$$

Again, since PF and PT are tangent to the sphere S from P,

$$\overline{PT} = \overline{PF},$$

giving

$$\frac{\sin \alpha}{\sin \beta} = \frac{\overline{PF}}{\overline{PQ}}.$$

Of course α and β are angles determined by the cone and the intersecting plane x; they are independent of the choice of the point P on the curve. Thus C is a conic with focus F, directrix d and eccentricity

$$e = \frac{\sin \alpha}{\sin \beta}.$$

Furthermore, if $0 < \alpha < \beta$, then $e < 1$ and the curve C is an ellipse; if $\alpha = \beta$, then $e = 1$ and the curve is a parabola; and if $\alpha > \beta$, then $e > 1$ and the curve is a hyperbola.

PROBLEMS

C 1. Show that if there is only one sphere inscribed in a cone and tangent to the intersecting plane, then the conic section is a parabola.

2. Show that if there are two spheres inscribed in a cone and tangent to the intersecting plane, and if these two spheres are in different nappes of the cone, then the conic section is a hyperbola.

3. Show that if there are two spheres inscribed in a cone and tangent to the intersecting plane at the same point, then the conic section is a circle.

4. Show that the intersection of a plane with a right circular cylinder is either a circle, an ellipse, or a pair of parallel lines.

REVIEW PROBLEMS

A *In Problems 1–6, sketch and discuss.*

    1. $y^2 = -6x$ **2.** $x^2 - 4y^2 = 4$

    3. $y^2 = 16x$ **4.** $4x^2 + 9y^2 = 36$

    5. $x^2 - y^2 + 9 = 0$ **6.** $4x^2 + 3y^2 = 48$

B 7. Find an equation(s) of the parabola(s) with vertex $(0, 0)$ and focus $(5, 0)$.

8. Find an equation(s) of the parabola(s) with vertex $(0, 0)$ and containing the points $(2, 4)$ and $(8, 8)$.

9. Find an equation(s) of the ellipse(s) with center $(0, 0)$, vertex $(10, 0)$, and focus $(6, 0)$.

10. Find an equation(s) of the ellipse(s) with center $(0, 0)$, vertex $(0, -3\sqrt{2})$, and containing the point $(2, -3)$.

11. Find an equation(s) of the hyperbola(s) with vertices $(0, \pm 8)$ and focus $(0, 10)$.

12. Find an equation(s) of the hyperbola(s) with asymptotes $y = \pm 3x/2\sqrt{2}$ and containing $(4, 3)$.

13. Find an equation of the line tangent to $4x^2 + 3y^2 = 16$ at $(1, 2)$.

C 14. The inner edge of a track has equation

$$\frac{x^2}{a^2} + \frac{y^2}{b^2} = 1.$$

The track is of width d. When asked to find an equation of the outer edge, a student gave the answer

$$\frac{x^2}{(a + d)^2} + \frac{y^3}{(b + d)^2} = 1.$$

Prove the student to be correct or incorrect.

6
TRANSFORMATION OF COORDINATES

6.1

TRANSLATION OF CONIC SECTIONS

The coordinate axes are something of an artificiality which we introduced on the plane in order to represent points and curves algebraically. Since the axes are of this nature, their placement is quite arbitrary. Thus we might prefer to move them in order to simplify some equation. Any change in the position of the axes may be represented by a combination of a **translation** and a **rotation**. A translation of the axes gives a new set of axes parallel to the old ones (see Figure 6.1a), while in a rotation, the axes are rotated about the origin (see Figure 6.1b).

Let us consider translation first. If the axes are translated in such a way that the origin of the new coordinate system is the point (h, k) of the old system (see Figure 6.2), then every point has two representations: (x, y) in the old coordinate

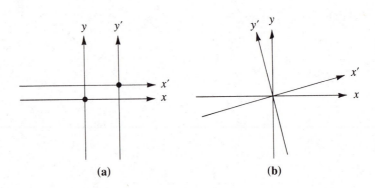

(a)   (b)

Figure 6.1

system, and (x', y') in the new. The relationship between the old and new co-ordinate system is easily seen from Figure 6.2 to be

$$x = x' + h \qquad \text{or} \qquad x' = x - h$$

and

$$y = y' + k \qquad \text{or} \qquad y' = y - k.$$

These equations (either set) are called **equations of translation.**

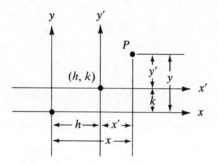

Figure 6.2

Furthermore, we can consider translation from a vector point of view. The vector $x\mathbf{i} + y\mathbf{j}$ can be used to represent the point (x, y). A graphical representation of this vector is a directed line segment with its tail at the origin and its head at (x, y). It is easily seen from Figure 6.3 that $\mathbf{u} = \mathbf{v} + \mathbf{w}$ or

$$x\mathbf{i} + y\mathbf{j} = (x' + h)\mathbf{i} + (y' + k)\mathbf{j}.$$

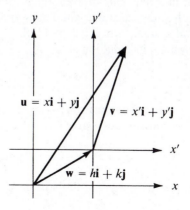

Figure 6.3

Thus

$$x = x' + h$$

and

$$y = y' + k.$$

Suppose we have a parabola with vertex at (h, k) and axis $y = k$ (see Figure 6.4).

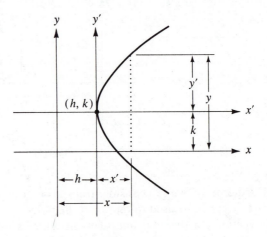

Figure 6.4

Let us put in a new pair of axes, the x' and y' axes, which are parallel to and in the same directions as the original axes and have their origin at the point (h, k) of the original system. Since the parabola's vertex is now at the origin of this coordinate system, its equation is

$$y'^2 = 4cx',$$

where $|c|$ is the distance from vertex to focus. Now the relationship between the old and new coordinates is given by the equations of translation

$$x' = x - h \quad \text{and} \quad y' = y - k.$$

Thus the equation of the parabola in the original coordinate system is

$$(y - k)^2 = 4c(x - h).$$

We can go through the same analysis for all three of our conics and obtain the following results.

Theorem 6.1 *A point (x, y) is on the parabola with focus $(h + c, k)$ and directrix $x = h - c$ if and only if it satisfies the equation*

$$(y - k)^2 = 4c(x - h).$$

A point (x, y) is on the parabola with focus $(h, k + c)$ and directrix $y = k - c$ if and only if it satisfies the equation

$$(x - h)^2 = 4c(y - k).$$

Theorem 6.2 *A point (x, y) is on the ellipse with center (h, k), vertices $(h \pm a, k)$, and covertices $(h, k \pm b)$ if and only if it satisfies the equation*

$$\frac{(x - h)^2}{a^2} + \frac{(y - k)^2}{b^2} = 1.$$

The foci are $(h \pm c, k)$, where $c^2 = a^2 - b^2$.
A point (x, y) is on the ellipse with center (h, k), vertices $(h, k \pm a)$, and covertices $(h \pm b, k)$ if and only if it satisfies the equation

$$\frac{(y - k)^2}{a^2} + \frac{(x - h)^2}{b^2} = 1.$$

The foci are $(h, k \pm c)$, where $c^2 = a^2 - b^2$.

Theorem 6.3 *A point (x, y) is on the hyperbola with center (h, k), vertices $(h \pm a, k)$, and foci $(h \pm c, k)$ if and only if it satisfies the equation*

$$\frac{(x - h)^2}{a^2} - \frac{(y - k)^2}{b^2} = 1,$$

where $b^2 = c^2 - a^2$.
A point (x, y) is on the hyperbola with center (h, k), vertices $(h, k \pm a)$, and foci $(h, k \pm c)$, if and only if it satisfies the equation

$$\frac{(y - k)^2}{a^2} - \frac{(x - h)^2}{b^2} = 1,$$

where $b^2 = c^2 - a^2$.

Suppose we now consider the equation

$$(y - k)^2 = 4c(x - h)$$

and carry out the indicated multiplications. The result is

$$y^2 - 2ky + k^2 = 4cx - 4ch$$
$$y^2 - 4cx - 2ky + (k^2 + 4ch) = 0.$$

This is in the form

$$y^2 + D'x + E'y + F' = 0,$$

where $D' = -4c$, $E' = -2k$, and $F' = k^2 + 4ch$. If we now multiply through by some number C ($C \neq 0$), we have

$$Cy^2 + Dx + Ey + F = 0.$$

We can consider the other five standard forms of Theorems 6.1–6.3 in the same way. We find that in every case we get an equation of the form

$$Ax^2 + Cy^2 + Dx + Ey + F = 0.$$

If we have a parabola, then either $A = 0$ or $C = 0$; if we have an ellipse, then A and C are both positive or both negative; if we have a hyperbola, then A and C have opposite signs. This is summarized in the following theorem.

Theorem 6.4   *Every conic with axis (or axes) parallel to or on a coordinate axis (or axes) may be represented by an equation of the form*

$$Ax^2 + Cy^2 + Dx + Ey + F = 0,$$

where A and C are not both zero. Furthermore, $AC = 0$ if the conic is a parabola, $AC > 0$ if it is an ellipse, and $AC < 0$ if it is a hyperbola.

The equations of Theorems 6.1–6.3 are called the **standard forms for conics**, while the equation of Theorem 6.4 is called the **general form**. It is a simple matter to go from the standard form to the general form; one merely carries out the indicated multiplications. We go from the general form to the standard form by completing the square, just as we did for a circle. Once we have the conic in standard form, it is a simple matter to carry out a translation.

Example 1 Sketch and discuss the parabola
$9y^2 + 36x - 6y - 23 = 0$.

Solution First we put the equation into standard form by completing the square on the y terms. To do this we first isolate the y terms on one side of the equation. Then we make the coefficient on y^2 one by dividing by 9. Finally we divide the resulting coefficient of y by 2 and square; this number is then added to both sides, making one side a perfect square.

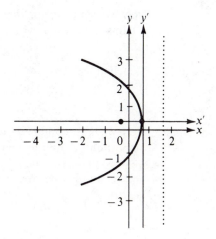

Figure 6.5

$$9y^2 - 6y = -36x + 23$$

$$y^2 - \frac{2}{3}y = -4x + \frac{23}{9}$$

$$y^2 - \frac{2}{3}y + \frac{1}{9} = -4x + \frac{24}{9}$$

$$\left(y - \frac{1}{3}\right)^2 = -4\left(x - \frac{2}{3}\right)$$

The equations of translation,

$$x' = x - \frac{2}{3} \quad \text{and} \quad y' = y - \frac{1}{3},$$

now give

$$y'^2 = -4x'.$$

In the $x'y'$ system, we have a parabola with vertex at the origin, axis the x' axis ($y' = 0$), focus at $(-1,0)$, directrix $x' = 1$, and a latus rectum of length 4. By using the equations of translation, we express the vertex, axis, and so on, in the xy coordinate system. Thus we have a parabola with vertex at $(2/3, 1/3)$, axis $y = 1/3$, focus at $(-1/3, 1/3)$, directrix $x = 5/3$, and a latus rectum of length 4. The parabola, with both sets of coordinate axes, is shown in Figure 6.5.

Note that the process of completing the square here is exactly the same one that we used in Section 4.1. If the coefficient on the square term is not one, we make it one by division. Then we complete the square by dividing the coefficient of the first degree term by two and squaring. Since this method of completing the square works only when the coefficient on the square term is one, the first step is essential. This is easily carried out when the x^2 and y^2 terms both have the same coefficient or when one of them is zero.

If they are nonzero and different, we cannot make them both one by dividing. In that case, instead of dividing, we factor. For example,

$$9x^2 + 18x = 9(x^2 + 2x).$$

Now we can complete the square inside the parentheses. This can be repeated on the y terms in the same way. This is illustrated in the following example.

Example 2   Sketch and discuss the hyperbola

$$9x^2 - 4y^2 - 18x - 24y - 63 = 0.$$

Solution   Here we must complete the square on both the x and the y terms. Thus we must make the coefficients of both x^2 and y^2 one. We do this by factoring.

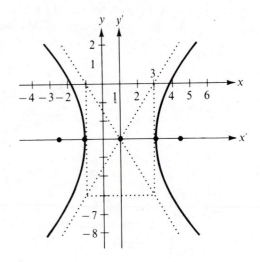

$$9x^2 - 18x - 4y^2 - 24y = 63$$

$$9(x^2 - 2x) - 4(y^2 + 6y) = 63$$

$$9(x^2 - 2x + 1) - 4(y^2 + 6y + 9)$$
$$= 63 + 9 \cdot 1 - 4 \cdot 9$$

$$9(x - 1)^2 - 4(y + 3)^2 = 36$$

$$\frac{(x - 1)^2}{4} - \frac{(y + 3)^2}{9} = 1$$

Now the equations of translation,

$$x' = x - 1$$

and

$$y' = y + 3,$$

Figure 6.6

give

$$\frac{x'^2}{4} - \frac{y'^2}{9} = 1.$$

In the $x'y'$ system, we have a hyperbola with center at $(0, 0)$, $a = 2$, $b = 3$, and $c^2 = a^2 + b^2 = 13$. Thus we have vertices $(\pm 2, 0)$, foci $(\pm \sqrt{13}, 0)$, asymptotes $y' = \pm 3x'/2$, and latera recta of length $2b^2/a = 2 \cdot 9/2 = 9$. Using the equations of translation, we see that in the xy system we have a hyperbola with center at $(1, -3)$, vertices $(3, -3)$ and $(-1, -3)$, foci $(1 \pm \sqrt{13}, -3)$, asymptotes $y + 3 = \pm 3(x - 1)/2$, or $3x - 2y - 9 = 0$ and $3x + 2y + 3 = 0$. The length of the latera recta is, of course, unchanged by the translation. The hyperbola with both coordinate systems is given in Figure 6.6.

While Theorem 6.4 tells us that any conic with axis (or axes) on or parallel to the coordinate axes can be represented in the form

$$Ax^2 + Cy^2 + Dx + Ey + F = 0,$$

it does not follow that any equation in this form represents a parabola, ellipse, or hyperbola. This point is demonstrated by the following example.

Example 3   Sketch and discuss $4x^2 + 3y^2 - 16x + 18y + 43 = 0$.

Solution   Since $A = 4$ and $C = 3$, $AC = 12 > 0$. The equation appears to represent an ellipse. But completing the square gives

$$4(x^2 - 4x) + 3(y^2 + 6y) = -43$$
$$4(x^2 - 4x + 4) + 3(y^2 + 6y + 9) = -43 + 4 \cdot 4 + 3 \cdot 9$$
$$4(x - 2)^2 + 3(y + 3)^2 = 0.$$

Since neither term on the left is negative, the sum can be zero only if both terms are zero. Thus $x = 2$ and $y = -3$. This equation is satisfied only by the point $(2, -3)$ as shown in Figure 6.7.

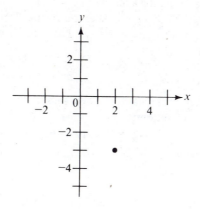

Figure 6.7

The foregoing example is called a **degenerate case of an ellipse,** since the equation has the form of an ellipse but does not actually represent an ellipse. It is com-

Conic	AC	Degenerate cases
Parabola	0	One line (two coincident lines) Two parallel lines* No graph*
Ellipse	+	Circle Point No graph*
Hyperbola	–	Two intersecting lines

parable to the degenerate cases of a circle, which we saw in Chapter 4. The degenerate cases of the three conics are given in the table.

In the previous examples we have been working from the equation of a conic section to its graph. Let us now consider the reverse problem; that is, starting with a description of a conic section, we shall find an equation for it.

Example 4 Find an equation(s) of the ellipse(s) with axes parallel to (or on) the coordinate axes and with vertex (3, 5) and covertex (1, 0).

Solution Since the axes are parallel to or on the coordinate axes and the vertex and covertex are the ends of the axes, we must have one of the situations shown in Figure 6.8. But (3, 5) and (1, 0) are given to be a vertex and covertex, respectively. This means that (3, 5) is an end of a major axis, and (1, 0) an end of a minor axis. Thus Figure 6.8a is the correct one. Now we can take the result directly from Figure 6.8a. Since the major axis is parallel to the y axis, we have the form

$$\frac{(y - k)^2}{a^2} + \frac{(x - h)^2}{b^2} = 1.$$

Furthermore, it is clear from the figure that the center is $(3, 0)$, $a = 5$, and $b = 2$. Thus the desired equation is

$$\frac{(y - 0)^2}{25} + \frac{(x - 3)^2}{4} = 1$$

or

$$25x^2 + 4y^2 - 150x + 125 = 0.$$

Note that there is no need to actually carry out the translation in the above example; the standard form of Theorem 6.2 gives us all we need.

---

*This is one of the cases mentioned in Section 5.1 in which an equation of the second degree does not represent a curve formed by the intersection of a plane with a right circular cone.

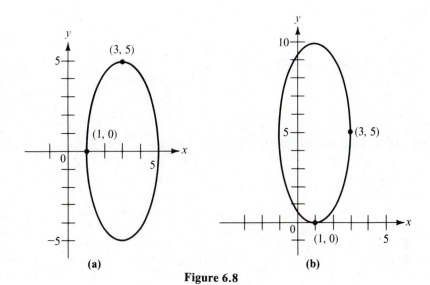

(a)

(b)

Figure 6.8

Example 5 Find an equation(s) of the hyperbola(s) containing $(7, -2)$, $(-1, -2)$, $(8, 1)$, and $(-2, -5)$.

Solution It is easier in this case to use the general form, $Ax^2 + Cy^2 + Dx + Ey + F = 0$, of Theorem 6.4 rather than one of the standard forms of Theorem 6.3. Since the four given points satisfy the equation of the conic section, we have the following four equations.

$$49A + 4C + 7D - 2E + F = 0$$
$$A + 4C - D - 2E + F = 0$$
$$64A + C + 8D + E + F = 0$$
$$4A + 25C - 2D - 5E + F = 0$$

Since there are four equations but five unknowns, we cannot solve for all of them; but let us solve for four of the unknowns in terms of the fifth. First let us subtract the second equation from the other three, eliminating F.

$$48A + 8D = 0$$
$$63A - 3C + 9D + 3E = 0$$
$$3A + 21C - D - 3E = 0$$

Adding the last pair, we get:

$$48A + 8D = 0$$
$$66A + 18C + 8D = 0$$

Finally we subtract the first from the second.

$$18A + 18C = 0$$

or

$$C = -A$$

Substituting back, we get:

$$D = -6A$$
$$E = -4A$$
$$F = -11A.$$

Thus the desired equation is

$$Ax^2 - Ay^2 - 6Ax - 4Ay - 11A = 0$$

or

$$x^2 - y^2 - 6x - 4y - 11 = 0.$$

PROBLEMS

In Problems 1–20, translate so that the center or vertex of the conic section is at the origin of the new coordinate system. Sketch the curve showing both the old and new axes.

A **1.** $(x - 3)^2 = 8(y - 2)$ **2.** $16(x - 3)^2 + 9(y + 1)^2 = 144$

3. $16(x + 3)^2 + 25y^2 = 400$ **4.** $(y - 4)^2 = 2x$

5. $9(x + 2)^2 + 4(y + 1)^2 = 36$ **6.** $9(x + 2)^2 - 16(y - 1)^2 = -144$

B **7.** $y^2 - 4x - 2y + 9 = 0$ **8.** $x^2 - 8x - 8y + 8 = 0$

9. $4x^2 + y^2 + 24x - 2y + 21 = 0$ **10.** $x^2 + 9y^2 - 10x + 36y + 52 = 0$

11. $9x^2 - 4y^2 + 90x + 32y + 125 = 0$ **12.** $9x^2 - 16y^2 + 72x + 96y + 144 = 0$

13. $9x^2 + 4y^2 - 72x + 16y + 160 = 0$ **14.** $4x^2 - y^2 - 40x + 6y + 91 = 0$

15. $4x^2 - 4x - 4y - 5 = 0$ **16.** $16x^2 + 36y^2 + 48x - 180y + 257 = 0$

17. $4x^2 - 16y^2 + 12x + 16y + 69 = 0$ **18.** $y^2 - 5y + 6 = 0$

19. $25x^2 + 4y^2 - 150x + 40y + 350 = 0$ **20.** $4x^2 - y^2 - 2x - y = 0$

In Problems 21–30, find an equation(s) of the conic section(s) described.

21. Parabola with focus $(3, 5)$ and directrix $x = -1$

22. Ellipse with vertices $(-1, 8)$ and $(-1, -2)$, containing $(1, 0)$

23. Hyperbola with vertices $(4, 1)$ and $(0, 1)$, and focus $(6, 1)$

24. Parabola with axis parallel to the y axis and containing $(0, 6)$, $(3, -6)$, and $(8, 14)$

25. Hyperbola with vertex $(6, -1)$ and asymptotes $3x - 2y - 6 = 0$ and $3x + 2y - 2 = 0$

26. Ellipse with covertices $(-5, 0)$ and $(1, 0)$ and having latera recta of length $9/2$

27. Ellipse with axes parallel to the coordinate axes and containing $(6, -1)$, $(-4, -5)$, $(6, -5)$, and $(-12, -3)$

28. Hyperbola with axes parallel to the coordinate axes and containing $(2, -2)$, $(-3, 8)$, $(-1, -1)$, and $(2, 8)$

29. Hyperbola with foci $(4, 0)$ and $(-6, 0)$ and eccentricity $5/2$

30. Ellipse with vertex $(8, -1)$, focus $(6, -1)$ and eccentricity $3/5$

C 31. If a parabola with a vertical axis contains the points (x_0, y_0), (x_1, y_1), and (x_2, y_2), show that its equation can be put into the form

$$\begin{vmatrix} x^2 & x & y & 1 \\ x_0^2 & x_0 & y_0 & 1 \\ x_1^2 & x_1 & y_1 & 1 \\ x_2^2 & x_2 & y_2 & 1 \end{vmatrix} = 0.$$

32. What happens to the determinant of Problem 31 if the three given points are collinear? [*Hint:* See Problem 25, page 108.]

6.2

TRANSLATION OF GENERAL EQUATIONS

The method of completing the square is simple to use, but it is rather limited in scope. It can be used only on second-degree equations with no xy term. If there is an xy term or if the equation is not of the second degree, another method, illustrated by the following examples can be used.

Example 1 Translate axes so that the constant and the x term of $x^2 - 2xy + y^2 + 4x - 6y + 10 = 0$ are eliminated.

Solution Since we have an xy term, completing the square will never give us one of the forms of Theorems 4.1–4.3. Thus we cannot determine h and k before translating. Therefore we use the equations of translation,

$$x = x' + h \qquad \text{and} \qquad y = y' + k,$$

and see what values of h and k are needed to eliminate the specified terms. Substituting the equations of translation into the given equation, we have

$$(x' + h)^2 - 2(x' + h)(y' + k) + (y' + k)^2 + 4(x' + h) - 6(y' + k) + 10 = 0$$
$$x'^2 + 2hx' + h^2 - 2x'y' - 2kx' - 2hy' - 2hk$$
$$+ y'^2 + 2ky' + k^2 + 4x' + 4h - 6y' - 6k + 11 = 0$$
$$x'^2 - 2x'y' + y'^2 + (2h - 2k + 4)x'$$
$$+ (-2h + 2k - 6)y' + (h^2 - 2hk + k^2 + 4h - 6k + 10) = 0$$

Since we want the constant term and the coefficient of x' to be zero, we must choose h and k so that

$$h^2 - 2hk + k^2 + 4h - 6k + 10 = 0$$
$$2h - 2k + 4 = 0.$$

Solving the second equation for h in terms of k and substituting into the first, we have

$$h = k - 2$$
$$(k - 2)^2 - 2(k - 2)k + k^2 + 4(k - 2) - 6k + 10 = 0$$
$$k^2 - 4k + 4 - 2k^2 + 4k + k^2 + 4k - 8 - 6k + 10 = 0$$
$$-2k + 6 = 0$$
$$k = 3$$
$$h = k - 2 = 1$$

Thus the equations of translation are

$$x = x' + 1 \qquad \text{and} \qquad y = y' + 3,$$

and the result in the new coordinate system is

$$x'^2 - 2x'y' + y'^2 - 2y' = 0.$$

The graph of the given equation (or equivalently of the new equation) showing both sets of coordinate axes is given in Figure 6.9.

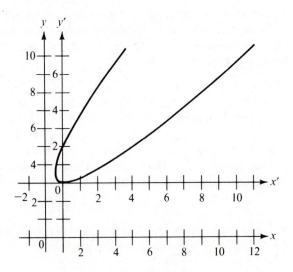

Figure 6.9

Example 2   Translate axes to eliminate the first degree terms of $x^2 - 4xy + 3x - 2y + 4 = 0$.

Solution   Substituting the equations of translation,

$$x = x' + h \quad \text{and} \quad y = y' + k,$$

into the given equation, we have

$$(x' + h)^2 - 4(x' + h)(y' + k) + 3(x' + h) - 2(y' + k) + 4 = 0$$
$$x'^2 + 2hx' + h^2 - 4x'y' - 4kx' - 4hy' - 4hk + 3x' + 3h - 2y' - 2k + 4 = 0$$
$$x'^2 - 4x'y' + (2h - 4k + 3)x' + (-4h - 2)y' + (h^2 - 4hk + 3h - 2k + 4) = 0.$$

Setting the coefficients of x' and y' equal to zero and solving, we have

$$2h - 4k + 3 = 0$$
$$-4h - 2 = 0$$

giving

$$h = -\frac{1}{2} \quad \text{and} \quad k = \frac{1}{2}.$$

Thus the equations of translation are

$$x = x' - \frac{1}{2} \quad \text{and} \quad y = y' + \frac{1}{2}$$

and the equation in the new coordinate system is

$$x'^2 - 4x'y' + \frac{11}{4} = 0$$

or

$$4x'^2 - 16x'y' + 11 = 0.$$

The result is shown graphically in Figure 6.10.

Both of these examples carried out translations on a general second degree equation

$$Ax^2 + Bxy + Cy^2 + Dx + Ey + F = 0.$$

Several things might be noted here. Except for multiplying through by a constant, as we did in Example 2 to eliminate a fraction, the translations did not change the values of A, B, and C. A, B, and C are said to be *invariant* under a translation. Of the remaining three coefficients, we may dictate the values of only two of them. This does not mean that we may dictate the values of *any* two of the three; we could not have eliminated both first-degree terms in Example 1.

It might also be noted that, when $B \neq 0$, we may *not* use AC to determine the type of conic section we have. In Example 1 it was seen that the graph is a parabola although AC is positive (not zero). In Example 2 we had a hyperbola although $AC = 0$. Thus the table on page 160 is only valid when $B = 0$.

The method that we have used here is more generally applicable—its use is not restricted to second degree equations.

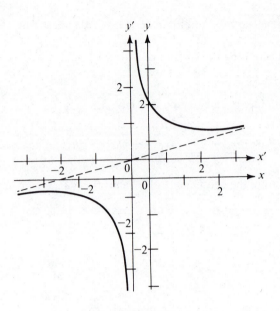

Figure 6.10

Example 3   Translate axes so that the constant and the x term of $y = x^3 - 5x^2 + 7x - 5$ are eliminated.

Solution   Since we do not know what values of h and k to choose, we simply use the equations of translation,

$$x = x' + h \qquad \text{and} \qquad y = y' + k,$$

and see what values of h and k are needed to eliminate the terms specified.

$$y' + k = (x' + h)^3 - 5(x' + h)^2 + 7(x' + h) - 5$$
$$y' = x'^3 + (3h - 5)x'^2 + (3h^2 - 10h + 7)x' + (h^3 - 5h^2 + 7h - 5 - k)$$

Now we must choose h and k so that

$$3h^2 - 10h + 7 = 0$$
$$h^3 - 5h^2 + 7h - 5 - k = 0.$$

The first of these two equations gives

$$h = 1 \qquad \text{or} \qquad h = 7/3.$$

Substituting these values into the second, we have

$$k = -2 \qquad \text{or} \qquad k = -86/27.$$

Using $h = 1$ and $k = -2$, we get

$$y' = x'^3 - 2x'^2.$$

Using $h = 7/3$ and $k = -86/27$, we get

$$y' = x'^3 + 2x'^2.$$

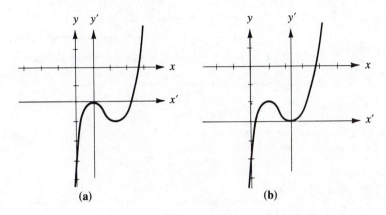

Figure 6.11

The graphs of both the cases described in Example 3 are given in Figure 6.11. While there are two different translations giving two different equations, both graphs are the same when referred to the original xy system. As an added bonus, this method has located the local maximum, $(1, -2)$, and minimum, $(7/3, -86/27)$. This method can also be used on second-degree equations with no xy term, but completing the square is so much simpler that most would prefer to use it.

PROBLEMS

B   *In Problems 1–14, translate to eliminate the terms indicated.*

1. $x^2 - 2xy + 4y^2 + 8x - 26y + 38 = 0$; first-degree terms
2. $2x^2 - xy - y^2 + 5x - 8y - 3 = 0$; first-degree terms
3. $x^2 + 4xy - y^2 - 2x - 14y - 3 = 0$; first-degree terms
4. $3x^2 + xy + y^2 - 16x - 10y + 30 = 0$; first-degree terms
5. $xy - 5x + 4y - 4 = 0$; first-degree terms
6. $x^2 + xy + 9x + 5y + 20 = 0$; first-degree terms
7. $y = x^3 - 6x^2 + 11x - 8$; constant, x^2 term

8. $y = x^3 - 3x + 6$; constant, x term

9. $y = x^3 - 3x^2 + 3x + 5$; constant, x term

10. $y = x^3 - 9x^2 + 24x + 3$; constant, x term

11. $y = x^4 - 8x^3 + 24x^2 - 28x + 7$; constant, x term

12. $y = x^4 - 10x^3 + 37x^2 - 120x + 138$; constant, x term

13. $x^2y - 2x^2 + 2xy + y - 4x - 6 = 0$; first-degree terms

14. $x^2y + x^2 + 2xy + x + y - 1 = 0$; second-degree terms

C   **15.** Prove that any translation on
$$Ax^2 + Bxy + Cy^2 + Dx + Ey + F = 0$$
leaves A, B, and C invariant.

16. Prove that any translation on $y = P(x)$, where $P(x)$ is a polynomial in x, leaves invariant the coefficient of the highest degree term of $P(x)$.

6.3

ROTATION

The second transformation of the axes that we wish to consider is a rotation of the axes about the origin (see Figure 6.1b). If the axes are rotated through an angle θ, then every point of the plane has two representations: (x, y) in the original coordinate system and (x', y') in the new coordinate system. Alternatively, every vector **v** in the plane has two representations: $\mathbf{v} = x\mathbf{i} + y\mathbf{j}$ in the original coordinate system and $\mathbf{v} = x'\mathbf{i}' + y'\mathbf{j}'$ in the new coordinate system (see Figure 6.12). In order

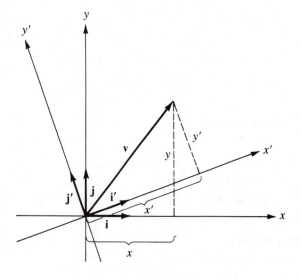

Figure 6.12

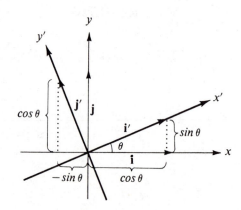

Figure 6.13

to find the relationships between the x and y of one coordinate system and the x' and y' of the other, let us consider the relationships of $\mathbf{i}$ and $\mathbf{j}$ with $\mathbf{i}'$ and $\mathbf{j}'$. Remembering that $\mathbf{i}$, $\mathbf{j}$, $\mathbf{i}'$, and $\mathbf{j}'$ are all unit vectors, we see from Figure 6.13 that

$$\mathbf{i}' = \cos\theta\mathbf{i} + \sin\theta\mathbf{j}$$
$$\mathbf{j}' = -\sin\theta\mathbf{i} + \cos\theta\mathbf{j}.$$

Thus

$$\mathbf{v} = x'\mathbf{i}' + y'\mathbf{j}'$$
$$= x'(\cos\theta\mathbf{i} + \sin\theta\mathbf{j}) + y'(-\sin\theta\mathbf{i} + \cos\theta\mathbf{j})$$
$$= (x'\cos\theta - y'\sin\theta)\mathbf{i} + (x'\sin\theta + y'\cos\theta)\mathbf{j}.$$

Since $\mathbf{v} = x\mathbf{i} + y\mathbf{j}$, we have

$$x = x'\cos\theta - y'\sin\theta$$

and

$$y = x'\sin\theta + y'\cos\theta,$$

called the **equations of rotation.**

An alternative method of finding the above equations of rotation without the use of vectors is the following. Recalling again that every point P of the plane has two representations, (x, y) in the original coordinate system and (x', y') in the new system, we see from Figure 6.14 that

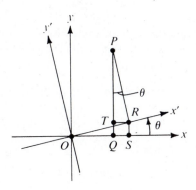

Figure 6.14

$$x = \overline{OQ}, \qquad x' = \overline{OR},$$
$$y = \overline{PQ}, \qquad y' = \overline{PR}.$$

Let us now consider the relations between x and y, and x' and y'. Noting first of all that

$$\sphericalangle ROQ = \sphericalangle RPQ = \theta,$$

we find from triangle ORS

$$\sin \theta = \frac{\overline{RS}}{\overline{OR}}, \qquad \cos \theta = \frac{\overline{OS}}{\overline{OR}},$$
$$\overline{RS} = \overline{OR} \sin \theta \qquad \overline{OS} = \overline{OR} \cos \theta$$
$$= x' \sin \theta, \qquad\qquad = x' \cos \theta,$$

and from triangle PRT

$$\sin \theta = \frac{\overline{TR}}{\overline{PR}}, \qquad \cos \theta = \frac{\overline{PT}}{\overline{PR}},$$
$$\overline{TR} = \overline{PR} \sin \theta \qquad \overline{PT} = \overline{PR} \cos \theta$$
$$= y' \sin \theta, \qquad\qquad = y' \cos \theta.$$

Now

$$x = \overline{OQ} \qquad\qquad y = \overline{PQ}$$
$$= \overline{OS} - \overline{QS} \qquad = \overline{TQ} + \overline{PT}$$
$$= \overline{OS} - \overline{TR} \qquad = \overline{RS} + \overline{PT}$$
$$= x' \cos \theta - y' \sin \theta; \qquad = x' \sin \theta + y' \cos \theta.$$

Thus we have the equations of rotation

$$x = x' \cos \theta - y' \sin \theta,$$
$$y = x' \sin \theta + y' \cos \theta.$$

Example 1   Find the new representation of

$$x^2 - xy + y^2 - 2 = 0$$

after rotating through an angle of 45°. Sketch the curve, showing both the old and new coordinate systems.

Solution   Since $\sin 45° = \cos 45° = 1/\sqrt{2}$, the equations of rotation are

$$x = \frac{x' - y'}{\sqrt{2}} \quad \text{and} \quad y = \frac{x' + y'}{\sqrt{2}}.$$

Substituting into the original equation, we have

$$\frac{(x' - y')^2}{2} - \frac{x' - y'}{\sqrt{2}} \cdot \frac{x' + y'}{\sqrt{2}} + \frac{(x' + y')^2}{2} - 2 = 0$$

$$\frac{x'^2 - 2x'y' + y'^2 - x'^2 + y'^2 + x'^2 + 2x'y' + y'^2}{2}$$

$$= 2$$

$$x'^2 + 3y'^2 = 4.$$

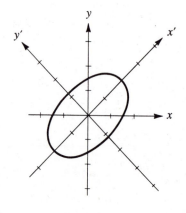

Figure 6.15

Figure 6.15 shows the final result.

Example 2   Find a new representation of $x^2 + 4xy - 2y^2 - 6 = 0$ after rotating through an angle $\theta = \text{Arctan } 1/2$. Sketch the curve, showing both the old and new coordinate systems.

Solution   Figure 6.16 shows that

$$\sin \theta = \frac{1}{\sqrt{5}} \quad \text{and} \quad \cos \theta = \frac{2}{\sqrt{5}},$$

giving equations of rotation

$$x = \frac{2x' - y'}{\sqrt{5}} \quad \text{and} \quad y = \frac{x' + 2y'}{\sqrt{5}}.$$

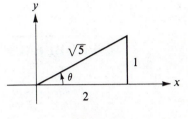

Figure 6.16

Substituting into the original equation, we have

$$\frac{(2x' - y')^2}{5} + 4\,\frac{2x' - y'}{\sqrt{5}}\cdot\frac{x' + 2y'}{\sqrt{5}} - 2\,\frac{(x' + 2y')^2}{5} - 6 = 0,$$

$$\frac{4x'^2 - 4x'y' + y'^2 + 8x'^2 + 12x'y' - 8y'^2 - 2x'^2 - 8x'y' - 8y'^2}{5} = 6,$$

$$2x'^2 - 3y'^2 = 6.$$

Figure 6.17 shows the final result. Note that Figure 6.16 can be used to determine the position of the new coordinate axes.

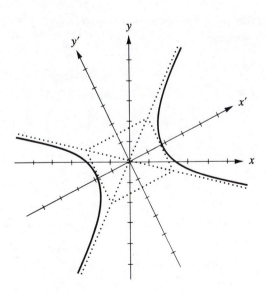

Figure 6.17

In both of these examples, we have seen that the given rotation has eliminated the xy term. Of course, not every rotation will do so—it must be specially chosen. We shall see in the next section how to choose θ to eliminate the xy term.

PROBLEMS

In Problems 1–10, find a new representation of the given equation after rotating through the given angle. Sketch the curve, showing both the old and new coordinate systems.

A   **1.** $2x + 3y = 6;\quad \theta = \text{Arctan } 3/2$ **2.** $3x - y = 5;\quad \theta = \text{Arctan } 3$

   3. $xy = 4;\quad \theta = 45°$ **4.** $2x^2 - xy + 2y^2 - 15 = 0;\quad \theta = 45°$

B   **5.** $x^2 - 2xy + y^2 + x + y = 0;\quad \theta = 45°$

   6. $31x^2 + 10\sqrt{3}xy + 21y^2 - 144 = 0;\quad \theta = 30°$

   7. $x^2 + 2\sqrt{3}xy + 3y^2 + 8\sqrt{3}x - 8y = 0;\quad \theta = 60°$

   8. $11x^2 - 50\sqrt{3}xy - 39y^2 + 576 = 0;\quad \theta = 60°$

   9. $8x^2 + 5xy - 4y^2 - 4 = 0;\quad \theta = \text{Arctan } 1/5$

   10. $6x^2 - 5xy - 6y^2 + 26 = 0;\quad \theta = \text{Arctan }(-1/5)$

In Problems 11–16, find a new representation of the given equation after rotating through the given angle.

   11. $3x^2 - 3xy - y^2 + 4 = 0;\quad \theta = \text{Arctan }(-1/2)$

   12. $4x^2 + 3xy - 5 = 0;\quad \theta = \text{Arctan } 1/2$

   13. $4x^2 + 3xy - 5 = 0;\quad \theta = 45°$

   14. $x^2 - 3xy + y^2 + 5 = 0;\quad \theta = 30°$

   15. $3x^2 - 3xy - y^2 + 4 = 0;\quad \theta = 60°$

   16. $x^2 - 5xy + 2 = 0;\quad \theta = \text{Arctan } 1/5$

C   **17.** Show that $x^2 + y^2 = 25$ is invariant under rotation through any angle.

   18. Show that a second form for the equations of rotation is

$$x' = x \cos \theta + y \sin \theta$$

and

$$y' = -x \sin \theta + y \cos \theta.$$

6.4

THE GENERAL EQUATION OF SECOND DEGREE

We have seen that any conic section with axes parallel to the coordinate axes can be represented by a second-degree equation with $B = 0$; furthermore, any second-degree equation with $B = 0$ represents a conic or degenerate conic with axes parallel to the coordinate axes. We now extend this concept to conic sections in any position. It is an easy matter to see that any conic can be represented by a second-degree equation, starting from our standard forms and translating and rotating.

Suppose, given a second-degree equation with $B \neq 0$, we rotate axes through an angle θ. If our assumption that this equation represents a conic or degenerate conic is correct, then a rotation of axes through some positive angle less than 90° should give us a conic with axis (or axes) on or parallel to the coordinate axes. Thus such a rotation should eliminate the xy term, and we shall assume throughout this discussion that $0° < \theta < 90°$ and

$$Ax^2 + Bxy + Cy^2 + Dx + Ey + F = 0.$$

Substituting the equations of rotation we have

$$A(x' \cos \theta - y' \sin \theta)^2 + B(x' \cos \theta - y' \sin \theta)(x' \sin \theta + y' \cos \theta)$$
$$+ C(x' \sin \theta + y' \cos \theta)^2 + D(x' \cos \theta - y' \sin \theta)$$
$$+ E(x' \sin \theta + y' \cos \theta) + F = 0.$$

After carrying out the multiplication and combining similar terms, we find that the coefficient of $x'y'$ is

$$(C - A)2 \sin \theta \cos \theta + B(\cos^2 \theta - \sin^2 \theta) = (C - A) \sin 2\theta + B \cos 2\theta.$$

We want this coefficient to be zero for the proper choice of θ. Let us set it equal to zero and see what θ should be.

$$(C - A) \sin 2\theta + B \cos 2\theta = 0$$

At this point we divide the argument into two cases.

Case I: If $A = C$, then

$$B \cos 2\theta = 0$$
$$\cos 2\theta = 0$$
$$2\theta = 90°$$
$$\theta = 45°.$$

Case II: If $A \neq C$, then

$$(A - C) \sin 2\theta = B \cos 2\theta$$

$$(A - C) \frac{\sin 2\theta}{\cos 2\theta} = B$$

$$(A - C) \tan 2\theta = B$$

$$(A - C) \frac{2 \tan \theta}{1 - \tan^2 \theta} = B \quad \text{(by a trigonometric identity)}$$

$$2(A - C) \tan \theta = B - B \tan^2 \theta$$

$$B \tan^2 \theta + 2(A - C) \tan \theta - B = 0$$

This is a quadratic equation in $\tan \theta$. Solving by the quadratic formula, we have

$$\tan \theta = \frac{2(C - A) \pm \sqrt{4(A - C)^2 + 4B^2}}{2B}$$

$$= \frac{(C - A) \pm \sqrt{(C - A)^2 + B^2}}{B}$$

Now we have two values of $\tan \theta$. Which one do we want? It is not difficult to see (refer to Problem 17) that the values of θ that we get from them must differ by an odd multiple of 90° and that the two values of θ have opposite signs. Thus either value should eliminate the xy term. Since we are assuming that $0° < \theta < 90°$, we want the positive value of $\tan \theta$. Since

$$\sqrt{(C - A)^2 + B^2} > \sqrt{(C - A)^2} = |C - A|$$

(that is to say, the radical in the numerator is always numerically larger than $C - A$), it follows that the sign of the numerator corresponds to the sign on the radical. By taking the sign on the radical to agree with the sign of B, we can be sure that the result is always positive. Once we have $\tan \theta$ it is a simple matter to find $\sin \theta$ and $\cos \theta$ and substitute them into the equations of rotation. Thus we are always able to rotate axes to eliminate the xy term. The resulting equation must then represent a conic or degenerate conic.

Theorem 6.5   *Any conic section can be represented by the second-degree equation*

$$Ax^2 + Bxy + Cy^2 + Dx + Ey + F = 0$$

where A, B, and C are not all zero. Any second-degree equation represents either a conic or a degenerate conic.

Let us sum up the results of the previous discussion. If $B \neq 0$, then the axes may be rotated to eliminate the xy term as follows: If $A = C$, then $\theta = 45°$. If $A \neq C$, then

$$\tan \theta = \frac{(C - A) \pm \sqrt{(C - A)^2 + B^2}}{B} ,$$

where the sign on the radical is taken to agree with the sign of B.

Example 1 Rotate axes to eliminate the xy term of $x^2 + 4xy - 2y^2 - 6 = 0$. Sketch, showing both sets of axes.

Solution

$$\tan \theta = \frac{(C - A) \pm \sqrt{(C - A)^2 + B^2}}{B}$$

$$= \frac{(-2 - 1) + \sqrt{(-2 - 1)^2 + 4^2}}{4}$$

$$= \frac{-3 + \sqrt{9 + 16}}{4}$$

$$= \frac{1}{2}$$

Using the triangle of Figure 6.18, we have

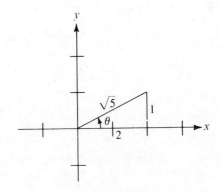

Figure 6.18

$$\sin \theta = \frac{1}{\sqrt{5}} \qquad \text{and} \qquad \cos \theta = \frac{2}{\sqrt{5}}.$$

Thus the equations of rotation are

$$x = \frac{2x' - y'}{\sqrt{5}} \qquad \text{and} \qquad y = \frac{x' + 2y'}{\sqrt{5}}$$

Substituting into the original equation (see Example 2 of the previous section), we have

$$2x'^2 - 3y'^2 = 6.$$

The sketch is given in Figure 6.17 on page 171.

Example 2   Rotate axes to eliminate xy term of

$$2x^2 - xy + 2y^2 - 2 = 0.$$

Sketch, showing both sets of axes.

Solution   Since $A = C, \theta = 45°$ and the equations of rotation are

$$x = \frac{x' - y'}{\sqrt{2}} \quad \text{and} \quad y = \frac{x' + y'}{\sqrt{2}}.$$

Substituting these into the original equation, we have the following.

$$2\frac{(x'-y')^2}{2} - \frac{x'-y'}{\sqrt{2}}\frac{x'+y'}{\sqrt{2}} + 2\frac{(x'+y')^2}{2} - 2 = 0$$

$$\frac{2x'^2 - 4x'y' + 2y'^2 - x' + y' + 2x'^2 + 4x'y' + 2y'^2}{2} = 2$$

$$3x'^2 + 5y'^2 = 4$$

$$\frac{x'^2}{4/3} + \frac{y'^2}{4/5} = 1$$

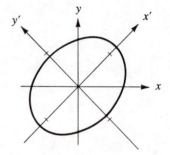

Figure 6.19

The sketch is given in Figure 6.19.

There is a method of determining which conic we have before we rotate the axes. It is based on the fact that certain expressions are invariant under rotation; that is, they have the same value before and after any rotation. Although there are several such expressions (see Problem 15), the one in which we are interested is $B^2 - 4AC$ for the equation

$$Ax^2 + Bxy + Cy^2 + Dx + Ey + F = 0.$$

If we substitute the equations of rotation,

$$x = x' \cos \theta - y' \sin \theta$$

and

$$y = x' \sin \theta + y' \cos \theta$$

into this equation, we get a new second-degree equation

$$A'x'^2 + B'x'y' + C'y'^2 + D'x' + E'y' + F' = 0$$

with

$$A' = A \cos^2\theta + B \sin \theta \cos \theta + C \sin^2\theta,$$

$$B' = -2A \sin \theta \cos \theta + B(\cos^2\theta - \sin^2\theta) + 2C \sin \theta \cos \theta,$$

and

$$C' = A \sin^2\theta - B \sin\theta \cos\theta + C \cos^2\theta.$$

From this, we find that the expression for $B'^2 - 4A'C'$ then simplifies to $B^2 - 4AC$. This gives the following theorem.

Theorem 6.6 *If the equation*

$$Ax^2 + Bxy + Cy^2 + Dx + Ey + F = 0$$

is transformed into the equation

$$A'x'^2 + B'x'y' + C'y'^2 + D'x' + E'y' + F' = 0$$

by rotating the axes, then

$$B^2 - 4AC = B'^2 - 4A'C'.$$

If we choose the angle of rotation properly, $B' = 0$ and the type of conic can be determined by looking at A' and C' (see the table on page 160). Thus we have the following results.

Theorem 6.7 *The equation*

$$Ax^2 + Bxy + Cy^2 + Dx + Ey + F = 0$$

represents a hyperbola, ellipse, or parabola (or a degenerate case of one of these) according to whether $B^2 - 4AC$ is positive, negative, or zero, respectively.

Let us apply this to the foregoing examples. In Example 1, the equation

$$x^2 + 4xy - 2y^2 - 6 = 0$$

gives

$$B^2 - 4AC = 16 - 4 \cdot 1 \cdot (-2) = 24,$$

indicating that the conic is a hyperbola, which we have seen to be the case. After rotation we have

$$2x'^2 - 3y'^2 = 6,$$

giving

$$B'^2 - 4A'C' = 0 - 4 \cdot 2 \cdot (-3) = 24.$$

In Example 2, the equation

$$2x^2 - xy + 2y^2 - 2 = 0$$

gives

$$B^2 - 4AC = 1 - 4 \cdot 2 \cdot 2 = -15.$$

This shows the conic to be an ellipse, which again is what we have already found. It might be noted that, after rotation, we get the result

$$\frac{3}{2}x'^2 + \frac{5}{2}y'^2 = 2$$

provided we do not multiply both sides by some constant. This again gives

$$B'^2 - 4A'C' = 0 - 4 \cdot \frac{3}{2} \cdot \frac{5}{2} = -15.$$

For the result

$$3x'^2 + 5y'^2 = 4,$$

we get a different value of $B'^2 - 4A'C'$, *which is still negative*. If, at some stage, we multiply both sides by some nonzero number k, $B^2 - 4AC$ is then multiplied by k^2. Since k^2 must be positive, the sign of $B^2 - 4AC$ is not changed, no matter what number k is.

This result is helpful here and will be used again in Section 7.4.

PROBLEMS

In Problems 1–14, rotate axes to eliminate the xy term. Sketch, showing both sets of axes.

B   **1.** $x^2 + xy + y^2 + 4\sqrt{2}x - 4\sqrt{2}y = 0$ **2.** $5x^2 + 6xy + 5y^2 - 8 = 0$

3. $7x^2 + 6xy - y^2 - 32 = 0$ **4.** $4x^2 + 4xy + y^2 + 8\sqrt{5}x - 16\sqrt{5}y = 0$

5. $8x^2 - 12xy + 17y^2 = 20$ **6.** $9x^2 + 8xy - 6y^2 = 70$

7. $5x^2 - 4xy + 8y^2 - 36 = 0$ **8.** $x^2 + 12xy + 6y^2 = 30$

9. $4x^2 + 12xy + 9y^2 + 8\sqrt{13}x + 12\sqrt{13}y - 65 = 0$

10. $6x^2 + 12xy + 11y^2 = 240$

11. $9x^2 - 6xy + y^2 - 12\sqrt{10}x - 36\sqrt{10}y = 0$

12. $x^2 + 8xy + 7y^2 - 36 = 0$

13. $8x^2 + 12xy - 8y^2 - 40 = 0$

14. $5x^2 - 6xy + 5y^2 = 72$

C   **15.** Given the equation

$$Ax^2 + Bxy + Cy^2 + Dx + Ey + F = 0,$$

which yields

$$A'x'^2 + B'x'y' + C'y'^2 + D'x' + E'y' + F' = 0$$

after rotation through the angle θ, show that $A' + C' = A + C$ for any value of θ; that is, $A + C$ is invariant under rotation.

16. It can easily be seen graphically that two conic sections have at most four points in common. But

$$2x^2 + xy - y^2 + 3y - 2 = 0$$

and

$$2x^2 + 3xy + y^2 - 6x - 5y + 4 = 0$$

have the five points $(1, 0), (-2, 3), (5, -4), (-6, 7)$, and $(10, -9)$ in common. Why?

17. Suppose

$$\tan \theta_1 = \frac{(C - A) + \sqrt{(C - A)^2 + B^2}}{B}$$

and

$$\tan \theta_2 = \frac{(C - A) - \sqrt{(C - A)^2 + B^2}}{B}.$$

Show that $\tan \theta_1 \cdot \tan \theta_2 = -1$. Use this result to show that $\tan \theta_1$ and $\tan \theta_2$ have opposite signs and θ_1 and θ_2 differ by an odd multiple of $90°$.

REVIEW PROBLEMS

In Problems 1–9, sketch and discuss.

A

1. $4x^2 + y^2 - 8x + 6y + 9 = 0.$

2. $y^2 - x + 2y + 4 = 0.$

3. $y^2 - 8x + 4y + 28 = 0.$

4. $9x^2 - 16y^2 + 36x - 128y - 364 = 0.$

5. $9x^2 + 25y^2 + 18x + 100y - 116 = 0.$

6. $9x^2 - 16y^2 + 18x - 16y - 139 = 0.$

7. $x^2 - 4x - 4y = 0.$

8. $3x^2 + 4y^2 + 30x - 16y + 91 = 0.$

9. $4x^2 - 9y^2 - 16x - 54y - 65 = 0.$

B

10. Find an equation(s) of the ellipse(s) with center $(-4, 1)$, axes parallel to the coordinate axes, and tangent to both coordinate axes.

11. Find an equation(s) of the parabola(s) with axis parallel to a coordinate axis, focus $(3, 5)$ and directrix $x = -1$.

12. Find an equation(s) of the hyperbola(s) with asymptotes $5x - 4y + 22 = 0$ and $5x + 4y - 18 = 0$ and containing the point $(32/5, -7)$.

13. Find an equation(s) of the parabola(s) with horizontal or vertical axis, vertex $(2, -3)$, and containing $(6, -1)$.

14. Find an equation(s) of the ellipse(s) with axes parallel to the coordinate axes, focus $(-1, -1)$, and covertex $(3, 2)$.

15. Find an equation(s) of the hyperbola(s) with vertices $(1, -1)$ and $(7, -1)$, and focus $(-1, -1)$.

16. Translate axes to eliminate the constant and second-degree terms of
$$y = x^3 + 6x^2 + 3x - 14.$$

17. Translate axes to eliminate the first- and third-degree terms of
$$y = x^4 - 16x^3 + 88x^2 - 192x + 140.$$

18. Rotate axes to eliminate the xy term of $3x^2 + 12xy - 2y^2 + 42 = 0$. Sketch, showing both sets of axes.

19. Rotate axes to eliminate the xy term of $2x^2 - \sqrt{3}\,xy + y^2 - 10 = 0$. Sketch, showing both sets of axes.

20. Rotate axes to eliminate the xy term of $4x^2 - 4xy + y^2 + \sqrt{5}\,x + 2\sqrt{5}\,y - 10 = 0$. Sketch, showing both sets of axes.

C   **21.** Show that $2x^2 - 2xy + y^2 - 9 = 0$ is an equation of an ellipse. This equation is a quadratic equation in y. Rearranging the terms, we have $y^2 - 2xy + (2x^2 - 9) = 0$, which can be solved by the quadratic formula to give $y = x \pm \sqrt{9 - x^2}$. Sketch $y = x$ and $y = \pm\sqrt{9 - x^2}$ (square both sides first in the latter equation) on the same set of axes. For each value of x, add the y coordinates for these two curves to get points on the original curve. Use this method to find the graph of the given curve.

7

CURVE SKETCHING

7.1

INTERCEPTS AND ASYMPTOTES

In the first chapter, we sketched the graph of an equation by the tedious process of point-by-point plotting—a method that sometimes causes one to overlook some "interesting" portions of the graph or to sketch certain portions incorrectly. Suppose, for example, you are asked to sketch the graph of

$$y = \frac{10x(x + 8)}{(x + 10)^2}.$$

The methods of Chapter 1 might lead you to the graph of Figure 7.1. A better sketch of the graph is given in Figure 7.2. While the earlier method produced cor-

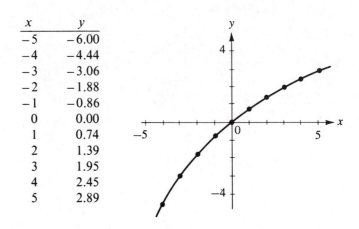

x	y
−5	−6.00
−4	−4.44
−3	−3.06
−2	−1.88
−1	−0.86
0	0.00
1	0.74
2	1.39
3	1.95
4	2.45
5	2.89

Figure 7.1

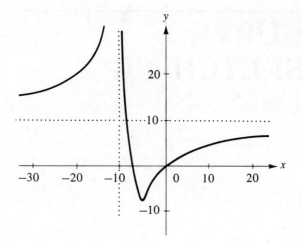

Figure 7.2

rect results for the portion we were sketching, it provided no means for determining which portions of the curve are most "interesting."

Let us consider one more example. Suppose we want to graph

$$y = \frac{2x(2x - 1)}{4x - 1}.$$

x	y
-4	-4.24
-3	-3.23
-2	-2.22
-1	-1.20
0	0.00
1	0.67
2	1.71
3	2.73
4	3.73
5	4.74

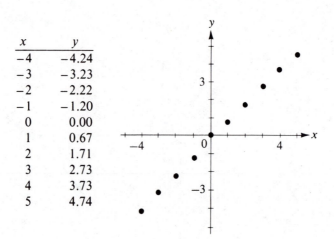

Figure 7.3

The methods of Chapter 1 lead us to the set of points shown in Figure 7.3. Now, what does the graph look like? How would you join the points? You might join them as indicated in Figure 7.4a. The correct graph is shown in Figure 7.4b. These examples demonstrate the need for better methods of sketching curves. We begin by considering intercepts and asymptotes.

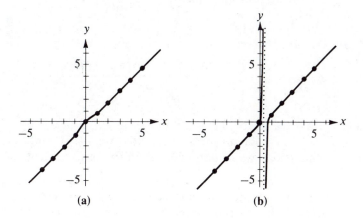

(a)   (b)

Figure 7.4

The **intercepts** of a curve are simply the points of the curve that lie on the coordinate axes; those on the x axis are the x intercepts, while those on the y axis are the y intercepts (the origin is both an x intercept and a y intercept). We determine the x intercepts (if any) by setting y equal to zero and solving for x; similarly, the y intercepts are found by setting x equal to zero and solving for y.

Example 1   Find the intercepts of $\dfrac{x^2}{4} + \dfrac{y^2}{9} = 1$.

Solution   When $y = 0$, $x^2 = 4$ and $x = \pm 2$. When $x = 0$, $y^2 = 9$ and $y = \pm 3$. Thus, the intercepts are $(2, 0)$, $(-2, 0)$, $(0, 3)$, and $(0, -3)$.

An equation often encountered is one of the type $y = P(x)$ or $y = P(x)/Q(x)$, where $P(x)$ and $Q(x)$ are polynomials having no common factor. If $P(x)$ can be factored to give the form

$$P(x) = c(x - a_1)^{n_1}(x - a_2)^{n_2} \cdots (x - a_k)^{n_k},$$

where $c, a_1, \ldots, a_k$ are real numbers, then the x intercepts are $(a_1, 0)$, $(a_2, 0), \ldots,$ $(a_k, 0)$. The y intercept (an equation in this form has at most one) is still found by setting x equal to zero.

Example 2   Find the intercepts of $y = (x + 1)^2(x - 3)$.

Solution   The x intercepts can be taken from the two factors: $(-1, 0)$ from $(x + 1)^2$ and $(3, 0)$ from $(x - 3)$. When $x = 0$,

$$y = 1^2(-3) = -3.$$

Thus the y intercept is $(0, -3)$.

Example 3   Find the intercepts of

$$y = \frac{(x - 2)^2(x + 1)}{(x - 3)(x - 1)^2}.$$

Solution   From the factors $(x - 2)^2$ and $(x + 1)$, we get $(2, 0)$ and $(-1, 0)$. When $x = 0$, $y = -4/3$, and so the y intercept is $(0, -4/3)$. Note that the factors of the denominator have no part in determining the x intercepts.

Let us now turn to **asymptotes**. We encountered asymptotes earlier when we studied the hyperbola (see Section 5.4). Let us consider some other examples. In Figure 7.2, the lines $x = -10$ and $y = 10$ are asymptotes. We see that portions of the curve approach $x = -10$ and $y = 10$. This is the main feature to be considered in determining asymptotes. Note that the curve contains the point $(-25/3, 10)$ of the line $y = 10$. This does not prevent $y = 10$ from being an asymptote. A curve *can* have one or more (even infinitely many) points in common with its asymptote; however, a line is not an asymptote of itself, nor does $y = |x|$ have an asymptote. In Figure 7.4b, the line $x = 1/4$ is an asymptote. Again we see that portions of the curve approach this line.

Although it is possible for any line to be an asymptote, we shall consider only horizontal and vertical asymptotes here (slant asymptotes are considered later in this chapter). First we take up vertical asymptotes. In Figure 7.2 you can see that, as x approaches -10 from either side, y approaches no definite number but gets larger and larger. As x approaches $1/4$ from the right in Figure 7.4b, y gets large and negative; and, as x approaches $1/4$ from the left, y gets large and positive. To find vertical asymptotes, we are not concerned with whether y gets large and positive or large and negative. We are interested only in determining values of x for which y gets large in absolute value. If the equation is in the form

$$y = \frac{P(x)}{Q(x)},$$

then, as x approaches a, y gets large in absolute value if $Q(x)$ approaches zero and $P(x)$ does not. Thus we need determine only the values of x which make $Q(x) = 0$ and $P(x) \neq 0$.

Example 4   Determine the vertical asymptotes of

$$y = \frac{(x + 1)(x - 3)}{(2x - 1)(x + 2)^2}.$$

Solution   The denominator is zero when either one of its two factors is zero.

$$2x - 1 = 0 \qquad \text{gives} \qquad x = 1/2.$$
$$x + 2 = 0 \qquad \text{gives} \qquad x = -2.$$

Since neither value of x gives zero for the numerator, $x = 1/2$ and $x = -2$ are the vertical asymptotes. Figure 7.5 shows the graph with the vertical asymptotes. The method of sketching the graph is deferred until the next section.

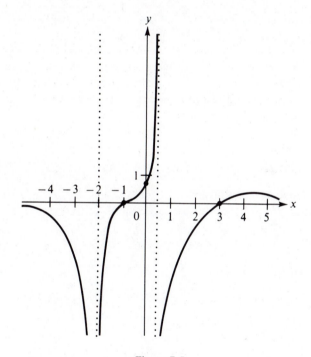

Figure 7.5

Let us now consider horizontal asymptotes. Of course, if the given equation is in the form

$$x = \frac{P(y)}{Q(y)},$$

or can easily be put into that form, we can simply use the methods given for vertical asymptotes. We merely reverse the role of the x and y here. Unfortunately, it is often difficult or impossible to solve for x as a function of y (consider the equation of Example 4), so another method must be found.

If $y = k$ is a horizontal asymptote for $y = f(x)$, then the distance between a point of the graph of $y = f(x)$ and the line $y = k$ must approach zero as x gets large in absolute value. Thus we must investigate the behavior of y as x gets large in one direction or the other; if y approaches the number k ($y \rightarrow k$) as $|x|$ gets large, then $y = k$ is a horizontal asymptote.

Example 5   Determine the horizontal asymptote of

$$y = \frac{x^2 - 4}{x^2 + 3x}.$$

Solution   As x gets large and positive, both the numerator and denominator are also getting large and positive. This fact alone tells us nothing about what number the quotient is approaching. Suppose we alter the equation by dividing both numerator and denominator by x^2. Then

$$y = \frac{x^2 - 4}{x^2 + 3x} = \frac{1 - 4/x^2}{1 + 3/x}.$$

Now, as x gets large,

$$\frac{4}{x^2} \rightarrow 0 \qquad \text{and} \qquad \frac{3}{x} \rightarrow 0.$$

Thus

$$y = \frac{1 - 4/x^2}{1 + 3/x} \rightarrow 1,$$

and $y = 1$ is the horizontal asymptote. Similarly, as x gets large and negative, $y \rightarrow 1$. Thus the asymptote $y = 1$ is approached by the curve in both directions. Note in Figure 7.6 that the graph crosses the horizontal asymptote at $(-4/3, 1)$.

In finding the asymptote in the preceding example, we first divided both numerator and denominator by the highest power of x (x^2 in this case) in the given expression. This trick often helps in finding asymptotes.

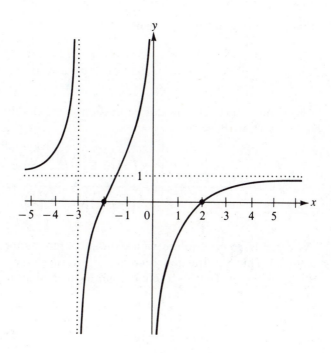

Figure 7.6

Example 6   Determine the horizontal asymptote of

$$y = \frac{(x + 1)(x - 3)}{(2x - 1)(x + 2)^2}.$$

Solution   If we multiplied out the numerator, the highest power of x would be x^2; in the denominator it would be x^3. Thus we shall divide the numerator and denominator by x^3 (do *not* divide the numerator by x^2 and the denominator by x^3; the result would *not* equal y).

$$y = \frac{(x + 1)(x - 3)}{(2x - 1)(x + 2)^2} = \frac{\left(\dfrac{x + 1}{x}\right)\left(\dfrac{x - 3}{x}\right)\left(\dfrac{1}{x}\right)}{\left(\dfrac{2x - 1}{x}\right)\left(\dfrac{x + 2}{x}\right)^2} = \frac{\left(1 + \dfrac{1}{x}\right)\left(1 - \dfrac{3}{x}\right)\left(\dfrac{1}{x}\right)}{\left(2 - \dfrac{1}{x}\right)\left(1 + \dfrac{2}{x}\right)^2}$$

As x gets large, all of the expressions with x in the denominator approach zero and

$$y \rightarrow \frac{(1 + 0)(1 - 0)(0)}{(2 - 0)(1 + 0)^2} = 0.$$

Thus $y = 0$ is the only horizontal asymptote. The graph was given in Figure 7.5. Again note that the graph crosses the horizontal asymptote at $(-1, 0)$ and $(3, 0)$.

Example 7   Find the horizontal asymptote of

$$y = \frac{x(x + 1)(x - 2)}{(x - 4)(x + 2)}.$$

Solution   The highest power of x in this expression is x^3. Dividing numerator and denominator by x^3, we have

$$y = \frac{x(x + 1)(x - 2)}{(x - 4)(x + 2)} = \frac{\left(\frac{x}{x}\right)\left(\frac{x + 1}{x}\right)\left(\frac{x - 2}{x}\right)}{\left(\frac{x - 4}{x}\right)\left(\frac{x + 2}{x}\right)\left(\frac{1}{x}\right)} = \frac{(1)\left(1 + \frac{1}{x}\right)\left(1 - \frac{2}{x}\right)}{\left(1 - \frac{4}{x}\right)\left(1 + \frac{2}{x}\right)\left(\frac{1}{x}\right)}.$$

As x gets large in absolute value, the numerator approaches 1 and the denominator, 0. Thus the fraction becomes arbitrarily large, rather than leveling off to some number k. There is, therefore, no horizontal asymptote. The graph is shown in Figure 7.7.

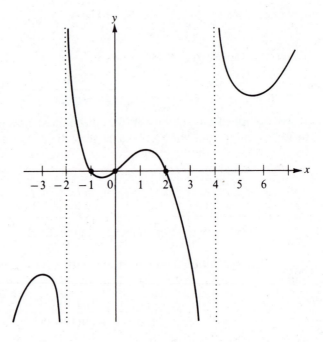

Figure 7.7

PROBLEMS

A *In Problems 1–16, find the intercepts.*

1. $y = x^2 + 3x$

2. $y = x^2 - x - 2$

3. $y = (x + 1)(x^2 - 1)$

4. $y = (2x - 1)^2(3x + 2)^3$

5. $y = (4x + 1)(x - 2)(2x + 3)^2$

6. $y = (2x - 1)^3(3x + 2)^2(x - 3)$

7. $y = (x - 1)(x^2 + 1)$

8. $y = (2x + 3)(x^2 + x + 1)$

9. $y = (3x - 1)^2(x^2 + 2)^3(2x + 1)^4$

10. $y = (2x - 5)(x^2 - x + 1)^3(x^2 + 2)$

11. $y = \dfrac{x}{x + 1}$

12. $y = \dfrac{(x + 1)(x - 2)}{(3x + 1)^2}$

13. $y = \dfrac{(x - 3)^2}{2x + 1}$

14. $y = \dfrac{1}{3x + 2}$

15. $y = \dfrac{2}{x^2}$

16. $y = \dfrac{(x + 1)^2(x^2 + 1)}{x^2}$

B *In Problems 17–34, find all horizontal and vertical asymptotes.*

17. $y = (x + 1)(x - 2)$

18. $y = (4x + 3)(x - 2)$

19. $y = \dfrac{1}{x - 1}$

20. $y = \dfrac{4x - 2}{x + 1}$

21. $y = \dfrac{x}{x + 3}$

22. $y = \dfrac{(x + 1)^2}{2x(x - 2)}$

23. $y = \dfrac{2x(x - 2)}{(x + 1)^2}$

24. $y = \dfrac{(x + 1)(x - 3)^2}{(x + 2)^2}$

25. $y = \dfrac{(2x - 3)(x - 2)}{(x + 1)(x - 3)^2}$

26. $y = \dfrac{(4x - 7)(x - 1)^2}{(x + 1)(x + 2)(x + 3)}$

27. $y = \dfrac{(x + 1)^2}{x^2 + 1}$

28. $y = \dfrac{(2x + 1)^2(x - 2)^2}{x(4x - 3)}$

29. $y = \dfrac{(3x + 2)^3(x - 4)}{(2x + 3)^2(x + 1)^3}$

30. $y = \dfrac{(2x + 1)^2(x - 3)^3}{x(2x - 3)^2}$

31. $y = x - \sqrt{x^2 + 1}$
    [*Hint:* Rationalize the numerator.]

32. $y = x + \sqrt{x^2 + 1}$

33. $y = 2x - \sqrt{4x^2 + 3}$

34. $y = 3x + \sqrt{9x^2 - 1}$

C   **35.** If $y = \dfrac{a_n x^n + a_{n-1} x^{n-1} + \cdots + a_1 x + a_0}{b_m x^m + b_{m-1} x^{m-1} + \cdots + b_1 x + b_0}$, where $a_n \neq 0$ and $b_m \neq 0$, what can
be said about horizontal asymptotes in case

    a. $n < m$?   **b.** $n = m$?   **c.** $n > m$?

36. A certain chemical decomposition proceeds according to the formula

$$x = \frac{15}{15 + t} \ (t \geq 0),$$

where t is the time in minutes and x is the proportion of undecomposed chemical; that
is, x is the ratio of the weight of undecomposed chemical to the weight of the chemical
when $t = 0$. Show that the chemical never decomposes entirely. How long does it take
for 99% of the chemical to decompose?

37. Under certain conditions the size of a colony of bacteria is related to time by

$$x = 40,000 \, \frac{2t + 1}{t + 1} \ (t \geq 0),$$

where x is the number of bacteria and t is the time in hours. To what number does the
size of the colony tend over a long period of time? In how many hours will the colony
have achieved 95% of its ultimate size?

7.2

SYMMETRY, SKETCHING

Another characteristic that helps in sketching a curve is symmetry. There are two
types: symmetry about a line and symmetry about a point. If a curve is symmetric
about a line, then one-half of it is the mirror image of the other half, with the mirror as the line of symmetry. More precisely, for every point P of the curve, on one
side of the line there is another point P' of the curve such that PP' is perpendicular to the line of symmetry and is bisected by it. An example of this type of symmetry occurs with the graph of $y = 1/x^2$, in which the y axis is the line of symmetry (see Figure 7.8).

    A curve is symmetric about a point O if for every point $P \neq O$ of the curve,
there corresponds a point P' such that PP' is bisected by the point O. The origin
is the point of symmetry of the graph of $y = 1/x$ (see Figure 7.9).

    While there is no restriction on the lines or points that may be lines or points of
symmetry, we shall consider here only symmetry about the axes and about the
origin. We begin with symmetry about the y axis. If a curve is symmetric about
the y axis, then, corresponding to every point $P = (x, y)$ on the curve, there is a
point $P' = (-x, y)$ (see Figure 7.8) with the same y coordinate and an x coordi-

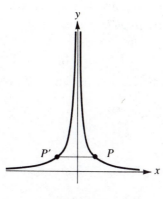

Figure 7.8

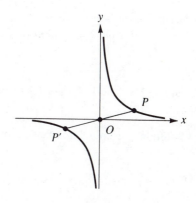

Figure 7.9

nate that is the negative of the x coordinate of P. In this situation, we get the same value for y whether we substitute a positive number x into the equation or its negative, $-x$.

Theorem 7.1   *If every x in an equation is replaced by $-x$ and the resulting equation is equivalent to the original (has the same graph), then its graph is symmetric about the y axis.*

Example 1   Use Theorem 7.1 to show that $y = 1/x^2$ is symmetric about the y axis.

Solution   Replacing x by $-x$, we have

$$y = \frac{1}{(-x)^2}.$$

Since $(-x)^2 = x^2$, we see that the substitution has produced an equation equivalent to the original equation, proving symmetry about the y axis.

An argument similar to the one preceding Theorem 7.1 can be used to obtain the following theorem.

Theorem 7.2   *If every y in an equation is replaced by $-y$ and the resulting equation is equivalent to the original, then its graph is symmetric about the x axis.*

If a curve is symmetric about the origin, then, for every point $P = (x, y)$ on the curve, there is a point $P' = (-x, -y)$ (see Figure 7.9). This is the statement of the next theorem.

Theorem 7.3   *If every x in an equation is replaced by $-x$ and every y by $-y$ and the resulting equation is equivalent to the original, then its graph is symmetric about the origin.*

Example 2   Use Theorem 7.3 to show that $y = 1/x$ is symmetric about the origin.

Solution   Replacing x by $-x$ and y by $-y$ gives

$$-y = \frac{1}{-x}.$$

This is equivalent to the original equation, since we get the original equation if we multiply both sides by -1.

Before using what we have observed about intercepts, asymptotes, and symmetry to sketch a curve, we shall look at a characteristic of curves represented by equations of the form

$$y = \frac{P(x)}{Q(x)},$$

where $P(x)$ and $Q(x)$ are polynomials in reduced form; that is, $P(x)$ and $Q(x)$ have no common factors. The factors of $P(x)$ determine the x intercepts, and the factors of $Q(x)$ determine the vertical asymptotes of the curve. These are the only two places at which y can change from positive to negative or from negative to positive. This is not to say that the value of y *must* change there—only that it cannot do so elsewhere. We can easily determine whether or not the change occurs at a given intercept or asymptote by considering the exponent on the factor that produces it.

Theorem 7.4   *Given an equation of the form*

$$y = \frac{P(x)}{Q(x)}$$

in reduced form, if $(x - a)^n$ (where n is a positive integer) is a factor of either $P(x)$ or $Q(x)$ and if $(x - a)^{n+1}$ is a factor of neither, then
   (a)   the graph crosses the x axis at $x = a$ if and only if n is odd, and
   (b)   the graph stays on the same side of the x axis at $x = a$ if and only if n is even.

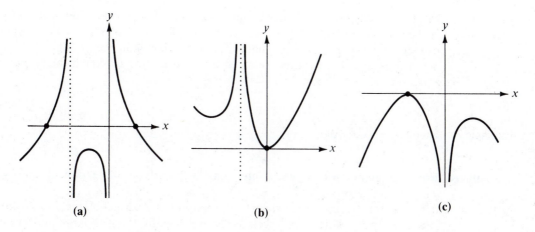

(a)   (b)   (c)

Figure 7.10

The expression "the graph crosses the x axis" does not necessarily imply that the graph has a point in common with the x axis. It means that the graph is above (or below) the x axis for $c < x < a$ and below (or above) for $a < x < d$ for some c and d. Of course the graph does contain a point of the x axis at an x intercept, but it "crosses the x axis" at an asymptote by "hopping" over and not touching the axis. This is illustrated in Figure 7.10a, in which the graph crosses the x axis at both intercepts and both asymptotes.

Similarly the graph can contain a point of the x axis and still stay on the same side of it. This is illustrated in Figures 7.10b and 7.10c, in which the graph stays on the same side of the x axis at the intercept and asymptote in each case.

Although the following discussion does not constitute a proof of this theorem, it serves to show why the theorem works. Let us consider the case in which $(x - a)^n$ is a factor of $P(x)$ (a similar argument can be used for the other case). Then

$$y = \frac{P(x)}{Q(x)} = \frac{R(x)}{Q(x)} (x - a)^n.$$

For all values of x at and "near" $x = a$, $R(x)/Q(x)$ is either positive throughout or negative throughout, not making any sign change. But

$$x - a < 0 \quad \text{for} \quad x < a,$$
$$x - a = 0 \quad \text{for} \quad x = a,$$

and

$$x - a > 0 \quad \text{for} \quad x > a.$$

In other words, $x - a$ changes sign at $x = a$. If n is odd, then $(x - a)^n$ also changes sign; thus y changes sign at $x = a$. If n is even, then $(x - a)^n$ is positive whether $x < a$ or $x > a$; that is, $(x - a)^n$ does not change sign and y does not change sign. Let us now use all of this information to sketch the graph of an equation.

Example 3   Sketch $y = (x - 3)(x + 1)^2$.

Solution   From the numerator, we get x intercepts $(3, 0)$ with an odd exponent and $(-1, 0)$ with an even exponent. If $x = 0$, then $y = -3$, which gives $(0, -3)$. Since the denominator is 1, there are no vertical asymptotes. As x gets large and positive, y gets large and positive; as x gets large and negative, y gets large and negative, which means that there is no horizontal asymptote. It is easy to see that no symmetry exists about either axis or the origin. Summing up, we have:

Intercepts: $(3, 0)$ odd, $(-1, 0)$ even, $(0, -3)$

No asymptotes: as $x \rightarrow + \infty, y \rightarrow + \infty$

as $x \rightarrow - \infty, y \rightarrow - \infty$

No symmetry

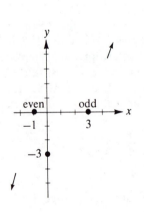

Figure 7.11

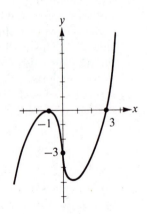

Figure 7.12

All of this is indicated in Figure 7.11. Let us sketch the graph, starting at the far left and working to the right (this choice is quite arbitrary; we might just as well go from right to left or start in the middle and work outward). We keep in mind that the curve must go through all intercepts and that y is a *function* of x; that is, it is single-valued. Since $y \to -\infty$ as $x \to -\infty$, we start in the lower left-hand corner. Going to the right, we first reach the intercept $(-1, 0)$. Since it is an even intercept, the graph merely touches the x axis but does not cross it. Next, the graph goes through $(0, -3)$ and then turns back up in order to go through $(3, 0)$. Since $(3, 0)$ is an odd intercept, the graph crosses the x axis there and proceeds upward. The result is given in Figure 7.12.

Note that we put the lowest point of the "dip" at approximately $x = 1$. How did we know to put it there? We didn't. We made no attempt to locate it—we simply guessed. Without further work, the best we can say is that it is between $x = -1$ and $x = 3$. Furthermore, how do we know that the graph does not have some extra "turns" and "wiggles" and perhaps look like Figure 7.13? Again, we

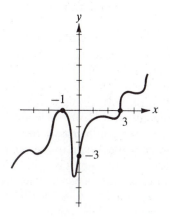

Figure 7.13

don't. As a general rule, unless there is some special reason to put in some extra "turn" or "wiggle," we shall leave it out. This rule will not necessarily give us the correct graph every time, but there is no point in needlessly complicating the situation. These methods give only a general idea of the graph. If you take a course in calculus, you will see how derivatives may be used to answer the questions raised here.

Note that, with the exception of the three intercepts, we have not plotted a single point! Yet we have some idea (within the restrictions noted above) of the main features of the curve. With a little practice, you should be able to sketch such curves quite quickly, and thus achieve the principal aim here.

Example 4   Sketch $y = \dfrac{(2x - 1)(x + 2)^2}{(x + 1)^2(x - 3)}$.

Solution   Intercepts: $(1/2, 0)$, odd; $(-2, 0)$, even; $(0, 4/3)$

Asymptotes:

From the denominator: $x = -1$, even; $x = 3$, odd

$$y = \frac{(2x - 1)(x + 2)^2}{(x + 1)^2(x - 3)} = \frac{\left(\dfrac{2x - 1}{x}\right)\left(\dfrac{x + 2}{x}\right)^2}{\left(\dfrac{x + 1}{x}\right)^2 \left(\dfrac{x - 3}{x}\right)}$$

$$= \frac{\left(2 - \dfrac{1}{x}\right)\left(1 + \dfrac{2}{x}\right)^2}{\left(1 + \dfrac{1}{x}\right)^2 \left(1 - \dfrac{3}{x}\right)}$$

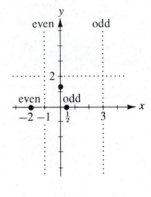

Figure 7.14

As $x \to \pm\infty$, $y \to \dfrac{2 \cdot 1}{1 \cdot 1} = 2$

Thus $y = 2$ is the horizontal asymptote.

No symmetry.

All of this information is summarized in Figure 7.14. If we begin sketching at one end or the other, we have the problem of not knowing whether the curve is approaching the asymptote from above or below. Similar problems exist at the vertical asymptotes and x intercepts. Suppose, then, we start at $(0, 4/3)$. Going to the right, we first come to $(1/2, 0)$. Since it is an odd intercept, the graph crosses the x axis there and then goes down to the vertical asymptote $x = 3$ (it cannot go up, since it cannot cross the x axis anywhere between $x = 1/2$ and $x = 3$). Since this asymptote is also odd, the graph now jumps above the x axis. Finally it comes down to the horizontal asymptote $y = 2$.

Going back to $(0, 4/3)$ and proceeding to the left, we see that the graph must go up to the vertical asymptote $x = -1$ (remember there is nothing to prevent the graph from crossing a horizontal asymptote). Since $x = -1$ is an even asymptote, the curve stays above the x axis. It must then proceed down to the intercept $(-2, 0)$. This is also even, so the graph again remains above the x axis, finally going up to the horizontal asymptote. Thus, we have the graph indicated in Figure 7.15.

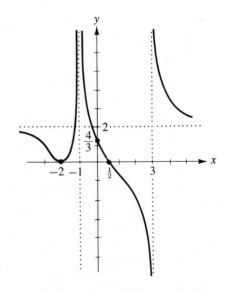

Figure 7.15

PROBLEMS

A In Problems 1–10, check for symmetry about both axes and the origin.

1. $y = x^4 - x^2$

2. $y = x^3 - x$

3. $y = x^3 - x^2$

4. $\dfrac{x^2}{4} + \dfrac{y^2}{9} = 1$

5. $y^2 = \dfrac{x + 1}{x}$

6. $y^3 = \dfrac{x + 1}{x}$

7. $xy = 1$

8. $x^2 y^2 = 1$

9. $y = \dfrac{x}{x^2 + 1}$

10. $y = \dfrac{(x + 1)(x - 1)}{x^2}$

B In Problems 11–30, use the methods of this and the preceding section to sketch the graph. Do not plot the graph point by point.

11. $y = (x + 1)(x - 3)$

12. $y = (x + 2)(x - 1)^2$

13. $y = x^2 - 5x - 6$

14. $y = x^3 + x^2 - 2x$

15. $y = x^4 - x^2$

16. $y = x^3 - x$

17. $y = \dfrac{x + 1}{x}$

18. $y = \dfrac{x - 2}{x + 2}$

19. $y = \dfrac{(2x + 1)(x - 1)^2}{(x - 2)(x + 1)^2}$

20. $y = \dfrac{x - 3}{(x + 1)(x - 2)}$

21. $y = \dfrac{(x + 2)(x - 4)}{x - 1}$

22. $y = \dfrac{x - 1}{(x + 2)(x - 4)}$

23. $y = \dfrac{(x + 2)^2(x - 4)}{(x - 1)^2}$

24. $y = \dfrac{x}{x^2 + 1}$

25. $y = \dfrac{x^2 + 1}{x}$

26. $xy = 2x + 1$

27. $x^2 y = 2x + 1$

28. $x^2 y - y = x^2$

29. $x^2 y - y = x^3$

30. $x^2 y - y = x$

C 31. Show that if a graph has any two of the three types of symmetry—about the x axis, about the y axis, about the origin—then it must have the third.

32. Given an example of a curve with exactly two lines of symmetry.

33. Can a graph have two points of symmetry?

34. Give an example of a curve with infinitely many lines of symmetry.

35. Show that if two perpendicular lines are lines of symmetry of a given curve, then their point of intersection is a point of symmetry.

36. An even function f is one for which $f(-x) = f(x)$; an odd function is one for which $f(-x) = -f(x)$. What can we say concerning the symmetry of the graph of an even function? Of an odd function?

37. A graph (with at least one point not on the x axis) is symmetric about the x axis. Can it be the graph of a function? Explain.

38. Show that if every x in an equation is replaced by $2k - x$ and the resulting equation is equivalent to the original, then its graph is symmetric about the line $x = k$.

7.3

RADICALS AND THE DOMAIN OF THE EQUATION

Recall that two things can keep us from getting a value for y when we substitute a value of x into an equation: a zero in the denominator and an even root of a negative number. A zero in the denominator gives a vertical asymptote (provided the numerator is not also zero for the same value of x). Even roots of negative numbers simply cause gaps in the domain of the equation.

Example 1 Sketch $y = \dfrac{2x}{\sqrt{x^2 - 4}}$.

Solution Using the previous methods for determining intercepts, asymptotes, and symmetry, we have the following.

Intercepts: $(0, 0)$ odd

Asymptotes: $x = 2, \quad x = -2$

The radical is equivalent to the one-half power, which is neither odd nor even. We have a special problem in finding the horizontal asymptotes. The highest power of x in the numerator is clearly x. The highest power in the denominator appears to be x^2. But it is under the radical; so the highest power is really $(x^2)^{1/2} = x$. Thus we shall want to divide the numerator and denominator by x. But we shall want to put the x under the radical in the denominator, which leads to further complications. The symbol $\sqrt{}$ means the *nonnegative* square root. Thus $x = \sqrt{x^2}$ is true only when $x \geq 0$; when $x < 0$, $\sqrt{x^2} = -x$ (note that, since x itself is negative, $-x$ is positive), and we have two cases to consider.

$$\frac{2x}{\sqrt{x^2 - 4}} = \frac{\dfrac{2x}{x}}{\sqrt{\dfrac{x^2 - 4}{x^2}}} = \frac{2}{\sqrt{1 - \dfrac{4}{x^2}}} \qquad \text{(when } x > 0)$$

$$\frac{2x}{\sqrt{x^2 - 4}} = \frac{\frac{2x}{-x}}{\sqrt{\frac{x^2 - 4}{x^2}}} = \frac{-2}{\sqrt{1 - \frac{4}{x^2}}} \qquad \text{(when } x < 0\text{)}$$

Thus,

$$\text{as } x \to +\infty, \quad y = \frac{2x}{\sqrt{x^2 - 4}} = \frac{2}{\sqrt{1 - \frac{4}{x^2}}} = 2$$

and

$$\text{as } x \to -\infty, \quad y = \frac{2x}{\sqrt{x^2 - 4}} = \frac{-2}{\sqrt{1 - \frac{4}{x^2}}} = -2,$$

giving two horizontal asymptotes: $y = 2$, which is approached on the right, and $y = -2$, which is approached on the left. Replacing x by $-x$ and y by $-y$ gives

$$-y = \frac{2(-x)}{\sqrt{(-x)^2 - 4}} = \frac{-2x}{\sqrt{x^2 - 4}},$$

which is equivalent to the original equation. Thus we have symmetry about the origin.

Finally, $\sqrt{x^2 - 4}$ represents a real number only when $x^2 - 4 \geq 0$, which gives

$$x^2 \geq 4 \qquad \text{or} \qquad \begin{cases} x \geq 2 \\ x \leq -2. \end{cases}$$

But y is real for one additional value of x, namely, $x = 0$. If $x = 0$, y equals zero divided by a complex number, which is still zero. Thus the domain is

$$\{x \mid x \geq 2 \quad \text{or} \quad x \leq -2 \quad \text{or} \quad x = 0\}.$$

We see here that $(0, 0)$ is an isolated point of the graph (see Note below).

All of this information is represented graphically in Figure 7.16, which shows the intercept as an isolated point; the fact that it is odd is of no use. Note one thing more: Since $\sqrt{x^2 - 4}$ is never negative, y is positive whenever x is positive and negative whenever x is negative. This additional information makes it easy for us to sketch the curve (see Figure 7.17).

Note: The existence of an isolated point of this graph at the origin is open to controversy. On one hand are those who say that there is no point of the graph at $(0, 0)$. Their reasoning is as follows: Whenever we restrict ourselves to real func-

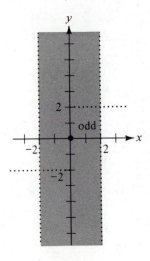

Figure 7.16

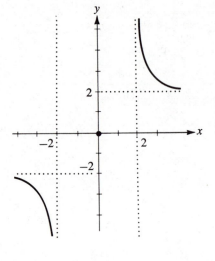

Figure 7.17

tions of real variables, we say, in effect, that imaginary numbers do not exist. Thus, instead of having zero over an imaginary number, we have zero divided by no number at all, which yields no number. By this line of reasoning, every "part" of the equation must be real in order to yield a valid result. On the other hand are those who maintain that the result is independent of the means of obtaining it. The mere fact that we go from one real number to another by way of imaginary numbers, they say, does not invalidate the result. Exactly the same controversy arose when Cardan published his solution of a cubic equation in 1545. His rule for the solutions of $x^3 = 15x + 4$ leads to

$$x = \sqrt[3]{\sqrt{2} + \sqrt{-121}} + \sqrt[3]{\sqrt{2} - \sqrt{-121}},$$

which simplifies to $x = 4$, the only positive root.* It was decided then that the excursion into complex numbers did not invalidate the result. While we shall take the latter point of view throughout this text, it is well to bear in mind the controversial nature of the problem.

*In this connection, see Carl B. Boyer, *A History of Mathematics,* (New York: John Wiley & Sons, Inc., 1968), pp 310–316.

Example 2   Sketch $y^2 = x^4 - x^2$.

Solution   To graph this equation, we use the following device: Since $y = \pm\sqrt{x^4 - x^2}$, we first graph $z = x^4 - x^2$ and then, from the values of z, get $y = \pm\sqrt{z}$. Graphing $z = x^4 - x^2 = x^2(x^2 - 1) = x^2(x + 1)(x - 1)$, we have the following.

Intercepts: $(0,0)$, even, $(1,0)$, odd, $(-1,0)$, odd.

No asymptotes.

Symmetry about the z axis (see Figure 7.18a).

We see on this graph that, for each value of x, we have a value of $z = x^4 - x^2$. Now let us find the corresponding values for $y = \pm\sqrt{z}$. But first, we note the following points to keep in mind.

1. $\sqrt{z} = z$ if $z = 0$ or $z = 1$
2. $\sqrt{z} > z$ if $0 < z < 1$
3. $\sqrt{z} < z$ if $z > 1$
4. $\sqrt{z}$ is not real if $z < 0$

The final result is given by the graph of Figure 7.18b. The origin is again an isolated point of this graph. It is convenient to sketch both graphs on the same pair of axes. This is done in Figure 7.18c.

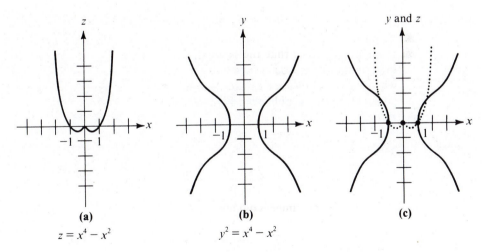

(a)	(b)	(c)
$z = x^4 - x^2$	$y^2 = x^4 - x^2$	

Figure 7.18

The same method could be used to sketch $y = \sqrt{x^4 - x^2}$. The only difference would be that we would have only the top half of the result in Figure 7.18b. We might also have used this method in Example 1, starting with

$$y^2 = \frac{4x^2}{x^2 - 4}.$$

In that case, we would have to be careful which branch we chose; we would have to choose the top portion when x is positive and the bottom portion when x is negative.

One final point. Let us recall that when we had an equation of the form

$$y = \frac{P(x)}{Q(x)},$$

we noted that x intercepts come from factors in the numerator and vertical asymptotes from factors in the denominator, *provided there is no value of x for which both numerator and denominator are zero.* In the examples we have been considering, this is equivalent to the provision that there is no factor common to both numerator and denominator. What happens if there *are* common factors? The answer is simple. You simply cancel the common factors and sketch the resulting equation. But remember that if you cancel the factor $x - a$, the original equation is not defined at $x = a$ (it gives 0/0) and there is no point on the graph with x coordinate a.

Example 3   Sketch $y = (x^2 - 1)/(x - 1)$.

Solution   Since the numerator and denominator have the common factor $x - 1$, we cancel them to get

$$y = x + 1,$$

which gives a straight line. But recall that the original equation gives no value of y when $x = 1$. Thus the point $(1, 2)$ should be deleted from the graph, as in Figure 7.19.

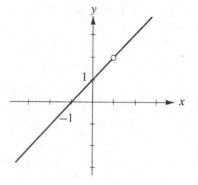

Figure 7.19

PROBLEMS

B   *Sketch the graphs of the following equations.*

1. $y = x\sqrt{x^2 - 1}$

2. $y = \dfrac{x}{\sqrt{x - 1}}$

3. $y = \dfrac{-x}{\sqrt{x^2 - 1}}$

4. $y = \dfrac{x - 1}{\sqrt{x(x + 1)}}$

5. $y = x + \sqrt{x^2 - 1}$

6. $y = x - \sqrt{x^2 - 1}$

7. $y^2 = \dfrac{x}{x + 1}$

8. $y^2 = \dfrac{x^2}{(x + 1)(x - 2)}$

9. $y^2 = \dfrac{2x}{(x - 1)^2}$

10. $y^2 = \dfrac{(x - 1)^2}{x}$

11. $y^2 = \dfrac{x(x + 1)}{(x - 2)^2}$

12. $y^2 = \dfrac{x(x + 1)^2}{x - 2}$

13. $y^2 = \dfrac{x(x - 1)}{(x + 1)^2}$

14. $y^2 = \dfrac{x(1 - x)}{(x + 1)^2}$

15. $y^2 = \dfrac{(x^2 - 1)^2}{x - 2}$

16. $y^2 = (1 - x)(3 - x)^2$

17. $y^2 = (x - 1)(x - 3)^2$

18. $y^2 = -(x - 1)(x - 3)^2$

19. $y = \dfrac{x^2 - 4}{x - 2}$

20. $y = \dfrac{x^2 + x}{x}$

21. $y = \dfrac{x^2 + x}{x^2}$

22. $y = \dfrac{x^3 + x^2}{x}$

23. $y = \dfrac{x(x + 1)^2}{(x - 1)(x + 1)^3}$

24. $y = \dfrac{2x(x - 1)}{x(x + 1)}$

25. $y = \dfrac{1 - (1 + h)^2}{h}$

26. $y = \dfrac{-1 - [(1 + h)^2 - 2(1 + h)]}{h}$

[*Hint:* Simplify the numerator.]

27. $y = \dfrac{2 - \dfrac{2 + h}{1 + h}}{h}$

28. $y = \dfrac{1 - \sqrt{1 + h}}{h}$

7.4

DIRECT SKETCHING OF CONICS

We have been able to sketch conics by putting them into a standard form in Chapters 5 and 6. But this was often quite tedious, especially when we had to rotate the axes. Let us see if we can determine some methods of sketching without going through the process of rotating axes.

Recall that any equation of second degree in x and y represents a conic or a degenerate conic. The type of conic can be determined by $B^2 - 4AC$, as indicated in Theorem 6.7 on page 177. Remember, however, that this test does not distinguish between the conics and their degenerate cases. The results are summarized in the following table.

Conic	$B^2 - 4AC$	Degenerate cases
Parabola	0	One line (two coincident lines) Two parallel lines No graph
Ellipse	–	Circle Point No graph
Hyperbola	+	Two intersecting lines

If we are dealing with a hyperbola, the greatest single aid in sketching the graph is determination of the asymptotes. If they are horizontal or vertical, the determination is relatively easy; so let us go to slant asymptotes. We shall consider two cases: the equation is linear in $y(C = 0)$ and the equation is quadratic in $y(C \neq 0)$. In either case, we first solve for y. Examples of each follow.

Example 1 Sketch $x^2 - xy - 3y - 1 = 0$ without rotating axes.

Solution First of all, $B^2 - 4AC = (-1)^2 - 4(1)(0) = 1$, indicating that the conic is a hyperbola or a degenerate case of one. Solving for y, we have

$$y = \frac{x^2 - 1}{x + 3}.$$

The methods of this chapter give intercepts $(\pm 1, 0)$ and $(0, -1/3)$, and vertical asymptote $x = -3$. There is no horizontal asymptote, but we know that there must be a second asymptote. To find it, we carry out the division.

$$y = \frac{x^2 - 1}{x + 3} = x - 3 + \frac{8}{x + 3}.$$

We now see that, for numerically large values of x, $8/(x + 3)$ is almost zero and y is very near $x - 3$. Thus the slant asymptote is

$$y = x - 3.$$

With this, we can easily sketch the hyperbola (see Figure 7.20).

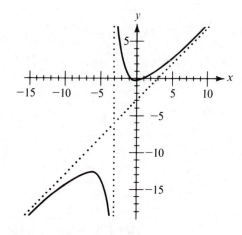

Figure 7.20

Example 2   Sketch $7x^2 + 6xy - y^2 - 32 = 0$ without rotating axes.

Solution   This equation is quadratic in y.

$$y^2 - 6xy + (32 - 7x^2) = 0.$$

Using the quadratic formula, we have

$$y = 3x \pm 4\sqrt{x^2 - 2}.$$

Again, for large values of x, $\sqrt{x^2 - 2}$ is almost $\sqrt{x^2}$, and y is very near $3x \pm 4x$. Thus, the slant asymptotes are (see Figure 7.21)

$$y = 7x \quad \text{and} \quad y = -x.$$

This and the intercepts give us a good idea of the curve.

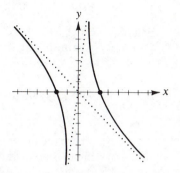

Figure 7.21

Slant asymptotes are not restricted to hyperbolas. In general, an equation of the form $y = P(x)/Q(x)$, where $P(x)$ and $Q(x)$ are polynomials, yields a slant asymptote whenever the degree of the numerator is one more than the degree of the denominator, since division gives a linear expression plus a remainder.

Another useful procedure in sketching conics (as well as other curves) is the method of addition of ordinates. Let us consider an example.

Example 3   Sketch $2x^2 - 2xy + y^2 - 9 = 0$ without rotating axes.

Solution   Since $B^2 - 4AC = -4$, the curve is an ellipse. Again, the equation is quadratic in y.

$$y^2 - 2xy + (2x^2 - 9) = 0$$

By the quadratic formula, we have

$$y = x \pm \sqrt{9 - x^2}.$$

Instead of trying to sketch this curve directly, let us sketch

$$y = x \quad \text{and} \quad y = \pm \sqrt{9 - x^2}.$$

By squaring both sides, we can put the second equation into the form

$$x^2 + y^2 = 9.$$

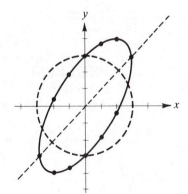

Figure 7.22

These two are easily sketched (see Figure 7.22).

For each value of x in the interval $[-3, 3]$, there is an ordinate on the line and one (or two) on the circle. Adding them, we have the ellipse of Figure 7.22.

Since values of x outside the interval $[-3, 3]$ give complex values of y in the equation $y = x \pm \sqrt{9 - x^2}$, there is no graph to the right of $x = 3$ or to the left of $x = -3$.

Example 4   Sketch $4x^2 - 4xy + y^2 - 5x + 2y + 1 = 0$ without rotating axes.

Solution   Since $B^2 - 4AC = 0$, the curve is a parabola. Again solving for y, we have

$$y = 2x - 1 \pm \sqrt{x}.$$

This gives the two equations

$$y = 2x - 1 \quad \text{and} \quad y = \pm \sqrt{x},$$

where the latter can be written $y^2 = x$. Sketching these two and adding ordinates, we have the result given in Figure 7.23. The line $y = 2x - 1$ is *not* the axis of the parabola.

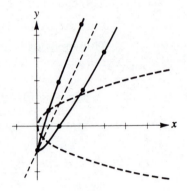

Figure 7.23

Addition of ordinates can be used for hyperbolas, as well as for ellipses and parabolas. One disadvantage of this method is that the two curves must be graphed relatively accurately, or the final result is likely to be extremely inaccurate.

One final word; it is *not* maintained that the methods of this section will *always* provide the simplest method of sketching conics. They are alternate methods that are useful in many cases.

PROBLEMS

In Problems 1–20, *sketch without rotating axes.*

A   **1.** $xy - x + y + 3 = 0$ **2.** $2xy - x - y - 2 = 0$
       3. $x^2 - xy - y - 4 = 0$ **4.** $x^2 - xy + x + 2y = 0$

B   **5.** $2x^2 - 2xy + y^2 - 1 = 0$ **6.** $5x^2 - 4xy + y^2 - 4 = 0$
       7. $x^2 - 2xy + y^2 - x = 0$ **8.** $x^2 - 2xy + y^2 + x - 2y + 2 = 0$
       9. $2xy - y^2 - 4 = 0$ **10.** $2x^2 - 2xy + y^2 + 4x - 4y - 5 = 0$
       11. $2xy - y^2 + 6x - 6y - 18 = 0$ **12.** $3x^2 - 4xy + y^2 - 4x + 2y + 5 = 0$
       13. $4x^2 + 4xy + y^2 - 3x + 2y + 1 = 0$ **14.** $x^2 - 2xy + y^2 - 12x + 8y + 24 = 0$
       15. $3x^2 + 2xy - y^2 + 10x + 2y + 8 = 0$ **16.** $x^2 - 2xy + y^2 - 2x + 2y - 3 = 0$
       17. $x^2 - xy - x - 2 = 0$ **18.** $xy - y^2 - y + 2 = 0$
       19. $10x^2 - 6xy + y^2 + 12x - 4y + 4 = 0$ **20.** $2xy + y^2 - 4 = 0$

C   **21.** Show that $\sqrt{x} + \sqrt{y} = \sqrt{a}$ is a portion of a parabola.

REVIEW PROBLEMS

A *In Problems 1–4, sketch the graphs of the equations as rapidly as possible without plotting points. Show all intercepts, asymptotes, and symmetry about either axis or the origin.*

1. $y = \dfrac{x^2 - 1}{x^2 + 1}$

2. $y = \dfrac{x^2(x - 3)}{(x^2 - 4)^2}$

3. $y = \dfrac{x^2 - 9}{x^2}$

4. $y = \dfrac{x^3 - 1}{x^2 + 3x}$

B **5.** $y = (x - 1)^2(x + 3)^3(2x - 3)$

6. $y = x + 1 - \dfrac{6}{x}$

7. $y = x - \sqrt{x^2 - 4}$

8. $y = \sqrt{\dfrac{(x - 1)(x + 2)}{x + 1}}$

9. $y^2 = \dfrac{x(x - 2)}{(x + 1)^2}$

10. $y^2 = \dfrac{2x - 1}{x + 2}$

11. $y = \dfrac{x^2 - 3x - 4}{x^2 + 4x + 3}$

12. $y = \dfrac{x^2}{x^2 - 2x}$

In Problems 13–16, sketch the graphs of the given conic sections without rotating axes.

13. $x^2 + xy - 2x + y = 0$

14. $x^2 - xy - y - 4 = 0$

15. $5x^2 - 4xy + y^2 - 4x + 2y - 8 = 0$

16. $4x^2 - 4xy + y^2 + 11x - 6y + 5 = 0$

8

TRANSCENDENTAL CURVES

8.1

TRIGONOMETRIC FUNCTIONS

We now turn to some of the nonalgebraic (or transcendental) functions, beginning with the trigonometric functions and their inverses. Although the measurement of angles in degrees is convenient for many purposes, in advanced mathematics, radian measure is the more natural way of measuring angles. Recall that the radian measure of an angle is the length s of the arc of a circle subtended by the angle (see Figure 8.1) divided by the radius r of the circle.

$$\theta = \frac{s}{r}$$

Since an angle of 360° subtends an arc whose length is the circumference of the circle, we see that 360° is equivalent to

$$\theta = \frac{s}{r} = \frac{2\pi r}{r} = 2\pi \text{ radians.}$$

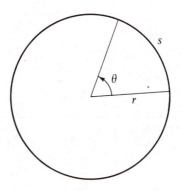

Figure 8.1

With this result, we can find a formula for the area of a sector (a pie-shaped region). If the central angle of a sector is 2π radians, the sector becomes a circle with area πr^2. Assuming that the area of a sector is proportional to the measure of the central angle, we compare the sector with central angle 2π and area πr^2 with one having central angle θ and area A. This gives the ratio

$$\frac{A}{\pi r^2} = \frac{\theta}{2\pi};$$

therefore the area of the sector is

$$A = \frac{1}{2} r^2 \theta.$$

Of course, θ is the **radian** measure of the angle. We shall use this result in the next section.

The graphs of the six common trigonometric functions are given in Figure 8.2. Note that all of them are periodic (repeating). To say that a function f has period p means that $f(x + p) = f(x)$ for all x; in other words, the function repeats itself after every p units. The period of $y = \tan x$ and $y = \cot x$ is π; the other

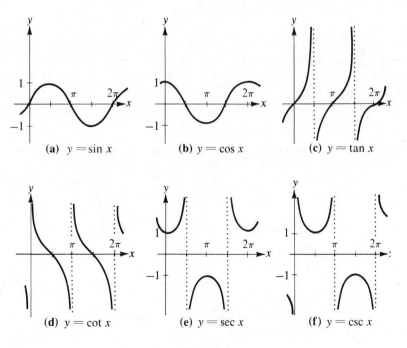

(a) $y = \sin x$ **(b)** $y = \cos x$ **(c)** $y = \tan x$

(d) $y = \cot x$ **(e)** $y = \sec x$ **(f)** $y = \csc x$

Figure 8.2

four functions have period 2π. Note also that four of them have vertical asymptotes, although there is no denominator to assume a value of zero. This is reasonable when we consider that all of them are defined as *ratios* of certain lengths. Let us see how these curves are altered by changing certain constants.

Example 1   Sketch $y = 3\sin 2x$.

Solution   First of all, note that $-1 \leq \sin x \leq 1$. Thus the factor 3 in $3\sin 2x$ changes this range by a factor of 3. It takes one complete cycle for whatever angle we are taking the sine of to go from 0 to 2π; that is, we have one complete cycle for

$$0 \leq 2x \leq 2\pi \qquad \text{or} \qquad 0 \leq x \leq \pi.$$

Thus, the 3 in $3\sin 2x$ triples the amplitude (or height) of the wave, while the 2 halves the period (or gives two complete cycles in the normal period of 2π). The result is shown in Figure 8.3.

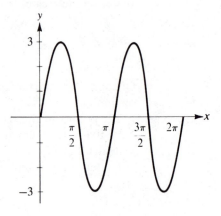

Figure 8.3

Example 2   Sketch $y = 4\cos\left(2x + \dfrac{\pi}{2}\right)$.

Solution   First let us write the equation in the form

$$y = 4\cos 2\left(x + \frac{\pi}{4}\right)$$

Now we see that the amplitude is 4 and the period is $2\pi/2 = \pi$. When $x = -\pi/4, \pi/4, 3\pi/4$, and so on, $x + \pi/4 = 0, \pi/2, \pi$, and so forth. Therefore $\pi/4$ has the effect of shifting the curve a distance $\pi/4$ to the left, as shown in Figure 8.4.

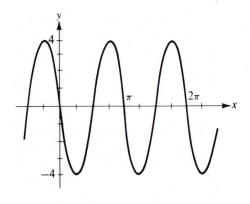

Figure 8.4

Example 3 Sketch $y = \cos x + \sin 2x$.

Solution The method of addition of ordinates (see Section 7.4) is quite useful for equations of this type. Sketching $y = \cos x$ and $y = \sin 2x$ and adding the ordinates, we have the result given in Figure 8.5.

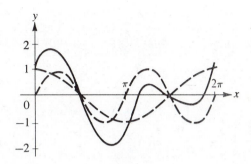

Figure 8.5

Example 4 Sketch $y = x + \sin x$.

Solution Perhaps you wonder how we can add x and $\sin x$, if x is an angle and $\sin x$ is a number. Actually both x and $\sin x$ are numbers. We take trigonometric functions not of angles, but of numbers. The numbers are simply the *measures* of angles. It is quite possible to consider trigonometric functions of numbers quite independently of any angular interpretations; but if we do want to impose such an interpretation, the value of x is the measure of an angle in **radians.** Again, addition of ordinates works very well; Figure 8.6 is self-explanatory.

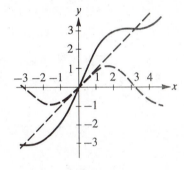

Figure 8.6

PROBLEMS

A

1. Express the following degree measures in radian measure: $45°$, $-210°$, $270°$, $30°$, $-180°$, $-60°$, $135°$, $150°$.

2. Express the following radian measures in degree measure: $\pi/3$, π, $3\pi/4$, $-\pi/2$, $5\pi/6$, $-2\pi/3$, $3\pi/2$, $10\pi/6$.

B

In Problems 3–26, sketch one complete cycle.

3. $y = 3 \cos x$

4. $y = 2 \sec x$

5. $y = -2 \sin x$

6. $y = 2 \tan 3x$

7. $y = 4 \sin \pi x$

8. $y = -2 \csc(\pi x/2)$

9. $y = 3 \cos 4x$

10. $y = 2 \sin(2x + \pi)$

11. $y = 3 \cos(2\pi x + \pi/2)$

12. $y = \tan(3x - \pi)$

13. $y = 2 \sec(4x - 2\pi)$

14. $y = -\cos(x - \pi/3)$

15. $y = -2 \sin(\pi/4 - x)$

16. $y = \sin x - \cos x$

17. $y = 2 \sin x + \sin 2x$

18. $y = \cos x - \sin 2x$

19. $y = 3 \cos x + \sin x$

20. $y = 4 \sin x + 2 \sin 2x - \sin 4x$

21. $y = 2 \sin x - \sin 2x + \dfrac{2}{3} \sin 3x$

22. $y = 1 - \cos x$

23. $y = 1 + 2 \sin x$

24. $y = 2 + \cos x$

25. $y = \sin^2 x$

26. $y = 2 - 3 \sin x$

In Problems 27–30, sketch the graph.

27. $y = x - \sin x$

28. $y = x^2 + \sin x$

29. $y = x \sin x$

30. $y = \dfrac{\sin x}{x}$

C   31. Sketch

$$y = \frac{\pi}{2} + 2 \sin x, \qquad y = \frac{\pi}{2} + 2 \sin x + \frac{2}{3} \sin 3x,$$

and

$$y = \frac{\pi}{2} + 2 \sin x + \frac{2}{3} \sin 3x + \frac{2}{5} \sin 5x$$

on the same coordinates. What do you think the graph of

$$y = \frac{\pi}{2} + 2\left(\sin x + \frac{1}{3} \sin 3x + \frac{1}{5} \sin 5x + \cdots\right)$$

looks like?

8.2

INVERSE TRIGONOMETRIC FUNCTIONS

Suppose we have the equation $y = \sin x$ and want to express x in terms of y. To do so, we introduce a new notation for the solution.

$$x = \arcsin y \qquad \text{or} \qquad x = \sin^{-1} y$$

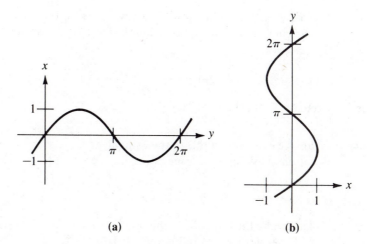

Figure 8.7

This is read: x is an inverse sine of y. Thus, $x = \arcsin y$ is equivalent to $y = \sin x$, or $y = \arcsin x$ is equivalent to $x = \sin y$. To graph $y = \arcsin x$, we merely graph $x = \sin y$. This is shown in Figure 8.7. Figure 8.7a shows the familiar sine curve with the labels on the axes reversed, while Figure 8.7b gives the same graph with the axes in their usual position. Figure 8.7b looks exactly like the graph of $y = \sin x$ with the x and y axes reversed. Note that arcsin x is *not* single-valued; one value of x gives many values of arcsin x. The remaining five trigonometric functions have inverses that are defined analogously.

Example 1   Sketch $y = 2 \arcsin 3x$.

Solution   We first convert this to the equivalent equation involving the sine.

$$\frac{y}{2} = \arcsin 3x$$

$$3x = \sin \frac{y}{2}$$

$$x = \frac{1}{3} \sin \frac{y}{2}$$

Graphing this by the methods of the previous section, we have Figure 8.8.

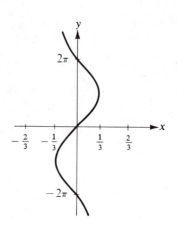

Figure 8.8

Note that the graph shown is a sine wave on the y axis. Furthermore, the 3 in the original equation gives an amplitude of $1/3$, while the 2 gives a period of $2 \cdot 2\pi = 4\pi$.

Example 2   Sketch $y = (1/3)\arccos 2x$.

Solution   Again we can convert this to the equivalent equation involving the cosine,

$$x = \frac{1}{2}\cos 3y,$$

and graph by the methods of the previous section. The result is given in Figure 8.9. This could also have been sketched by noting that we have a cosine wave on the y axis with amplitude $1/2$ and period $1/3 \cdot 2\pi = 2\pi/3$.

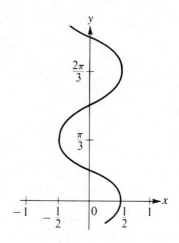

Figure 8.9

Example 3   Sketch $y = \pi/2 + \arctan x$.

Solution   In the tangent form, the equation is

$$x = \tan\left(y - \frac{\pi}{2}\right),$$

giving the graph of Figure 8.10. The $\pi/2$ has the effect of raising the curve a distance $\pi/2$.

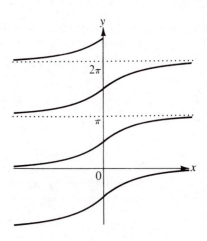

Figure 8.10

Example 4   Sketch $y = \text{arcsec}\,(x + 1)$.

Solution   In secant form we have

$$x = \sec y - 1,$$

which gives the graph of Figure 8.11. The 1 has the effect of shifting the graph one unit to the left.

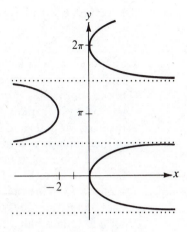

Figure 8.11

PROBLEMS

A   **1.** $y = \arccos x$ **2.** $y = \arctan x$ **3.** $y = \text{arccot}\, x$ **4.** $y = \text{arcsec}\, x$

B   **5.** $y = 2 \arcsin x$ **6.** $y = 3 \arccos x$

7. $y = 3 \,\text{arcsec}\, x$ **8.** $y = 4 \,\text{arccsc}\, x$ **9.** $y = \arcsin 3x$

10. $y = \arctan 2x$ **11.** $y = \arccos 4x$ **12.** $y = \text{arcsec}\, 2x$

13. $y = 3 \arcsin 2x$ **14.** $y = 2 \arccos 4x$ **15.** $y = 2 \arctan 3x$

16. $y = 4 \,\text{arcsec}\, 2x$ **17.** $y = -2 \arcsin \dfrac{x}{2}$ **18.** $y = -2 \arccos \dfrac{x}{3}$

19. $y = \dfrac{\pi}{2} + \arcsin x$ **20.** $y = \dfrac{\pi}{4} + \arccos x$ **21.** $y = -\dfrac{\pi}{4} + 2 \arcsin x$

22. $y = \dfrac{\pi}{4} - 2 \arctan x$ **23.** $y = \arcsin (x + 2)$ **24.** $y = \arccos (x - 1)$

25. $y = 3 \arcsin 2 (x - 1)$ **26.** $y = 2 \arctan (3x + 2)$

C   **27.** $y = \dfrac{\pi}{4} + 2 \arcsin 3 (x + 2)$ **28.** $y = \dfrac{\pi}{3} - 2 \,\text{arcsec}\, 4 (x - 2)$

29. $y = -\dfrac{2\pi}{3} + \dfrac{1}{2} \arccos \dfrac{x + 1}{3}$ **30.** $y = \dfrac{\pi}{4} + \dfrac{1}{3} \arctan \dfrac{2x + 1}{3}$

8.3

EXPONENTIAL AND LOGARITHMIC FUNCTIONS

Suppose we consider the graph of $y = 2^x$. By plotting a few points, we have the graph shown in Figure 8.12. The curve has the x axis as a horizontal asymptote, but it approaches this asymptote only at the left end. At the right end the graph increases very rapidly; there is, however, no vertical asymptote. The graph crosses the y axis at $(0, 1)$ and is always above the x axis. These are general characteristics of the graph of $y = a^x$, where $a > 1$. If $a < 1$, then the graph rises steeply on the left and approaches the x axis on the right. The graph of $y = 1^x$ is the horizontal line $y = 1$, since $1^x = 1$ for all x. The graphs of several such equations are given in Figure 8.13. Note that, for $a > 1$, the larger the base a, the more steeply the curve rises on the right and the more rapidly it approaches the x axis on the left; while, for $a < 1$, the smaller the base, the more steeply the curve rises on the left and the more rapidly it approaches the x axis on the right.

A base of zero gives the x axis without the origin (since 0^0 is meaningless), and negative bases are not considered at all, since powers of negative bases are not defined for any but integer exponents.

A number that is often used for a base is the number $e = 2.71828\ldots$. This base is especially useful in calculus, and, for this reason, is frequently encountered. The graph of $y = e^x$ lies between the graphs of $y = 2^x$ and $y = 3^x$, as in Figure 8.13.

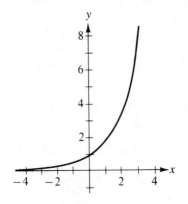

Figure 8.12

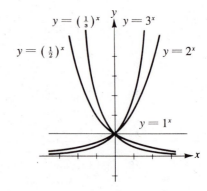

Figure 8.13

Example 1   Sketch the graph of $y = 2^{x-2}$.

Solution   This can be considered from two different points of view. One way is to consider the desired graph to be the graph of $y = 2^x$, translated two units to the right. Another is to note that

$$y = 2^{x-2} = 2^x \cdot 2^{-2} = \frac{2^x}{4}.$$

Thus the y coordinate of $y = 2^{x-2}$ is one-fourth of the corresponding y coordinate of $y = 2^x$. The result is given in Figure 8.14.

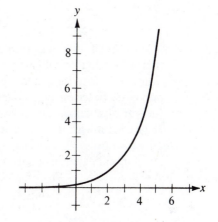

Figure 8.14

Example 2   Sketch the graph of $y = 2^{2x}$.

Solution   Since

$$y = 2^{2x} = (2^2)^x = 4^x,$$

we have the graph shown in Figure 8.15.

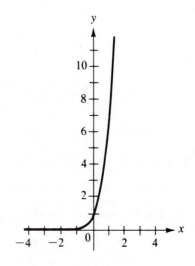

Figure 8.15

Note the distinction between $y = 2^x$ and $y = x^2$. The first, with a constant base and variable exponent, is an exponential function such as we have been discussing. The other, with a variable base and constant exponent, is a power function; its graph is a parabola.

Let us now consider the logarithm, which we define in the customary way.

Definition *The **logarithm**, base a ($a > 0$, $a \neq 1$), of the number x ($x > 0$) is the number y such that $a^y = x$. Thus*

$$y = \log_a x \qquad means \qquad x = a^y.$$

We see that the logarithm and exponential functions are inverses of each other; that is, $y = \log_a x$, which is equivalent to $x = a^y$, is simply the exponential function $y = a^x$ with the x and y reversed. Thus, we may use our knowledge of the exponential function to sketch the graph of a logarithm. The graph of $y = \log_a x$ for various values of a is given in Figure 8.16. This figure is the same as Figure 8.13, with the x and y axes interchanged. Note that logarithms are defined only for positive values of x—there is no graph to the left of the y axis. The y axis is a vertical asymptote and $y = \log_a x$ increases slowly for $a > 1$ and decreases slowly for $a < 1$. All contain the point $(1, 0)$.

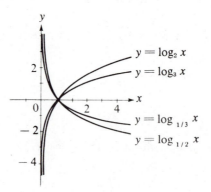

Figure 8.16

Example 3 Sketch the graph of $y = \log_2 (x + 2)$.

Solution The expression $x + 2$ merely translates the graph of $y = \log_2 x$ two units to the left. That is, there is a vertical asymptote when $x + 2 = 0$ or $x = -2$, and the graph crosses the x axis when $x + 2 = 1$ or $x = -1$. Thus we have the graph shown in Figure 8.17.

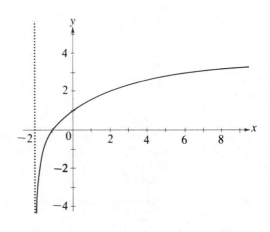

Figure 8.17

Example 4   Sketch the graph of $y = \log_4(-x)$.

Solution   The logarithm is defined only when $-x$ is positive, or when x is negative. The result is the mirror image of $y = \log_4 x$ reflected in the y axis (see Figure 8.18).

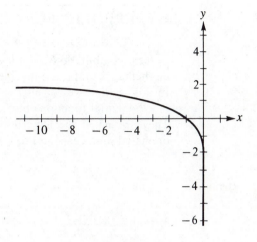

Figure 8.18

Two bases that are frequently used are 10 and e. The following abbreviations are frequently encountered and are used throughout this book.

$$\log x = \log_{10} x$$
$$\ln x = \log_e x$$

PROBLEMS

Sketch the graph of each of the following equations.

A   **1.** $y = 2^{x+1}$ **2.** $y = 2^{x-1}$ **3.** $y = 3^{-x}$ **4.** $y = e^x$

B   **5.** $y = 4^{x-3}$ **6.** $y = 3^{1-x}$ **7.** $y = 3^{2x}$ **8.** $y = 4^{-2x}$

9. $y = 2^{2x+1}$ **10.** $y = 5^{2x-1}$ **11.** $y = \left(\frac{1}{2}\right)^{x+1}$ **12.** $y = \left(\frac{1}{3}\right)^{x-2}$

13. $y = \left(\frac{1}{4}\right)^{2x+1}$ **14.** $y = \left(\frac{1}{2}\right)^{2-x}$ **15.** $y = 2^{|x|}$ **16.** $y = 2^{x^2}$

17. $y = \log_2(x+1)$ **18.** $y = \log_2(x-1)$ **19.** $y = \log_3(x+4)$

20. $y = \log_4(x-3)$ **21.** $y = \log_3(-x)$ **22.** $y = \ln x$

23. $y = \log_2(2x)$ **24.** $y = \log_3(9x)$ **25.** $y = \log_2(4x-1)$

26. $y = \log_4(4x+1)$ **27.** $y = \log_2|x|$ **28.** $y = \log_2 x^2$

29. $y = \log_{1/2}(x+1)$ **30.** $y = \log_{1/3}(x-2)$

8.4

HYPERBOLIC FUNCTIONS

Certain combinations of functions occur frequently enough in applications of mathematics that it is convenient to set them apart by giving them special names. In this section, we consider the hyperbolic functions, which are defined in terms of exponential functions but are like the trigonometric functions in many ways. They are called the hyperbolic sine, hyperbolic cosine, and so on, and are abbreviated sinh, cosh, and so forth, respectively.

Definition 1. $\sinh x = \dfrac{e^x - e^{-x}}{2}$ 2. $\cosh x = \dfrac{e^x + e^{-x}}{2}$

3. $\tanh x = \dfrac{\sinh x}{\cosh x}$ 4. $\coth x = \dfrac{\cosh x}{\sinh x}$

5. $\operatorname{sech} x = \dfrac{1}{\cosh x}$ 6. $\operatorname{csch} x = \dfrac{1}{\sinh x}$

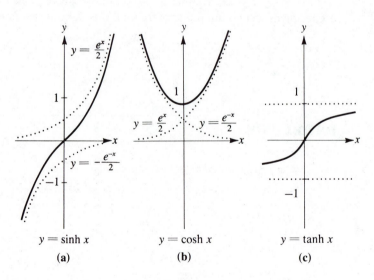

$y = \sinh x$

(a)

$y = \cosh x$

(b)

$y = \tanh x$

(c)

Figure 8.19

The first two hyperbolic functions are easily graphed by addition of ordinates, and the remaining four by division of ordinates. For example, since

$$\sinh x = \frac{e^x}{2} - \frac{e^{-x}}{2},$$

we graph $y = e^x/2$ and $y = -e^{-x}/2$ and add the ordinates. This is given in Figure 8.19a. The graph of $y = \tanh x$ is found by noting that

$$\tanh x = \frac{\sinh x}{\cosh x}.$$

For each value of x, we divide $y_1 = \sinh x$ by $y_2 = \cosh x$ to find $\tanh x$. The result is given in Figure 8.19c.

Example 1 Sketch the graph of $y = \sinh (x/2)$.

Solution Note that the substitution of $x = 1$ into this equation is equivalent to the substitution of $x = 1/2$ into $y = \sinh x$. Thus the graph of $y = \sinh (x/2)$ is like that of $y = \sinh x$, but "stretched" horizontally so that each point is twice as far from the y axis. This is given in Figure 8.20.

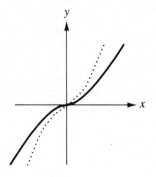

Figure 8.20

Example 2 Sketch the graph of $y = 2 \cosh x$.

Solution The effect of the 2 here is to stretch the graph of $y = \cosh x$ vertically so that each point is twice as far from the x axis. This is given in Figure 8.21.

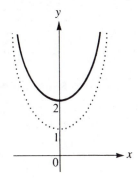

Figure 8.21

The inverses of hyperbolic functions can be considered in the same way as those of trigonometric functions. They are called the inverse hyperbolic sine, and so forth, and are represented by $\sinh^{-1}$, and so forth.

Definition $y = \sinh^{-1} x$ *means* $x = \sinh y$

The other five inverse hyperbolic functions are defined similarly.

Again the graphs of the inverses can be determined from those of the corresponding hyperbolic functions.

Example 3   Sketch the graph of $y = \sinh^{-1} x$.

Solution   Since this is equivalent to $x = \sinh y$, its graph is the same as the graph of $y = \sinh x$ with the x and y reversed. The result is given in Figure 8.22.

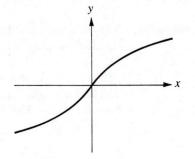

Figure 8.22

PROBLEMS

Sketch the graph of each of the following equations.

A   **1.** $y = \coth x$ **2.** $y = \operatorname{sech} x$ **3.** $y = \operatorname{csch} x$ **4.** $y = \sinh 2x$

   5. $y = \dfrac{1}{2} \cosh x$ **6.** $y = \tanh \dfrac{x}{3}$ **7.** $y = 2 \sinh x$ **8.** $y = 3 \cosh x$

B   **9.** $y = 2 \cosh 3x$ **10.** $y = 2 \tanh 3x$ **11.** $y = \sinh |x|$

   12. $y = \sinh x^2$ **13.** $y = \sinh (x - 1)$ **14.** $y = \cosh (x + 2)$

   15. $y = \tanh |x + 1|$ **16.** $y = 2 \cosh (x - 3)$ **17.** $y = \cosh^{-1} x$

   18. $y = \tanh^{-1} x$ **19.** $y = 2 \sinh^{-1} x$ **20.** $y = \sinh^{-1} 2x$

   21. Express $\tanh x$ in terms of exponential functions.

C   **22.** Show that $\tanh^{-1} x = \dfrac{1}{2} \ln \dfrac{1 + x}{1 - x}$, $|x| < 1$.

   23. Show that $\sinh^{-1} x = \ln (x + \sqrt{x^2 + 1})$.

REVIEW PROBLEMS

Sketch the graph of each of the following equations.

A

1. $y = 2 \sin 3x$

2. $y = -\cos 2x$

3. $y = 2 \arccos 3x$

4. $y = \dfrac{1}{3} \arcsin x$

5. $y = 2^{x-2}$

6. $y = \sinh 2x$

7. $y = \cosh \dfrac{x}{3}$

8. $y = 3 \sinh x$

B

9. $y = 2 \sin \left(x - \dfrac{\pi}{6} \right)$

10. $y = 3 \sin 2x + 2 \cos 4x$

11. $y = 2 \sin \left(2x + \dfrac{\pi}{2} \right)$

12. $y = \dfrac{x}{2} + \cos 2x$

13. $y = \dfrac{\pi}{4} - \arctan x$

14. $y = \dfrac{\pi}{3} + \dfrac{1}{3} \arctan \dfrac{2x - 1}{2}$

15. $y = 3^{2x+1}$

16. $y = \left(\dfrac{1}{2} \right)^{x-1}$

17. $y = \log_2 (x + 3)$

18. $y = \log_3 (3x + 1)$

19. $y = 2 \cosh^{-1} x$

20. $y = \operatorname{sech}^{-1} x$

9

POLAR COORDINATES AND PARAMETRIC EQUATIONS

9.1

POLAR COORDINATES

Up to now, a point in the plane has been represented by a pair of numbers, (x, y), which represent (for perpendicular axes) the distances of the point from the y and x axes, respectively. Another way of representing points is by *polar coordinates*. In this case, we need only one axis (the *polar axis*) and a point on it (the *pole*). These correspond to the x axis and the origin of the rectangular coordinate system. Normally we shall include the y axis, even though it is not necessary to do so.

Before considering points in polar coordinates, let us recall that an angle in the standard position has its vertex at the origin (or pole) and its initial side on the positive end of the x axis (or polar axis). The terminal side is another ray (or half-line) with the origin as its end point. The ray with the same end point and on the same line as the terminal side is called the ray opposite the terminal side. For example, the terminal side of a 90° angle in standard position is the positive end

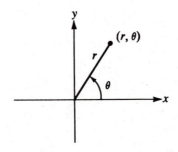

Figure 9.1

of the y axis together with the origin; the ray opposite the terminal side is the negative end of the y axis together with the origin.

A point P is represented, in polar coordinates, by an ordered pair of numbers (r, θ) (see Figure 9.1). It is determined in the following way: first find the terminal side of the angle θ in standard position; if $r \geq 0$, then P is on this terminal side and at a distance r from the pole; if $r < 0$, then P is on the ray opposite the terminal side and at a distance $|r|$ from the pole. A few points are given with their polar coordinates in Figure 9.2.

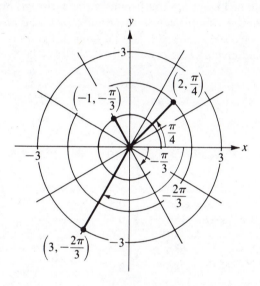

Figure 9.2

It might be noted that while the terminal side of the angle $-\pi/3$ is in the fourth quadrant, $(-1, -\pi/3)$ is in the second quadrant. The quadrant that a point is in is *not* determined by the signs of the two polar coordinates, as it is with rectangular coordinates. It is determined by the size of θ and the sign of r. If r is positive, the point is in whatever quadrant θ is in; if r is negative, the point is in the opposite quadrant.

Polar coordinates present one problem we did not have with rectangular coordinates—a point has more than one representation. For example: $(2, \pi/2)$ and $(-2, -\pi/2)$ represent the same point. In fact, if (r, θ) is one representation of a point, then $(r, \theta + \pi n)$, where n is an even integer, and $(-r, \theta + \pi n)$, where n is an odd integer, are representations of the same point. Furthermore, $(0, \theta)$ is the pole for any choice of θ.

9.2

GRAPHS IN POLAR COORDINATES

Equations in polar coordinates can be graphed by point-by-point plotting, as we graphed rectangular coordinates.

Example 1   Graph $r = \sin \theta$.

Solution   Note in Figure 9.3 that we have the entire graph for $0 \le \theta < \pi$. The remaining values of θ simply repeat the graph a second time, since $(0,0) = (0, \pi)$, $(0.5, \pi/6) = (-0.5, 7\pi/6)$, and so forth. Of course, values of θ outside the range $0 \le \theta \le 2\pi$ would give no new points. As we shall see later, this is a circle.

θ			r
0	$=$	$0°$	0.00
$\pi/6$	$=$	$30°$	0.50
$\pi/3$	$=$	$60°$	0.87
$\pi/2$	$=$	$90°$	1.00
$2\pi/3$	$=$	$120°$	0.87
$5\pi/6$	$=$	$150°$	0.50
π	$=$	$180°$	0.00
$7\pi/6$	$=$	$210°$	-0.50
$4\pi/3$	$=$	$240°$	-0.87
$3\pi/2$	$=$	$270°$	-1.00
$5\pi/3$	$=$	$300°$	-0.87
$11\pi/6$	$=$	$330°$	-0.50
2π	$=$	$360°$	0.00

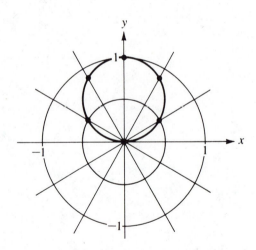

Figure 9.3

This method of point-by-point plotting is quite cumbersome here, as it was in the case of rectangular coordinates. One way to simplify the proceedings is to represent the table of values of r and θ by means of a graph. This may sound as if we are going in circles—we can get the graph from a table of values of r and θ that is represented by a graph. Actually, this is not so bad as it sounds. We shall represent the table by a graph in *retangular coordinates*.

For example, the table of values of r and θ, used in Example 1 can be represented by the graph shown in Figure 8.2a on page 209. Of course, Figure 8.2a represents the graph of $y = \sin x$; we merely replace the symbols x and y by θ and r.

Example 2 Graph $r = 1 + \cos \theta$.

Solution We can easily graph this equation in rectangular coordinates by using addition of ordinates. The result is given in Figure 9.4. Now we can read off values of r and θ just as we would from a table. As θ increases from 0 to $\pi/2$, r goes from 2 to 1. This gives the portion of the curve shown in Figure 9.5a. As θ goes from $\pi/2$ to π, r goes from 1 down to 0 (shown in Figure 9.5b). As θ goes from π to $3\pi/2$, r goes from 0 back up to 1 (as in Figure 9.5c); and finally, as θ goes from $3\pi/2$ to 2π, we see in Figure

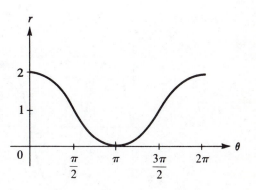

Figure 9.4

9.5d that θ goes from 1 to 2. The same path is traced for values of θ beyond 2π or less than 0. Putting all of this together, we have the desired graph, shown in Figure 9.5e. This curve is called a *cardioid,* which means "heart-shaped." It is a special case of a more general curve called a *limaçon* (French for snail), which has the form $r = a + b \sin \theta$ or $r = a + b \cos \theta$. If $|a| = |b|$, then we have a cardioid. See Problems 17–20 for the graphs of limaçons.

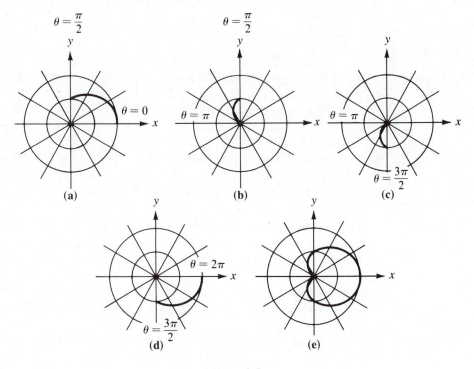

Figure 9.5

Example 3   Graph $r = \sin 2\theta$.

Solution

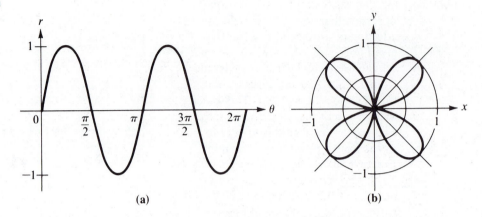

(a)   **(b)**

Figure 9.6

The graph is given in rectangular coordinates in Figure 9.6a. This is then put on the polar graph shown in Figure 9.6b. Note that for θ in the range $\pi/2 < \theta < \pi$, r is negative. Thus, instead of giving the loop in the second quadrant, it gives the one in the fourth quadrant. Similarly, for θ in the range $3\pi/2 < \theta < 2\pi$, r is negative. This gives the loop in the second quadrant. The resulting curve is called a four-leafed rose.

Example 4   Graph $r^2 = \sin 2\theta$.

Solution   Graphing in rectangular coordinates by the methods of Section 7.3, we have the result given in Figure 9.7a. There are a couple of things of interest here. First, $r^2 = \sin 2\theta$ has two values of r for each θ in the ranges $0 < \theta < \pi/2$ and $\pi < \theta < 3\pi/2$, while it has no value at all for $\pi/2 < \theta < \pi$ and $3\pi/2 < \theta < 2\pi$. Since it has two values in the range $0 < \theta < \pi/2$, we get both loops for $0 \le \theta \le \pi/2$, shown in Figure 9.7b. Similarly, we get both loops a second time for $\pi \le \theta \le 3\pi/2$. Because there is no value of r for $\pi/2 < \theta < \pi$ and $3\pi/2 < \theta < 2\pi$, there are no points of the graph in the second or fourth quadrants. This is called a lemniscate.

There are tests for symmetry in polar coordinates which are somewhat like those in rectangular coordinates. Suppose, for example, that for each point (r, θ) on a given curve there corresponds another point $(r, -\theta)$ on the same curve. Then (see Figure 9.8a) the curve is symmetric about the x axis. Thus if θ is replaced by $-\theta$ and the result is equivalent* to the original equation, then the graph is symmetric about the x axis. Figures 9.8b and 9.8c illustrate conditions leading to symmetry

---

*Two equations are equivalent if any point that satisfies one of them also satisfies the other. To determine equivalence we use algebraic or trigonometric identities, add any expression to both sides of one equation, or multiply both sides of an equation by a nonzero constant in order to make that equation identical to the other.

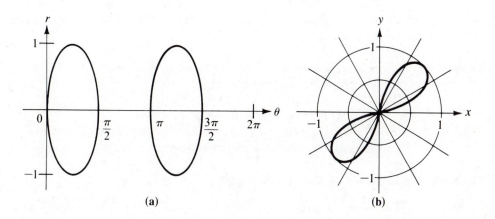

(a)   (b)

Figure 9.7

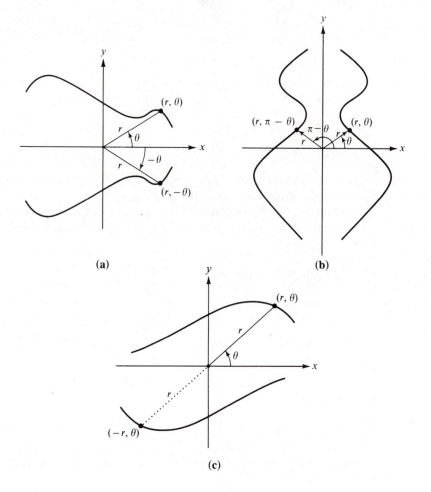

(a)   (b)

(c)

Figure 9.8

about the y axis and the pole, respectively. These tests are summarized in the following theorem.

Theorem 9.1 (a) *If θ is replaced by $-\theta$ and the result is equivalent to the original equation, then the graph is symmetric about the x axis.*

(b) *If θ is replaced by $\pi - \theta$ and the result is equivalent to the original equation, then the graph is symmetric about the y axis.*

(c) *If r is replaced by $-r$ and the result is equivalent to the original equation, then the graph is symmetric about the pole.*

Example 5 Test $r = 1 + \cos \theta$ for symmetry.

Solution The three tests are represented schematically below (the arrow is used to represent "is replaced by").

(a) $\begin{cases} r \to r \\ \theta \to -\theta \end{cases}$ (b) $\begin{cases} r \to r \\ \theta \to \pi - \theta \end{cases}$ (c) $\begin{cases} r \to -r \\ \theta \to \theta \end{cases}$

$\quad\ \ r = 1 + \cos(-\theta) \qquad\ \ r = 1 + \cos(\pi - \theta) \qquad -r = 1 + \cos \theta$

$\qquad\ = 1 + \cos \theta \qquad\qquad\ = 1 - \cos \theta$

Since $r = 1 + \cos(-\theta)$ is equivalent to $r = 1 + \cos \theta$, we have symmetry about the x axis. The other two tests are negative; however, *this in itself is not enough to say that we do not have the other two types of symmetry* (we shall consider this in more detail in the next example). Nevertheless the graph of $r = 1 + \cos \theta$ (given in Figure 9.5e) indicates that we do have symmetry only about the x axis.

We see from this example that, while symmetry is sometimes an aid in graphing an equation, the graph can also be an aid in determining the presence or absence of symmetry. This is especially true in polar coordinates, because negative results in the tests of Theorem 9.1 do not necessarily imply a lack of symmetry, as the next example makes evident.

Example 6 Test $r = \sin 2\theta$ for symmetry.

Solution (a) $\begin{cases} r \to r \\ \theta \to -\theta \end{cases}$ (b) $\begin{cases} r \to r \\ \theta \to \pi - \theta \end{cases}$ (c) $\begin{cases} r \to -r \\ \theta \to \theta \end{cases}$

$\quad\ \ r = \sin(-2\theta) \qquad\ \ r = \sin(2\pi - 2\theta) \qquad -r = \sin 2\theta$

$\qquad = -\sin 2\theta \qquad\qquad = -\sin 2\theta$

All three tests are negative. Yet we can see from Figure 9.6b that all three types of symmetry are present.

The reason for the rather strange behavior of the last example can be traced directly to the fact that one point has many different representations in polar coordinates. For example, $(r, -\theta) = (-r, \pi - \theta) = (r, 2\pi - \theta)$, etc. Thus there are many tests for symmetry about the x axis. The equalities above lead to the three tests:

$$\begin{cases} r \to r \\ \theta \to -\theta \end{cases} \quad \begin{cases} r \to -r \\ \theta \to \pi - \theta \end{cases} \quad \begin{cases} r \to r \\ \theta \to 2\pi - \theta. \end{cases}$$

If any one of these gives an equation that is equivalent to the original, there is symmetry about the x axis. Notice that while the first and third of these three tests give negative results in the equation of Example 6, the second gives a positive result. This is sufficient to assure us that there is symmetry about the x axis. If they all give negative results, nothing can be concluded, since there are still other possible tests. The student can easily devise other tests for all three types of symmetry. The result of all of this is that we must be content with tests for symmetry which do not guarantee a lack of symmetry when the test is negative.

Because of the multiplicity of the foregoing tests and the indecisiveness of negative results, you might prefer to rely upon the graphs of an equation. For example, Figure 9.6b suggests that we have all three types of symmetry for $r = \sin 2\theta$. Then the symmetry of Figure 9.6a assures us that we really do have the suspected symmetry. Note, however, that a particular type of symmetry in the rectangular coordinate graph does not necessarily imply the same type of symmetry in polar coordinates. Although the rectangular coordinate graph of $r^2 = \sin 2\theta$ (i.e., of $y^2 = \sin 2x$) shows symmetry about the x axis, the polar graph does not (see Figure 9.7). Furthermore, the polar graph exhibits symmetry about the pole (or origin), while the rectangular coordinate graph does not. But the symmetry of the loops in rectangular coordinates does imply symmetry of the loops in polar coordinates and thus symmetry about the pole.

PROBLEMS

A   1. Plot the following points: $(1, \pi/3)$, $(2, 45°)$, $(0, 30°)$, $(-2, 90°)$, $(-1, 3\pi/4)$, $(2, 300°)$.

*2. Give an alternate polar representation with $0° \le \theta < 180°$: $(4, 330°)$, $(-2, 420°)$, $(1, 210°)$, $(0, 283°)$, $(-3, 270°)$, $(2, 240°)$.

*3. Give an alternate polar representation with $r \ge 0$ and $0 \le \theta < 2\pi$: $(-4, 2\pi/3)$, $(3, -\pi/3)$, $(0, 7\pi/2)$, $(-1, 11\pi/6)$, $(-2, 13\pi/6)$, $(-2, 3\pi/4)$.

4. Give an alternate polar representation: $(1, 30°)$, $(-2, \pi)$, $(4, 210°)$, $(0, \pi/3)$, $(-1, 30°)$, $(2, \pi/2)$.

B   *In Problems 5–34, sketch the graph of the given equation.*

5. $r = \cos \theta$ (circle)

6. $r = 2 \sin \theta$ (circle)

7. $r = 1 - \cos \theta$ (cardioid)

8. $r = 1 + \sin \theta$ (cardioid)

9. $r = 1 - \sin \theta$ (cardioid)

10. $r = \sin \theta - 1$ (cardioid)

11. $r = \cos 2\theta$ (four-leafed rose)

12. $r = \sin 4\theta$ (eight-leafed rose)

13. $r = \sin 3\theta$ (three-leafed rose)

14. $r = \cos 3\theta$ (three-leafed rose)

15. $r = \cos 5\theta$ (five-leafed rose)

16. $r = \sin 6\theta$ (twelve-leafed rose)

*Answers are given for the first three parts of Problems 2 and 3 rather than for none of 2 and all of 3.

17. $r = 1 + 2 \sin \theta$ (limaçon)

18. $r = 1 - 2 \cos \theta$ (limaçon)

19. $r = 2 + \cos \theta$ (limaçon)

20. $r = 2 + 3 \sin \theta$ (limaçon)

21. $r = \tan \theta$

22. $r = \sec \theta$

23. $r^2 = \sin \theta$

24. $r^2 = \cos 3\theta$ (lemniscate)

25. $r^2 = \cos 4\theta$

26. $r^2 = \sin^2 \theta$

27. $r^2 = 1 + \cos \theta$

28. $r^2 = 1 - \sin \theta$

29. $r = \theta$ (spiral of Archimedes)

30. $r = 1/\theta$ (hyperbolic spiral)

31. $r^2 = \theta^2$

32. $r = \dfrac{2}{1 - \cos \theta}$ (parabola)

33. $r = \dfrac{2}{1 - 2 \cos \theta}$ (hyperbola)

34. $r = \dfrac{2}{2 - \cos \theta}$ (ellipse)

C **35.** Find two tests for symmetry about the y axis that are different from the one given in this section.

36. Find two tests for symmetry about the pole that are different from the one given in this section.

9.3

POINTS OF INTERSECTION

Suppose we have a pair of equations in polar form that we solve simultaneously to obtain pairs of numbers satisfying both equations—that is, points of intersection of the two curves.

Example 1 Find the points of intersection of $r = 1$ and $r = 2 \sin \theta$.

Solution Eliminating r from this pair of equations, we get

$$\sin \theta = \frac{1}{2}, \quad \text{or} \quad \theta = \frac{\pi}{6}, \frac{5\pi}{6},$$

giving the points $(1, \pi/6)$ and $(1, 5\pi/6)$. The graphs of these two curves, showing the two points of intersection are given in Figure 9.9.

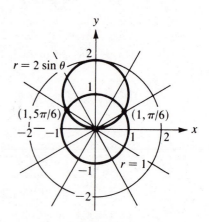

Figure 9.9

While the solutions of a pair of simultaneous equations must be points of inter-section of the curves represented by the equations, some points of intersection can-not be found in this way. The reason is that they have different representations on the two curves. Thus *we must graph both curves* to be sure that we have found all points of intersection.

Example 2   Find the points of intersection of
$r = \sin \theta$ and $r = \cos \theta$.

Solution   Eliminating r between the two equa-tions, we have

$$\sin \theta = \cos \theta.$$

If we divide by $\cos \theta$, then

$$\tan \theta = 1$$

and

$$\theta = \frac{\pi}{4} + \pi \cdot n.$$

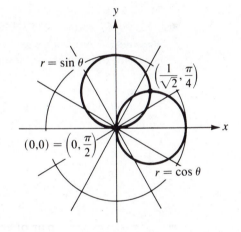

In the range $0 \le \theta < 2\pi$, we have $(1/\sqrt{2}, \pi/4)$ and $(-1/\sqrt{2}, 5\pi/4)$. But these are different representations for the same point. Thus we have found only one point of intersection. As we can see from Figure 9.10, there are really two points of intersection—the one we found and the pole.

Figure 9.10

The pole has many different repre-sentations. On the curve $r = \sin \theta$, it is represented by $(0, \pi n)$; on $r = \cos \theta$, it is represented by $(0, \pi/2 + \pi n)$. Thus, while the pole is common to both curves, it does not have a common representation that satisfies both equations. So we cannot find this point of intersection by finding simultaneous solutions of the two equations. This point has been identified in Figure 9.10 by both $(0, 0)$ and $(0, \pi/2)$ to show that there is no one representation satisfying both equations.

One convenient way to think of the phenomenon of this last example is to imagine the two curves as paths traced by points as θ increases uniformly with time. From this point of view, the curves both go through the origin; but they do so at different times.

Example 3 Find all points of intersection of
$r = \cos 2\theta$ and $r = \sin \theta$.

Solution

$$\cos 2\theta = \sin \theta$$
$$1 - 2\sin^2 \theta = \sin \theta$$
$$2\sin^2 \theta + \sin \theta - 1 = 0$$
$$(2\sin \theta - 1)(\sin \theta + 1) = 0$$

$$\sin \theta = \frac{1}{2}, \quad \sin \theta = -1$$

$$\theta = \frac{\pi}{6}, \quad \frac{5\pi}{6}, \quad \frac{3\pi}{2}$$

Thus, we have the points $(1/2, \pi/6)$, $(1/2, 5\pi/6)$, and $(-1, 3\pi/2)$. In addition, we can see from Figure 9.11 that the pole is a point of intersection; it may be represented by $(0, \pi/4)$ satisfying the equation $r = \cos 2\theta$ or by $(0, 0)$ satisfying $r = \sin \theta$. It might also be noted that the point $(-1, 3\pi/2)$ can also be written $(1, \pi/2)$, but this form satisfies only $r = \sin \theta$.

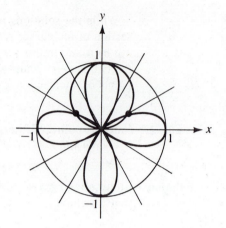

Figure 9.11

PROBLEMS

Find all points of intersection of the given curves.

A **1.** $r = \sqrt{2}, r = 2\cos \theta$ **2.** $r = \sqrt{3}, r = 2\sin \theta$

   3. $r = 2, r = \sin \theta + 2$ **4.** $r = 1, r = 2\cos 2\theta$

   5. $r = \cos \theta, r = 1 - \cos \theta$

B **6.** $r = \cos \theta, r = 1 + \sin \theta$

   7. $r = \sin 2\theta, r = \sin \theta$ **8.** $r = \sin 2\theta, r = \sqrt{2}\cos \theta$

   9. $r = \sec \theta, r = \csc \theta$ **10.** $r = \sec \theta, r = \tan \theta$

   11. $r = 3\cos \theta + 4, r = 3$ **12.** $r = \sin 2\theta, r = \cos 2\theta$

   13. $r = 2(1 + \cos \theta), r(1 - \cos \theta) = 1$ **14.** $r = 1 - \sin \theta, r(1 - \sin \theta) = 1$

   15. $r = 1 - \sin \theta, r = 1 - \cos \theta$ **16.** $r^2 = \sin \theta, r^2 = \cos \theta$

17. $r^2 = \cos \theta, r^2 = \sec \theta$ **18.** $r = 2 \cos \theta + 1, r = 2 \cos \theta - 1$

19. $r^2 = \sin \theta, r = \sin \theta$ **20.** $r^2 = \sin \theta, r = \cos \theta$

9.4

RELATIONSHIPS BETWEEN RECTANGULAR AND POLAR COORDINATES

There are some simple relationships between rectangular and polar coordinates. These can be found easily by a consideration of Figure 9.12. They are given in the following theorem.

Theorem 9.2   *If (x, y) and (r, θ) represent the same point in rectangular and polar coordinates, respectively, then*

$$x = r \cos \theta$$
$$y = r \sin \theta$$
$$r^2 = x^2 + y^2$$
$$\tan \theta = \frac{y}{x}$$

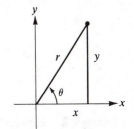

Figure 9.12

The last two, which may be solved for r and θ, would give us expressions involving $\pm$ and arctan. Thus, we prefer to leave them in their present form.

With these, we can now change from one coordinate system to the other.

Example 1   Give a rectangular coordinate representation for the point having polar coordinates $(2, \pi/6)$.

Solution
$$
\begin{aligned}
x &= r \cos \theta & y &= r \sin \theta \\
&= 2 \cos \pi/6 & &= 2 \sin \pi/6 \\
&= 2 \cdot \frac{\sqrt{3}}{2} & &= 2 \cdot \frac{1}{2} \\
&= \sqrt{3} & &= 1
\end{aligned}
$$

Thus the point $(2, \pi/6)$ in polar coordinates is the point $(\sqrt{3}, 1)$ in rectangular coordinates (see Figure 9.13).

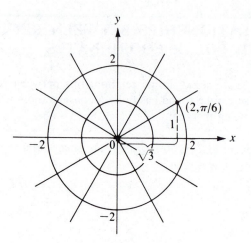

Figure 9.13

Example 2   Express in polar coordinates the point having rectangular coordinates $(4, -4)$.

Solution

$$r^2 = x^2 + y^2 \qquad \tan\theta = \frac{y}{x}$$

$$= 16 + 16 \qquad\qquad = \frac{-4}{4}$$

$$= 32 \qquad\qquad\quad = -1$$

$$r = \pm 4\sqrt{2} \qquad\quad \theta = \frac{3\pi}{4} + \pi n$$

We have a choice for both r and θ. The values of r and θ cannot be selected independently; the value we choose for one will limit the available choices for the other. In this case, the point $(4, -4)$ is in the fourth quadrant. Thus, we may choose either a fourth-quadrant angle and a positive r or a second-quadrant angle and a negative r. Thus (see Figure 9.14) the point $(4, -4)$ has any one of the following polar coordinate representations:

$$(4\sqrt{2}, 7\pi/4) = (-4\sqrt{2}, 3\pi/4) = (4\sqrt{2}, -\pi/4), \text{ etc.}$$

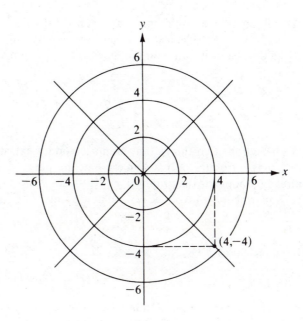

Figure 9.14

Of course we can use these equations to find a polar equation corresponding to one in rectangular coordinates, and vice versa.

Example 3   Express $y = x^2$ in polar coordinates.

Solution

$$y = x^2$$
$$r \sin \theta = r^2 \cos^2 \theta$$
$$\sin \theta = r \cos^2 \theta \quad \text{or} \quad r = 0$$
$$r = \frac{\sin \theta}{\cos^2 \theta}$$
$$r = \sec \theta \tan \theta$$

Since $r = 0$ represents only the pole and it is included in $r = \sec \theta \tan \theta$, we may drop $r = 0$. The result is

$$r = \sec \theta \tan \theta.$$

Example 4   Express $r = 1 - \cos \theta$ in rectangular coordinates.

Solution   First, multiply through by r.

$$r^2 = r - r \cos \theta$$

At this point, we could make the substitutions $r^2 = x^2 + y^2$, $r\cos\theta = x$, and $r = \pm\sqrt{x^2 + y^2}$. The last is rather bothersome, since it involves a $\pm$. In order to avoid this, let us isolate r on one side of the equation and square.

$$r = r^2 + r\cos\theta$$
$$r^2 = (r^2 + r\cos\theta)^2$$
$$x^2 + y^2 = (x^2 + y^2 + x)^2$$

We have done two things that might introduce extraneous roots: (1) Multiplying by r may introduce only a single point, the pole, to the graph. Since the pole is already a point of the graph of $r = 1 - \cos\theta$, no new point is introduced here. (2) Squaring may introduce several new points. The equation

$$r^2 = (r^2 + r\cos\theta)^2$$

is equivalent to

$$r = \pm(r^2 + r\cos\theta).$$

Now $r = r^2 + r\cos\theta$ is equivalent to our original equation, $r = 1 - \cos\theta$, while $r = -(r^2 + r\cos\theta)$ is equivalent to $r = -1 - \cos\theta$. Thus

$$x^2 + y^2 = (x^2 + y^2 + x)^2$$

is equivalent to $r = 1 - \cos\theta$ together with $r = -1 - \cos\theta$. But $r = 1 - \cos\theta$ and $r = -1 - \cos\theta$ have the same graph. Thus we have introduced no new points by squaring.

PROBLEMS

A

1. The following points are given in polar coordinates. Give the rectangular coordinate representation of each: $(1, \pi)$, $(\sqrt{3}, \pi/3)$, $(-1, 3\pi)$, $(\sqrt{2}, 3\pi/4)$, $(2\sqrt{3}, 5\pi/3)$, $(-3, 7\pi/6)$, $(0, 5\pi/4)$, $(4, 0)$, $(-2, 7\pi/4)$.

2. The following points are given in rectangular coordinates. Give a polar coordinate representation of each: $(\sqrt{2}, -\sqrt{2})$, $(-1, \sqrt{3})$, $(4, 0)$, $(-1, -1)$, $(0, -2)$, $(0, 0)$, $(-2\sqrt{3}, 2)$, $(-3, 1)$, $(4, 3)$, $(-2, 4)$.

In Problems 3–12, express the given equation in polar coordinates.

3. $x = 2$ 4. $y = 5$

5. $x^2 + y^2 = 1$ 6. $x^2 - y^2 = 4$

7. $x = y^2$ 8. $y = x^3$

9. $x + 2y - 4 = 0$ **10.** $x = y$

11. $y = 3x$ **12.** $y^2 = x^3$

In Problems 13–18, *express the given equation in rectangular coordinates.*

13. $r = a$ **14.** $\theta = \pi/4$

15. $\theta = \pi/3$ **16.** $r = 2\sin\theta$

17. $r = 4\cos\theta$ **18.** $r = \sin 2\theta$

B *In Problems* 19–24, *express the given equation in polar coordinates.*

19. $(x + y)^2 = x - y$ **20.** $x^2 + y^2 - 2x = 0$

21. $x^2 + y^2 - 2x - 2y + 1 = 0$ **22.** $x^2 + 9y^2 = 9$

23. $xy = 1$ **24.** $y = \dfrac{x}{x + 1}$

In Problems 25–34, *express the given equation in rectangular coordinates.*

25. $r = \cos 2\theta$ **26.** $r = 1 - \cos\theta$

27. $r = 3 + 2\sin\theta$ **28.** $r^2 = \sin\theta$

29. $r^2 = 1 + \sin\theta$ **30.** $r^2 = \sin 2\theta$

31. $r = \dfrac{1}{1 - \cos\theta}$ **32.** $r = \dfrac{1}{1 + \sin\theta}$

33. $r = 2\sin\theta + 3\cos\theta$ **34.** $r = \sec\theta$

9.5

CONICS IN POLAR COORDINATES

We found earlier that the equations of conic sections (in rectangular coordinates) have very simple forms if the center or vertex is at the origin and the axes are the coordinate axes. There are, however, three different forms corresponding to the three different types of conics. We find that conics can be easily represented in polar coordinates if a focus is at the origin and one axis is a coordinate axis. Furthermore, the same type of equation represents all three types of conics if we use the unifying concept of eccentricity.

Recall that any conic can be determined by a single focus, the corresponding directrix, and the eccentricity. If P is a point on the conic, then the distance from P

to the focus divided by the distance from P to the directrix equals the eccentricity. The particular conic we get depends upon the eccentricity; the eccentricity is a positive number and

> if $e < 1$, the conic is an ellipse,
>
> if $e = 1$, the conic is a parabola,
>
> if $e > 1$, the conic is a hyperbola.

This is illustrated in Figure 9.15, where we have an ellipse, a parabola, and a hyperbola, all having the same focus and directrix. In this case the ellipse has eccentricity 1/2, and the hyperbola has eccentricity 2.

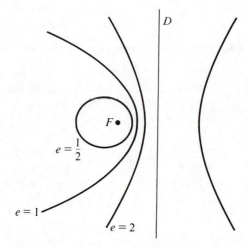

Figure 9.15

If $P = (r, \theta)$ is a point on a conic with focus 0, directrix $x = p$ (p positive), and eccentricity e, then (see Figure 9.16)

$$\frac{\overline{OP}}{\overline{PD}} = e, \qquad \text{or} \qquad \frac{|r|}{|p - r\cos\theta|} = e.$$

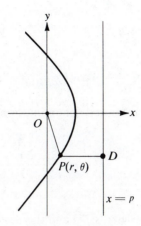

Figure 9.16

There are now two cases to consider.

$$\frac{r}{p - r\cos\theta} = e \qquad \text{and} \qquad \frac{r}{p - r\cos\theta} = -e$$

Either of these yields an equation of the desired conic (see Problems 21 and 22); however, the first yields the commonly-used form. Solving for r in this equation, we have

$$r = \frac{ep}{1 + e\cos\theta}.$$

If the directrix is $x = -p$ (p positive), then the equation is

$$r = \frac{ep}{1 - e\cos\theta}.$$

If the directrix is $y = \pm p$ (p positive), then the equation is

$$r = \frac{ep}{1 \pm e\sin\theta}.$$

Theorem 9.3 *The conic section with focus at the origin, directrix $x = \pm p$ (p positive), and eccentricity e has polar equation*

$$r = \frac{ep}{1 \pm e\cos\theta};$$

if the directrix is $y = \pm p$ (p positive), it has equation

$$r = \frac{ep}{1 \pm e \sin \theta}.$$

Example 1   Describe $r = 6/(4 + 3\cos\theta)$.

Solution   Dividing numerator and denominator by 4, we have

$$r = \frac{\dfrac{3}{2}}{1 + \dfrac{3}{4}\cos\theta} = \frac{\dfrac{3}{4}\cdot 2}{1 + \dfrac{3}{4}\cos\theta}.$$

Thus the eccentricity is 3/4 and the directrix is $x = 2$. The conic is an ellipse with focus at the origin, directrix $x = 2$, and eccentricity 3/4.

Example 2   Sketch $r = 15/(2 - 3\cos\theta)$.

Solution   Dividing by 2, we have

$$r = \frac{\dfrac{15}{2}}{1 - \dfrac{3}{2}\cos\theta} = \frac{\dfrac{3}{2}\cdot 5}{1 - \dfrac{3}{2}\cos\theta}.$$

Thus we have a hyperbola with focus at the origin, eccentricity 3/2, and directrix $x = -5$. The vertices are on the x axis, one between the focus and directrix and the other to the left of the directrix. When $\theta = 0, r = -15$; when $\theta = \pi, r = 3$. Thus the vertices are $(-15, 0)$ and $(3, \pi)$. When $\theta = \pi/2$ or $3\pi/2, r = 15/2$. Thus, the ends of one of the latera recta are $(15/2, \pi/2)$ and $(15/2, 3\pi/2)$. This information is enough to give a reasonably accurate picture of the hyperbola. If the asymptotes are desired, they can best be found by considering some of the above points in rectangular coordinates. Thus the vertices are $(-3, 0)$ and $(-15, 0)$, and the center is $(-9, 0)$, giving $a = 6$ and $c = 9$. We can now use the equation

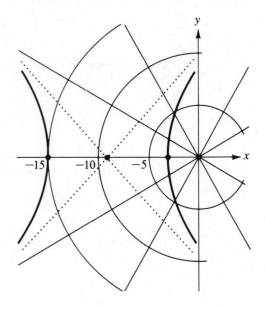

Figure 9.17

$$b^2 = c^2 - a^2$$

to find $b^2 = 45$ or $b = 3\sqrt{5}$. Once we have this, the asymptotes are easily found (see Figure 9.17).

Example 3   Find a polar equation of the parabola with focus at the origin and directrix $y = -4$.

Solution   The equation is in the form

$$r = \frac{ep}{1 - e\sin\theta},$$

since the directrix is a horizontal line below the focus. Furthermore, $e = 1$, since the conic is a parabola, and directrix $y = -4$ gives $p = 4$. Thus the equation is

$$r = \frac{4}{1 - \sin\theta}.$$

PROBLEMS

A   *In Problems 1–8, state the type of conic and give a focus and its corresponding directrix and the eccentricity.*

1. $r = \dfrac{4}{1 + 2\cos\theta}$

2. $r = \dfrac{12}{1 - 3\sin\theta}$

3. $r = \dfrac{4}{3 + 2\sin\theta}$

4. $r = \dfrac{5}{4 - 4\cos\theta}$

5. $r = \dfrac{3}{1 + \sin\theta}$

6. $r = \dfrac{10}{5 - 2\cos\theta}$

7. $r(3 + 2\sin\theta) = 6$

8. $r(2 - 4\cos\theta) = 5$

In Problems 9–14, find a polar equation of the conic with focus at the origin and the given eccentricity and directrix.

9. Directrix: $x = 5$; $e = 2/3$

10. Directrix: $y = -3$; $e = 2$

11. Directrix: $y = 2$; $e = 1$

12. Directrix: $x = -4$; $e = 1$

13. Directrix: $x = 5$; $e = 5/4$

14. Directrix: $y = 3$; $e = 3/4$

B   *In Problems 15–20, sketch the given conic.*

15. $r = \dfrac{2}{1 + \cos\theta}$

16. $r = \dfrac{16}{5 - 3\cos\theta}$

17. $r = \dfrac{16}{4 - 5\sin\theta}$

18. $r(3 - 5\cos\theta) = 9$

19. $r(13 + 12\sin\theta) = 25$

20. $r(3 + 3\sin\theta) = 4$

21. Sketch $r = \dfrac{-2}{1 - \cos\theta}$. Compare it with the conic of Problem 9 (see the following problem).

22. Show that the conic section with focus at the origin, directrix $x = p$ (p positive), and eccentricity e has polar equation

$$r = \frac{-ep}{1 - e\cos\theta}.$$

23. Suppose, in the equation $r = \dfrac{ep}{1 + e\cos\theta}$, $e \to 0$ and $p \to +\infty$ in such a way that ep remains constant. What happens to the shape of the conic? What happens to the equation of the conic?

C **24.** A comet has a parabolic orbit with the sun at the focus. When the comet is 100,000,000 miles from the sun, the line joining the sun and the comet makes an angle of 60° with the axis of the parabola (see Figure 9.18). How close to the sun will the comet get?

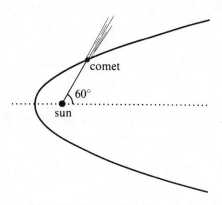

Figure 9.18

25. A satellite has an elliptical orbit with the earth at one focus. At its closest point it is 100 miles above the surface of the earth; at its farthest point, 500 miles. Find a polar equation of its path. (Take the radius of the earth to be 4000 miles.)

26. Find a polar equation of a circle with center (k, α) and radius a by using the law of cosines (see Figure 9.19). What is the result when $k = 0$?

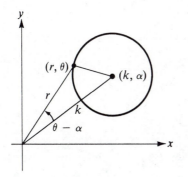

Figure 9.19   Figure 9.20

27. By using the trigonometry of right triangles, show that the line PQ (Figure 9.20) can be represented by the equation

$$x \cos \alpha + y \sin \alpha - p = 0.$$

This is called the **normal form** of the line, since it is expressed in terms of the polar coordinates of the point Q, which is the intersection of the original line and another perpendicular (or normal) to it and through the origin (see Problem 38, Section 3.3).

28. By using the identity

$$\sin^2 \alpha + \cos^2 \alpha = 1,$$

show that $Ax + By + C = 0$ can be put into the normal form by dividing through by $\pm \sqrt{A^2 + B^2}$ (see Problem 27 for the normal form).

29. Show that the distance from the point (x_1, y_1) to the line $Ax + By + C = 0$ is

$$d = \frac{|Ax_1 + By_1 + C|}{\sqrt{A^2 + B^2}}.$$

This result was found without the use of polar coordinates in Theorem 3.6, page 74. [*Hint:* Put the original line and the one parallel to it and through (x_1, y_1) into the normal form (see Problems 27 and 28).]

9.6

PARAMETRIC EQUATIONS

Up to now all of the equations we have dealt with have been in the form

$$y = f(x) \qquad \text{or} \qquad F(x, y) = 0.$$

In either case, a direct relationship between x and y is given. Sometimes, however, it is more convenient to express both x and y in terms of a third variable, called a parameter. That is,

$$x = f(t) \qquad \text{and} \qquad y = g(t).$$

Each value of the parameter t gives a value of x and a value of y.

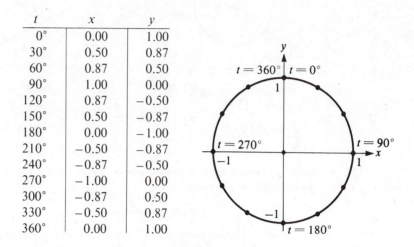

t	x	y
0°	0.00	1.00
30°	0.50	0.87
60°	0.87	0.50
90°	1.00	0.00
120°	0.87	−0.50
150°	0.50	−0.87
180°	0.00	−1.00
210°	−0.50	−0.87
240°	−0.87	−0.50
270°	−1.00	0.00
300°	−0.87	0.50
330°	−0.50	0.87
360°	0.00	1.00

Figure 9.21

For instance, if $t = 0°$ in the parametric equations

$$x = \sin t \qquad \text{and} \qquad y = \cos t,$$

we see that $x = 0$ and $y = 1$. Thus the point $(0, 1)$ is a point of the graph. Note that we still have just the x and y axes; t does not appear on the graph. Continuing with this process gives the results shown in Figure 9.21. Of course, we could continue with values of t beyond 360°, but we would simply go over the same points again. Although the value of t need not appear anywhere on the graph, we have labeled several points with their corresponding values of t. Once the points are plotted, they are joined in the order of increasing (or decreasing) values of t.

The result seems to resemble a circle. How can we be *sure* it is a circle? If we had a single equation in x and y, we could easily see by the form of the equation whether or not we have a circle. Let us try to eliminate the parameter t between the equations $x = \sin t$ and $y = \cos t$.

$$\sin^2 t + \cos^2 t = 1$$
$$x^2 + y^2 = 1$$

We now see that we have a circle with center at the origin and radius 1.

Elimination of the parameter not only assures us that this particular curve is a circle, it also gives us a basis for sketching more rapidly than can be done by point-by-point plotting. However, we must be careful with the domain of the resulting equation. Let us illustrate this with some examples and see how the domain of $F(x, y) = 0$ plays an important role in sketching the graph.

Example 1 Graph the following two pairs of parametric equations by eliminating the parameter.

$$x = t, y = t \quad \text{and} \quad x = t^2, y = t^2$$

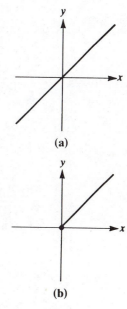

Solution Elimination of the parameter gives $y = x$ in both cases. But the graphs are not the same, as the domains in the two cases will show. The domain can be determined from the first of the two parametric equations in each case. In the first case, $x = t$ and, since there is no restriction on t, there is none on x; the domain is the set of all real numbers. In the second case, $x = t^2$. The domain of $y = x$ is the range of $x = t^2$, which is $\{x \mid x \geq 0\}$. Thus we have a restricted domain here that we did not have in the first case. The graphs are given in Figure 9.22.

(a)

(b)

Figure 9.22

Example 2 Eliminate the parameter between $x = t + 1$ and $y = t^2 + 3t + 2$ and sketch the graph.

Solution Solving $x = t + 1$ for t, we have

$$t = x - 1.$$

If this is substituted into $y = t^2 + 3t + 2$, then

$$y = (x - 1)^2 + 3(x - 1) + 2$$
$$= x^2 + x.$$

Note that there is no restriction on x; the domain of $y = x^2 + x$ is the set of all real numbers. It is now a simple matter to sketch the curve; it is given in Figure 9.23.

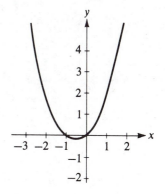

Figure 9.23

Eliminating the parameter is not always so simple as in these examples. Occasionally it is difficult or impossible. In such cases, the curve must be plotted point by point as we did with $x = \sin t$, $y = \cos t$. Closely related to parametric equations are vector-valued functions. They are represented in the following way.

$$\mathbf{f}(t) = f_1(t)\mathbf{i} + f_2(t)\mathbf{j}$$

Thus, when a value of t is substituted into the equation, the function takes on a vector value. For example, if

$$\mathbf{f}(t) = t\mathbf{i} + t^2\mathbf{j},$$

then

$$\mathbf{f}(1) = \mathbf{i} + \mathbf{j}$$

and

$$\mathbf{f}(2) = 2\mathbf{i} + 4\mathbf{j}.$$

Recall that, in graphing vectors, we graph only representatives. Thus, in graphing vector functions, let us graph representatives of the vectors, each having its tail at the origin. Thus,

$$\mathbf{f}(t) = t\mathbf{i} + t^2\mathbf{j}$$

has the graphical representation shown in Figure 9.24a. Normally we shall omit the directed line segments and show only their heads, as in Figure 9.24b. Thus, the result is equivalent to graphing the curve represented parametrically by

$$x = t \quad\text{and}\quad y = t^2.$$

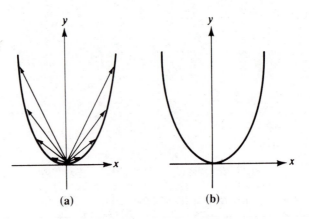

(a)   (b)

Figure 9.24

Example 3   Sketch the curve $\mathbf{f}(t) = (t + 2)\mathbf{i} + (t^2 + 7t + 12)\mathbf{j}$.

Solution   This is equivalent to the parametric equations

$$x = t + 2 \quad \text{and} \quad y = t^2 + 7t + 12.$$

Eliminating the parameter, we have

$$y = x^2 + 3x + 2.$$

Its graph is given in Figure 9.25.

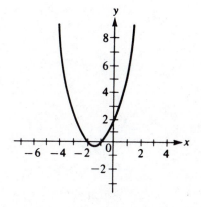

Figure 9.25

Let us consider a line determined by two of its points, $A = (x_1, y_1)$ and $B = (x_2, y_2)$. The directed line segments $\overrightarrow{OA}$ (where O is the origin) and $\overrightarrow{AB}$ represent vectors **u** and **v**, respectively (see Figure 9.26). Now suppose $P = (x, y)$ is any point on the line, and **w** is the vector represented by $\overrightarrow{OP}$. Since $\overrightarrow{AB}$ and $\overrightarrow{AP}$ have the same directions (or opposite directions if P is above A),

$$\overrightarrow{AP} = r\overrightarrow{AB}$$

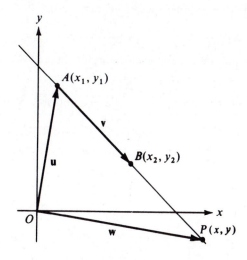

Figure 9.26

for some number r. (Note that r is negative if $\overrightarrow{AB}$ and $\overrightarrow{AP}$ have opposite directions.) Then

$$\overrightarrow{OP} = \overrightarrow{OA} + r\overrightarrow{AB}.$$
$$\mathbf{w} = \mathbf{u} + r\mathbf{v}.$$

But

$$\mathbf{w} = x\mathbf{i} + y\mathbf{j},$$
$$\mathbf{u} = x_1\mathbf{i} + y_1\mathbf{j},$$
$$\mathbf{v} = (x_2 - x_1)\mathbf{i} + (y_2 - y_1)\mathbf{j},$$

which gives

$$x\mathbf{i} + y\mathbf{j} = x_1\mathbf{i} + y_1\mathbf{j} + r[(x_2 - x_1)\mathbf{i} + (y_2 - y_1)\mathbf{j}]$$
$$= [x_1 + r(x_2 - x_1)]\mathbf{i} + [y_1 + r(y_2 - y_1)]\mathbf{j}.$$

Thus we have a vector-valued function of r which represents points on the given line. By tracing the argument backward, we see that for each value of r there is a vector which represents a point on the line. Thus the above vector-valued function represents the given line. The corresponding parametric form for the line is

$$x = x_1 + r(x_2 - x_1)$$
$$y = y_1 + r(y_2 - y_1).$$

These are the familiar point-of-division formulas that we saw on page 11. This use of the point-of-division formulas as a parametric representation for a line is a convenient one; it is one of the principal ways we shall have for representing lines in three-dimensional space (see page 280).

Example 4   Find a parametric representation for the line through $(1, 5)$ and $(-2, 3)$.

Solution   Letting $(1, 5)$ and $(-2, 3)$ be the first and second points, respectively, of

$$x = x_1 + r(x_2 - x_1)$$

and

$$y = y_1 + r(y_2 - y_1),$$

we then have

$$x = 1 - 3r$$

and

$$y = 5 - 2r.$$

The choice of $(1, 5)$ and $(-2, 3)$ as the first and second points is quite arbitrary; we could have reversed the designation. In that case the parametric representation would be

$$x = -2 + 3s$$

and

$$y = 3 + 2s.$$

While the two representations appear to have little in common, it is easily seen that they represent the same line. We get point $(1, 5)$ when $r = 0$ or when $s = 1$; we get $(-2, 3)$ when $r = 1$ or $s = 0$. In fact, whatever point we get for a given value of r, we get the same point for $s = 1 - r$. By using other points on the line, we can get still more parametric representations. Thus a line does not have a unique parametric representation.

PROBLEMS

A   *In Problems 1–6, eliminate the parameter and sketch the curve.*

1. $x = t^2 + 1, y = t + 1$

2. $x = t^2 + t - 2, y = t + 2$

3. $x = t - 1, y = t^2 - 2t$

4. $x = 2t^2 + t - 3, y = t - 1$

5. $\mathbf{f}(t) = (t - 1)\mathbf{i} + t^2\mathbf{j}$

6. $\mathbf{f}(t) = (t^2 + 1)\mathbf{i} + t^2\mathbf{j}$

In Problems 7–12, give equations in parametric form for the line through the given pair of points.

7. $(1, 5), (3, 1)$

8. $(4, 2), (-1, 3)$

9. $(2, 5), (-1, 2)$

10. $(4, 1), (-8, 3)$

11. $(2, 3), (5, 3)$

12. $(-3, 2), (-3, 5)$

B   *In Problems 13–24, eliminate the parameter and sketch the curve.*

13. $x = t^2 + t, y = t^2 - t$

14. $x = t^2 + 1, y = t^2 - 1$

15. $x = t^3, y = t^2$

16. $x = a \cos \theta, y = b \sin \theta$

17. $x = 2 + \cos \theta, y = -1 + \sin \theta$

18. $x = 3 - \cos \theta, y = 2 + 4 \sin \theta$

19. $x = 3 + \cosh \theta, y = 2 + \sinh \theta$

20. $x = 4 + 2 \cosh \theta, y = 1 - 4 \sinh \theta$

21. $\mathbf{f}(t) = t^2\mathbf{i} + t^3\mathbf{j}$

22. $\mathbf{f}(t) = (3t + 1)\mathbf{i} + (t^3 - 1)\mathbf{j}$

23. $\mathbf{f}(t) = \cos t\mathbf{i} + \sin t\mathbf{j}$

24. $\mathbf{f}(t) = t^2\mathbf{i} + e^t\mathbf{j}$

In Problems 25–30, sketch the curve.

25. $x = e^t, y = \sin t$

26. $x = \theta - \sin \theta, y = 1 - \cos \theta$

27. $x = \cos \theta + \theta \sin \theta, y = \sin \theta - \theta \cos \theta; \theta \geq 0$

28. $x = a \cos^3 \theta, y = a \sin^3 \theta$

29. $x = 2a \cos \theta - a \cos 2\theta, y = 2a \sin \theta - a \sin 2\theta$

30. $x = t - a \tanh \dfrac{t}{a}, y = a \operatorname{sech} \dfrac{t}{a}$

C **31.** Show that the parametric representation

$$x = x_1 + r(x_2 - x_1) \qquad \text{and} \qquad y = y_1 + r(y_2 - y_1)$$

of the line through (x_1, y_1) and (x_2, y_2) is equivalent to the two-point form of the line given on page 64.

32. Sketch each of the following parametric equations and note the similarities and differences.

a. $x = t, y = t^2$

b. $x = \sqrt{t}, y = t$

c. $x = e^t, y = e^{2t}$

d. $x = \sin t, y = 1 - \cos^2 t$

33. Sketch each of the following parametric equations and note the similarities and differences.

a. $x = t, y = \sqrt{t^2 - 1}$

b. $x = \sqrt{t}, y = \sqrt{t - 1}$

c. $x = \sec t, y = \tan t$

d. $x = \cosh t, y = \sinh t$

9.7

PARAMETRIC EQUATIONS OF A LOCUS

The principal advantage of parametric equations is in the determination of equations of a locus. It is frequently simpler to relate x and y to some third variable than to relate them to each other directly. One example of this is the determination of the path of a projectile. The position, measured by the distance above or below a fixed reference point, of an object thrown vertically upward or downward is given by

$$y = -16t^2 + v_0t + y_0,$$

where v_0 is the initial velocity and y_0 is the initial position. But if the object is thrown in any nonvertical direction, then its position must have both a horizontal and a vertical component. Of course these can be determined by reference to a pair of axes. However, it is easier to relate the horizontal and vertical components of the position to time than to relate them directly to each other.

As soon as a projectile is released, it becomes a falling object. Thus the vertical motion is governed by the laws of falling bodies, and the y coordinate is easily related to time by the above formula with v_0 replaced by v_y, the vertical component of the initial velocity. The x coordinate is even easier to deal with. Because the force of gravity is the *only* force acting upon the projectile (we neglect the resistance of the air), there is no force tending to change the horizontal velocity—it remains constant. Using the familiar formula, distance = (rate)(time), and adding the initial position, we have

$$x = v_x t + x_0,$$

where v_x is the horizontal component of the initial velocity.

Example 1   A gun is inclined to the horizontal at an angle of 30° and fired from ground level with an initial speed of 1500 feet/second. Determine the path of the bullet.

Solution   Let us place the axes so that the gun is at the origin. First we consider the initial velocity as a vector (see Figure 9.27) and break it down into its horizontal and vertical components. We have

$$\mathbf{v} = v_x \mathbf{i} + v_y \mathbf{j},$$

where

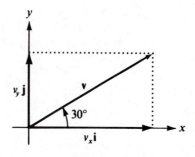

Figure 9.27

$$v_x = |\mathbf{v}| \cos 30° = 1500 \cdot \frac{\sqrt{3}}{2} = 750\sqrt{3}$$

and

$$v_y = |\mathbf{v}| \sin 30° = 1500 \cdot \frac{1}{2} = 750.$$

Now we consider the vertical component, which is a falling-body problem. Since the vertical component of the initial velocity is $\mathbf{v}_y = 750$,

$$y = -16t^2 + 750t + y_0.$$

By the placement of the axes, $y = 0$ when $t = 0$. Thus $y_0 = 0$ and

$$y = -16t^2 + 750t.$$

For the horizontal component, $v_x = 750\sqrt{3}$, giving

$$x = 750\sqrt{3}t + x_0.$$

Again by the placement of the axes, $x = 0$ when $t = 0$, which gives $x_0 = 0$ and

$$x = 750\sqrt{3}t.$$

Thus, in parametric form, the path of the projectile is given by

$$x = 750\sqrt{3}t \quad \text{and} \quad y = -16t^2 + 750t.$$

By eliminating the parameter, we see that the path is parabolic.

$$y = \frac{-4x^2}{421,875} + \frac{x}{\sqrt{3}}$$

In the past we have found equations of curves from a geometric description. Again, this can often be accomplished by relating the x and y coordinates of a point on the curve to some third variable.

Example 2   Find parametric equations for the set of all points P which are determined as illustrated in Figure 9.28.

Solution   Since the ray OA is determined by the angle θ, θ is a convenient parameter. The x coordinate of P is the x coordinate of A; thus

$$x = a \cos \theta.$$

Similarly, the y coordinate of P is the y co-ordinate of B.

$$y = b \sin \theta$$

Thus we have the curve in parametric form. It might be noted here that the parameter can easily be eliminated to give

$$\frac{x^2}{a^2} + \frac{y^2}{b^2} = 1.$$

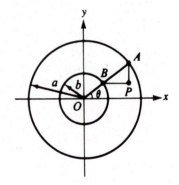

Figure 9.28

Example 3   A wheel of radius a is rolling along a line. Find the path traced by a point on the circumference if the line is the x axis and the point starts at the origin.

Solution Suppose we use the angle θ (see Figure 9.29) as the parameter. Since the wheel is rolling along the x axis,

$$\overline{OT} = \overset{\frown}{PT} = a\theta.$$

Thus C is the point $(a\theta, a)$. Furthermore, from triangle CPQ,

$$\overline{PQ} = a \sin \theta \quad \text{and} \quad \overline{CQ} = a \cos \theta.$$

Thus,

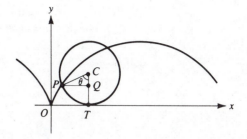

Figure 9.29

$$x = \overline{OT} - \overline{PQ} \qquad y = \overline{CT} - \overline{CQ}$$
$$= a\theta - a \sin \theta \qquad = a - a \cos \theta$$
$$= a(\theta - \sin \theta) \qquad = a(1 - \cos \theta).$$

This curve is called a cycloid.

PROBLEMS

A

1. A gun that is inclined to the horizontal at an angle of 60° is fired from ground level with an initial speed of 2000 feet/second. Determine the path of the bullet.

2. A gun is clamped in a horizontal position 8 ft above the ground and fired with an initial speed of 1600 feet/second. Determine the path of the bullet. Where does it hit the ground?

3. A ball is thrown upward from ground level at a 30° angle of inclination. Given that it is thrown at 32 feet/second, find the path of the ball. Where does it hit the ground?

4. A ball is thrown downward and inclined to the horizontal at an angle of 60° from a building 300 feet high. Find the path of the ball when it is thrown at 16 feet/second. How far from the base of the building does it hit the ground?

B

5. A cannon, inclined at an angle θ to the horizontal, is fired from ground level with speed v_0. Determine the path of the projectile.

6. Find parametric equations for the set of all points P determined as shown in Figure 9.30. Eliminate the parameter. [*Hint:* Consider triangles OAR and OSQ.]

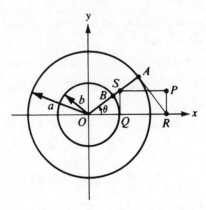

Figure 9.30

7. Suppose, in Example 3, the point starts at $(0, 2a)$. Find parametric equations for the curve. It is the curve traced by P' in Figure 9.31.

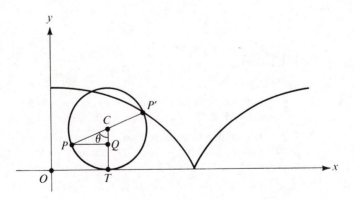

Figure 9.31

8. Find the path traced by a point P a distance b from the center of the circle of Example 3 if P starts at $(0, a - b)$. The circle of radius a rolls along the x axis, but P is on a second circle of radius b.

9. Sketch the curve of Problem 8 if

    a. $a = 2$ and $b = 1$ **b.** $a = 1$ and $b = 2$.

10. In Figure 9.32, $\overline{BA} = \overline{BP} = \overline{BP}$. Find parametric equations for the set of all points P and P' determined as shown. Sketch. This curve is called a strophoid. [*Hint:* First find the coordinates of B.]

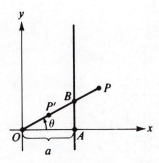

Figure 9.32

11. Find a polar representation for the curve of Problem 10.

12. Find parametric equations for the set of all points P and P' determined as shown in Figure 9.33. This curve is called a conchoid. [*Hint:* First find the coordinates of B.]

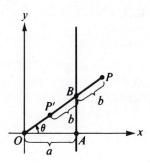

Figure 9.33

13. Find a polar equation for the curve of Problem 12.

14. Sketch the conchoid (see Problems 12 and 13) if

    a. $a = b = 1$ **b.** $a = 1$ and $b = 2$ **c.** $a = 2$ and $b = 1$.

C **15.** If a string which is wound on a spool is unwound while the string is kept taut (see Figure 9.34), the curve traced by the end of the string is called the involute of the circle. Find parametric equations for the involute of a circle of radius a.

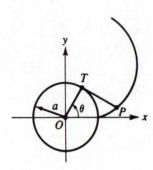

Figure 9.34

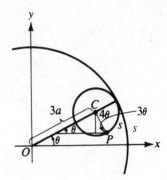

Figure 9.35

16. Find the path traced by a point P on the circumference of a circle of radius a which rolls inside a circle of radius $4a$ (see Figure 9.35). This curve is called a hypocycloid.

17. Find the path of the point P of Problem 16 if the smaller circle rolls outside the larger one. This curve is called an epicycloid.

18. What is the result if both of the circles of Problem 17 have radius a? This curve is called a cardioid.

REVIEW PROBLEMS

A *In Problems 1–6, sketch the graph of the given equation and indicate any symmetry about either axis or the pole.*

1. $r = 1 + 3 \cos \theta$ **2.** $r = \cos 4\theta$

3. $r = 1 + \sin 2\theta$ **4.** $r = e^{\theta}$

5. $r^2 = \sin 2\theta$ **6.** $r^2 = 1 + \cos \theta$

In Problems 7–12, find all points of intersection.

7. $r = 1, r = 1 + \cos \theta$ **8.** $r = 1 + \sin \theta, r = 1 - \cos \theta$

9. $r = \sin\theta$, $r = 1 + 2\sin\theta$

10. $r = 2\sin 2\theta$, $r = 1$

11. $r = \cos\theta$, $r = \cos 3\theta$

12. $r = 1 + \sin\theta$, $r = \cos\theta - 1$

13. **(a)** The following points are given in polar coordinates; give the rectangular coordinate representation of each: $(-1, \pi/2)$, $(2\sqrt{2}, 3\pi/4)$, $(6, 7\pi/6)$, $(-2\sqrt{3}, 2\pi/3)$.
    (b) The following points are given in rectangular coordinates; give a polar coordinate representation of each: $(-2, 2)$, $(-5, 0)$, $(1, -\sqrt{3})$, $(10, 4)$.

14. Express $x^2 + y^2 + 3y = 0$ in polar coordinates.

15. Give a polar equation for the ellipse with focus at the pole, the corresponding directrix $y = 3$, and eccentricity $2/3$.

16. Give a polar equation for the parabola with focus at the pole and directrix $x = -6$.

In Problems 17–20, eliminate the parameter and sketch the curve.

17. $x = t + 1$, $y = t^2 + 4t - 2$

18. $x = 1 - 3\cos\theta$, $y = 2 + 2\sin\theta$

19. $\mathbf{f}(t) = (t + 1)^2\mathbf{i} + t^2\mathbf{j}$

20. $\mathbf{f}(t) = e^t\mathbf{i} + \ln t\,\mathbf{j}$

21. Give a parametric representation of the line containing $(2, 3)$ and $(4, -5)$.

22. Give a parametric representation of the line containing $(-1, 4)$ and $(3, -3)$.

B 23. Express $r = 1 + \sin\theta$ in rectangular coordinates.

24. Express $r(3\cos\theta - 5) = 16$ in rectangular coordinates.

25. Sketch the conic $r = 2/(1 - \sin\theta)$. Identify the focus, directrix, and eccentricity.

26. Sketch the conic $r = 2/(1 - 2\cos\theta)$. Identify a focus, the corresponding directrix, and the eccentricity.

27. Eliminate the parameter and sketch $x = \sin\theta$, $y = \sin 2\theta$.

28. Sketch $\mathbf{f}(\theta) = \cos\theta\,(1 + \cos\theta)\mathbf{i} + \sin\theta\,(1 + \cos\theta)\mathbf{j}$.

29. Find parametric equations for the set of all points P determined as shown in Figure 9.36. [*Hint:* Use a polar equation for the circle.] Sketch. This curve is called the witch of Agnesi.

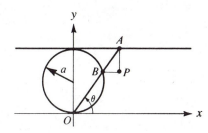

Figure 9.36

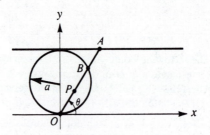

Figure 9.37

30. In Figure 9.37, $\overline{OP} = \overline{AB}$. Find parametric equations for the set of all points P determined as shown. [*Hint:* Use a polar equation for the circle.] Sketch. This curve is called the cissoid of Diocles.

10

SOLID ANALYTIC GEOMETRY

10.1

INTRODUCTION: THE DISTANCE AND POINT-OF-DIVISION FORMULAS

So far we have dealt almost exclusively with plane figures. Let us now consider the geometry of solid figures. Forming the bridge between algebra and geometry is the assignment of numbers to points in space, similar to the assignment we made to points in a plane; a point in space, however, is represented by a set of three numbers rather than two. We begin with a set of three lines, called **axes**, concurrent at a point (the origin). The only requirement is that these three lines not be coplanar—that is, that they not all lie in the same plane. However, we shall consider only the case in which the axes are mutually perpendicular. The three axes, labeled x, y, and z, with a scale on each, determine a set of three numbers, called **coordinates**, associated uniquely with any point in space. Since any pair of intersecting lines determines a plane, the three pairs of axes determine three **coordinate planes**, which we shall call the **xy plane**, the **xz plane**, and the **yz plane** (see Figure 10.1a. The x *coordinate* of a point P in space is the number associated with the point on the x axis that is the intersection of the x axis and the plane through P parallel to the yz plane. The y and z *coordinates* of P are defined in a similar fashion by considering the points of intersection of the y and z axes and with planes through P parallel to the xz and xy coordinate planes, respectively (see Figure 10.1b).

The coordinate planes separate space into eight **octants**. Although we shall not number all of them, the one in which all three coordinates are positive is called the **first octant**. Note that points of the xy plane have z coordinate 0, points of the xz plane have y coordinate 0, and points of the yz plane have x coordinate 0. Similarly, points of the x axis have y and z coordinates 0, and so on. Of course the origin has all of these coordinates 0.

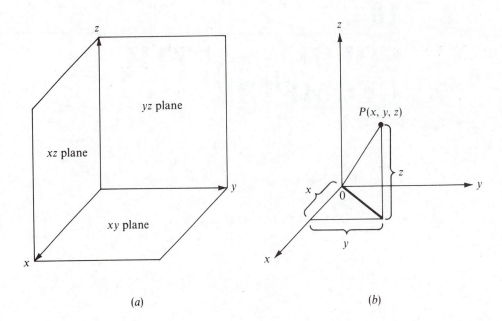

$$(a) \qquad\qquad\qquad\qquad (b)$$

Figure 10.1

The two basic geometric representatives of the axes are given in Figure 10.2; 10.2a shows a **right-hand system**, while 10.2b shows a **left-hand system**. Graphs of equations in the two systems are mirror images of each other. Since we shall normally represent space by a right-hand system, the axes will usually appear in the positions indicated in Figure 10.2a. This is sometimes represented by the right-hand rule illustrated in Figure 10.2c. If the index and second fingers of the right hand point in the direction of the x and y axes, respectively, then the thumb points in the direction of the z axis.

Many of the formulas of solid analytic geometry are simple extensions of plane analytic geometry. The one that follows is an example.

Theorem 10.1   *The distance between two points* (x_1, y_1, z_1) *and* (x_2, y_2, z_2) *is*

$$d = \sqrt{(x_1 - x_2)^2 + (y_1 - y_2)^2 + (z_1 - z_2)^2}.$$

Proof   Suppose we project the points P_1 and P_2 onto the xy plane, giving Q_1 and Q_2 (see Figure 10.3). Since Q_1 and Q_2 are in the xy plane, we can use our distance formula for the plane to get

$$d_1 = \overline{Q_1 Q_2} = \sqrt{(x_2 - x_1)^2 + (y_2 - y_1)^2}.$$

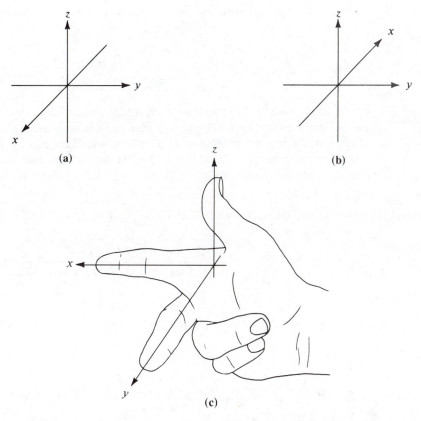

(a)   **(b)**

(c)

Figure 10.2

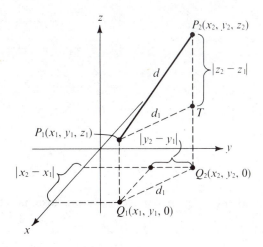

Figure 10.3

We now use the Pythagorean Theorem on triangle P_1TP_2 with right angle at T.

$$d^2 = \overline{P_1P_2} = \overline{P_1T}^2 + \overline{P_2T}^2$$
$$= d_1^2 + |z_2 - z_1|^2$$
$$= (x_2 - x_1)^2 + (y_2 - y_1)^2 + (z_2 - z_1)^2$$
$$d = \sqrt{(x_2 - x_1)^2 + (y_2 - y_1)^2 + (z_2 - z_1)^2}$$

If the line joining the two points is on or parallel to one of the coordinate planes, at least one of the three terms of this formula is zero, and it reduces to the plane case. Similarly, if the line joining the points is on or parallel to one of the axes, at least two terms are zero and the distance is the absolute value of the difference between the coordinates of the remaining pair.

Example 1   Find the distance between $(1, -2, 5)$ and $(-3, 6, 4)$ (see Figure 10.4).

Solution
$$d = \sqrt{(x_1 - x_2)^2 + (y_1 - y_2)^2 + (z_1 - z_2)^2}$$
$$= \sqrt{(1 + 3)^2 + (-2 - 6)^2 + (5 - 4)^2}$$
$$= \sqrt{16 + 64 + 1}$$
$$= \sqrt{81}$$
$$= 9$$

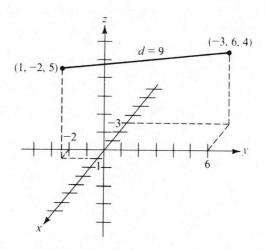

Figure 10.4

Another easy extension from two dimensions is the point-of-division formula.

Theorem 10.2   *If $P_1 = (x_1, y_1, z_1)$, $P_2 = (x_2, y_2, z_2)$, and P is a point such that $r = \overline{P_1P}/\overline{P_1P_2}$, then the coordinates of P are*

$$x = x_1 + r(x_2 - x_1),$$
$$y = y_1 + r(y_2 - y_1),$$

and

$$z = z_1 + r(z_2 - z_1).$$

The proof is similar to the one for the two-dimensional case, and it is left to the student.

Example 2   Find the point $1/3$ of the way from $(-2, 4, 1)$ to $(4, 1, 7)$.

Solution

$$x = x_1 + r(x_2 - x_1) = -2 + \frac{1}{3}(4 + 2) = 0$$

$$y = y_1 + r(y_2 - y_1) = 4 + \frac{1}{3}(1 - 4) = 3$$

$$z = z_1 + r(z_2 - z_1) = 1 + \frac{1}{3}(7 - 1) = 3$$

The desired point is $(0, 3, 3)$ (see Figure 10.5).

The following theorem is a direct corollary of the point-of-division formulas.

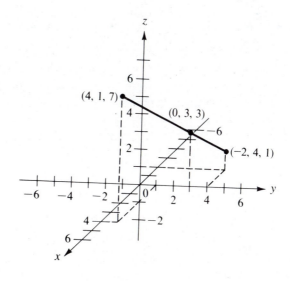

Figure 10.5

Theorem 10.3   *If $P_1 = (x_1, y_1, z_1)$ and $P_2 = (x_2, y_2, z_2)$, then the coordinates of the midpoint of the segment $P_1 P_2$ are*

$$x = \frac{x_1 + x_2}{2}, \qquad y = \frac{y_1 + y_2}{2}, \qquad z = \frac{z_1 + z_2}{2}.$$

Example 3   Find the midpoint of the segment with ends $(4, -3, 1)$ and $(-2, 5, 3)$.

Solution

$$x = \frac{x_1 + x_2}{2} = \frac{4 - 2}{2} = 1$$

$$y = \frac{y_1 + y_2}{2} = \frac{-3 + 5}{2} = 1$$

$$z = \frac{z_1 + z_2}{2} = \frac{1 + 3}{2} = 2$$

The point is $(1, 1, 2)$.

PROBLEMS

A   *In Problems 1–10, plot both points and find the distance between them.*

1. $(2, 5, 0), (-3, 1, 3)$
2. $(4, -2, 1), (2, 2, -3)$
3. $(5, 4, -1), (2, 0, -1)$
4. $(2, 5, 3), (-2, 4, -1)$
5. $(-5, 0, 2), (4, 1, -5)$
6. $(-2, 5, 1), (-2, 8, 4)$
7. $(3, -1, 4), (3, 4, 4)$
8. $(2, 5, 0), (5, 5, 0)$
9. $(4, 7, -1), (3, -1, 3)$
10. $(5, 2, 3), (4, 5, -1)$

In Problems 11–16, find the point P such that $\overline{AP}/\overline{AB} = r$.

11. $A = (4, 3, -2), B = (-5, 0, 4), r = 2/3$
12. $A = (5, 2, 3), B = (-5, 7, -2), r = 2/5$
13. $A = (-2, 0, 1), B = (10, 8, 5), r = 1/4$
14. $A = (5, 5, 3), B = (2, -4, 0), r = 1/3$
15. $A = (3, 1, 5), B = (-3, 4, 2), r = 2$
16. $A = (-2, 5, 1), B = (4, -1, 2), r = 3/2$

In Problems 17–20, *find the midpoint of segment AB.*

17. $A = (5, -2, 3), B = (-3, 4, 7)$
18. $A = (4, 3, 5), B = (-2, -1, 2)$
19. $A = (-3, 2, 0), B = (5, 4, 3)$
20. $A = (4, 3, -1), B = (4, 8, -3)$

B 21. Given $A = (5, -2, 3), P = (6, 0, 0)$, and $\overline{AP}/\overline{AB} = 1/3$, find B.
22. Given $B = (-4, 14, 4), P = (-1, 8, -4)$, and $\overline{AP}/\overline{AB} = 2/5$, find A.
23. Given $B = (6, 0, 9), P = (4, 1, 6)$, and $\overline{AP}/\overline{AB} = 3/4$, find A.
24. Given $A = (5, 3, -2), P = (1, 5, 2)$, and $\overline{AP}/\overline{AB} = 2/3$, find B.

In Problems 25–28, *find the unknown quantity.*

25. $A = (5, 1, 0), B = (1, y, 2), \overline{AB} = 6$
26. $A = (-2, 4, 3), B = (x, -4, 2), \overline{AB} = 9$
27. $A = (x, 4, -2), B = (-x, -6, 3), \overline{AB} = 15$
28. $A = (x, x, 5), B = (-1, -2, 0), \overline{AB} = 5\sqrt{2}$

29. The point $(-1, 5, 2)$ is a distance 6 from the midpoint of the segment joining $(1, 3, 2)$ and $(x, -1, 6)$. Find x.

30. The point $(1, -2, 9)$ is a distance $5\sqrt{5}$ from the midpoint of the segment joining $(1, y, 2)$ and $(5, -1, 6)$. Find y.

C 31. Prove Theorem 10.2.   32. Prove Theorem 10.3.

10.2

VECTORS IN SPACE

Vectors in three-dimensional space may be handled in much the same way as vectors in the plane. Vectors themselves, the sum and difference of two vectors, the absolute value of a vector, and scalar multiple of a vector are defined in the same way as they were in Chapter 2; and Theorem 2.2 (see page 41) holds for vectors in space as well as for vectors in the plane. The following theorems and definitions are the three-dimensional analogs of theorems and definitions of Chapter 2. The proofs of the theorems are simple extensions of the corresponding arguments in two dimensions.

Definition   *If* $O = (0, 0, 0)$, $X = (1, 0, 0)$, $Y = (0, 1, 0)$ *and* $Z = (0, 0, 1)$, *then the vectors represented by* $\overrightarrow{OX}$, $\overrightarrow{OY}$, *and* $\overrightarrow{OZ}$ *are* **i**, **j**, *and* **k**, *respectively, and are called* **basis vectors**.

Theorem 10.4   *Every vector in space can be written in the form*

$$a\mathbf{i} + b\mathbf{j} + c\mathbf{k}$$

*in one and only one way. The numbers a, b, and c are called the **first**, **second**, and **third components**, respectively, of the vector.*

Theorem 10.5   *If $\overrightarrow{AB}$, where $A = (x_1, y_1, z_1)$ and $B = (x_2, y_2, z_2)$, represents a vector $\mathbf{v}$ in space, then*

$$\mathbf{v} = (x_2 - x_1)\mathbf{i} + (y_2 - y_1)\mathbf{j} + (z_2 - z_1)\mathbf{k}.$$

Theorem 10.6 $(a_1\mathbf{i} + b_1\mathbf{j} + c_1\mathbf{k}) + (a_2\mathbf{i} + b_2\mathbf{j} + c_2\mathbf{k}) = (a_1 + a_2)\mathbf{i} + (b_1 + b_2)\mathbf{j} + (c_1 + c_2)\mathbf{k}$
$(a_1\mathbf{i} + b_1\mathbf{j} + c_1\mathbf{k}) - (a_2\mathbf{i} + b_2\mathbf{j} + c_2\mathbf{k}) = (a_1 - a_2)\mathbf{i} + (b_1 - b_2)\mathbf{j} + (c_1 - c_2)\mathbf{k}$

$$d(a\mathbf{i} + b\mathbf{j} + c\mathbf{k}) = da\mathbf{i} + db\mathbf{j} + dc\mathbf{k}$$

$$|a\mathbf{i} + b\mathbf{j} + c\mathbf{k}| = \sqrt{a^2 + b^2 + c^2}$$

Definition   *The **angle** between two nonzero vectors $\mathbf{u}$ and $\mathbf{v}$ is the smaller angle between the representatives of $\mathbf{u}$ and $\mathbf{v}$ having their tails at the origin.*

Theorem 10.7   *If $\mathbf{u} = a_1\mathbf{i} + b_1\mathbf{j} + c_1\mathbf{k}$ and $\mathbf{v} = a_2\mathbf{i} + b_2\mathbf{j} + c_2\mathbf{k}$ ($\mathbf{u} \neq \mathbf{0}$ and $\mathbf{v} \neq \mathbf{0}$) and if θ is the angle between them, then*

$$\cos\theta = \frac{a_1 a_2 + b_1 b_2 + c_1 c_2}{|\mathbf{u}||\mathbf{v}|}.$$

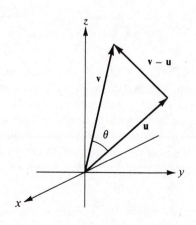

Figure 10.6

Proof   By the law of cosines (see Figure 10.6)

$$| \mathbf{v} - \mathbf{u} |^2 = | \mathbf{u} |^2 + | \mathbf{v} |^2 - 2 | \mathbf{u} | | \mathbf{v} | \cos \theta.$$

Since

$$\mathbf{v} - \mathbf{u} = (a_2 - a_1)\mathbf{i} + (b_2 - b_1)\mathbf{j} + (c_2 - c_1)\mathbf{k},$$

we have

$$(a_2 - a_1)^2 + (b_2 - b_1)^2 + (c_2 - c_1)^2$$
$$= a_1^2 + b_1^2 + c_1^2 + a_2^2 + b_2^2 + c_2^2 - 2 | \mathbf{u} | | \mathbf{v} | \cos \theta.$$

and it follows that

$$| \mathbf{u} | | \mathbf{v} | \cos \theta = a_1 a_2 + b_1 b_2 + c_1 c_2,$$

and

$$\cos \theta = \frac{a_1 a_2 + b_1 b_2 + c_1 c_2}{| \mathbf{u} | | \mathbf{v} |}.$$

Compare this proof with the proof of Theorem 2.6 on page 48.

Definition   *If* $\mathbf{u} = a_1 \mathbf{i} + b_1 \mathbf{j} + c_1 \mathbf{k}$ *and* $\mathbf{v} = a_2 \mathbf{i} + b_2 \mathbf{j} + c_2 \mathbf{k}$, *then the* **dot product** (*scalar product, inner product*) *of* $\mathbf{u}$ *and* $\mathbf{v}$ *is*

$$\mathbf{u} \cdot \mathbf{v} = a_1 a_2 + b_1 b_2 + c_1 c_2.$$

Theorems 2.7–2.10 (pages 50–52) still hold for three-dimensional vectors. They are restated here for convenience.

Theorem 10.8   *The vectors* $\mathbf{u}$ *and* $\mathbf{v}$ (*not both* $\mathbf{0}$) *are orthogonal* (*perpendicular*) *if and only if* $\mathbf{u} \cdot \mathbf{v} = 0$ (*the zero vector is taken to be orthogonal to every vector*).

Theorem 10.9   *If* $\mathbf{u}, \mathbf{v},$ *and* $\mathbf{w}$ *are vectors, then*

$$\mathbf{u} \cdot \mathbf{v} = \mathbf{v} \cdot \mathbf{u}$$

and

$$(\mathbf{u} + \mathbf{v}) \cdot \mathbf{w} = \mathbf{u} \cdot \mathbf{w} + \mathbf{v} \cdot \mathbf{w}.$$

Theorem 10.10   *If* $\mathbf{u}$ *and* $\mathbf{v}$ *are vectors and* θ *is the angle between them, then*

$$\mathbf{u} \cdot \mathbf{v} = | \mathbf{u} | | \mathbf{v} | \cos \theta$$

and

$$\mathbf{v} \cdot \mathbf{v} = |\mathbf{v}|^2.$$

Definition *Suppose the nonzero vectors* **u** *and* **v** *are represented by* $\overrightarrow{AB}$ *and* $\overrightarrow{AC}$, *respectively. Then the* **projection** *of* **u** *on* **v** *is the vector* **w** *represented by* $\overrightarrow{AB}$, *where D is on the line AC and BD* $\perp$ *AC (see Figure 2.15, page 52).*

Theorem 10.11 *If* **w** *is the projection of* **u** *on* **v**, *then*

$$|\mathbf{w}| = \frac{|\mathbf{u} \cdot \mathbf{v}|}{|\mathbf{v}|} \quad and \quad \mathbf{w} = \left(\frac{\mathbf{u} \cdot \mathbf{v}}{|\mathbf{v}|}\right)\frac{\mathbf{v}}{|\mathbf{v}|} = \frac{\mathbf{u} \cdot \mathbf{v}}{|\mathbf{v}|^2}\,\mathbf{v}.$$

Theorems 10.8–10.11 are not stated in terms of the dimensions of the vectors; their proofs are identical to those of Chapter 2. Proofs of Theorems 10.4–10.6 are left to the student (see Problems 27–29).

Example 1 Given $\mathbf{u} = 2\mathbf{i} + \mathbf{j} - 3\mathbf{k}$ and $\mathbf{v} = \mathbf{i} - 2\mathbf{j} - \mathbf{k}$, find $\mathbf{u} + \mathbf{v}$, $\mathbf{u} - \mathbf{v}$, and $\mathbf{u} \cdot \mathbf{v}$.

Solution
$$\mathbf{u} + \mathbf{v} = (2 + 1)\mathbf{i} + (1 - 2)\mathbf{j} + (-3 - 1)\mathbf{k} = 3\mathbf{i} - \mathbf{j} - 4\mathbf{k}$$
$$\mathbf{u} - \mathbf{v} = (2 - 1)\mathbf{i} + (1 + 2)\mathbf{j} + (-3 + 1)\mathbf{k} = \mathbf{i} + 3\mathbf{j} - 2\mathbf{k}$$
$$\mathbf{u} \cdot \mathbf{v} = 2 \cdot 1 + 1(-2) + (-3)(-1) = 3$$

Example 2 Give in component form the vector **v** that is represented by $\overrightarrow{AB}$, where $A = (4, 3, -1)$ and $B = (-1, 2, -3)$ (see Figure 10.7).

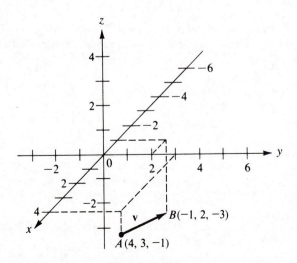

Figure 10.7

Solution $$\mathbf{v} = (-1 - 4)\mathbf{i} + (2 - 3)\mathbf{j} + (-3 + 1)\mathbf{k} = -5\mathbf{i} - \mathbf{j} - 2\mathbf{k}$$

Example 3   Find the end points of the representative $\overrightarrow{AB}$ of $\mathbf{v}$ if $\mathbf{v} = 2\mathbf{i} - 4\mathbf{j} + \mathbf{k}$ and $(2, -3, 5)$ is the midpoint of AB (see Figure 10.8).

Solution   If $A = (x_1, y_1, z_1)$ and $B = (x_2, y_2, z_2)$, then, by Theorem 10.5,

$$x_2 - x_1 = 2, \qquad y_2 - y_1 = -4, \qquad \text{and} \qquad z_2 - z_1 = 1.$$

By Theorem 10.3,

$$\frac{x_1 + x_2}{2} = 2, \qquad \frac{y_1 + y_2}{2} = -3, \qquad \text{and} \qquad \frac{z_1 + z_2}{2} = 5.$$

Solving simultaneously, we have

$$A = \left(1, -1, \frac{9}{2}\right) \qquad \text{and} \qquad B = \left(3, -5, \frac{11}{2}\right).$$

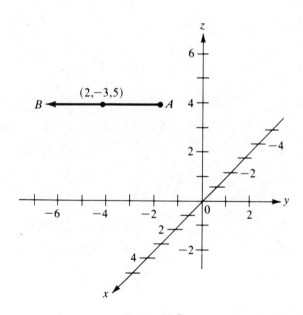

Figure 10.8

Example 4   Determine whether the vectors $\mathbf{u} = 3\mathbf{i} + \mathbf{j} - 2\mathbf{k}$ and $\mathbf{v} = 2\mathbf{i} - 4\mathbf{j} + \mathbf{k}$ are orthogonal.

Solution $$\mathbf{u} \cdot \mathbf{v} = (3)(2) + (1)(-4) + (-2)(1) = 0$$

Thus $\mathbf{u}$ and $\mathbf{v}$ are orthogonal.

Example 5   Find the projection **w** of $\mathbf{u} = 4\mathbf{i} - \mathbf{j} + \mathbf{k}$ upon $\mathbf{v} = 3\mathbf{i} + \mathbf{j} - 4\mathbf{k}$ (see Figure 10.9).

Solution $$\mathbf{u} \cdot \mathbf{v} = (4)(3) + (-1)(1) + (1)(-4) = 7$$

and

$$|\mathbf{v}| = \sqrt{3^2 + 1^2 + (-4)^2} = \sqrt{26}$$

Thus

$$\mathbf{w} = \left(\frac{\mathbf{u} \cdot \mathbf{v}}{|\mathbf{v}|}\right)\frac{\mathbf{v}}{|\mathbf{v}|}$$

$$= \frac{7}{\sqrt{26}}\frac{3\mathbf{i} + \mathbf{j} - 4\mathbf{k}}{\sqrt{26}}$$

$$= \frac{21}{26}\mathbf{i} + \frac{7}{26}\mathbf{j} - \frac{14}{13}\mathbf{k}.$$

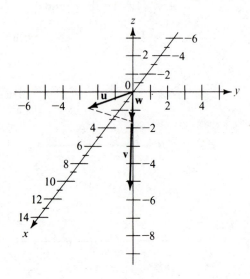

Figure 10.9

PROBLEMS

A   *In Problems 1–4, give in component form the vector* **v** *that is represented by* $\overrightarrow{AB}$.

1. $A = (2, 3, -5), \quad B = (-4, 1, 2)$ **2.** $A = (3, -2, 4), \quad B = (5, 4, -1)$

3. $A = (5, 0, -2), \quad B = (2, -4, 1)$ **4.** $A = (2, -3, 8), \quad B = (2, 5, 2)$

In Problems 5–8, give the unit vector in the direction of **v**.

5. $\mathbf{v} = 4\mathbf{i} + \mathbf{j} - 2\mathbf{k}$ **6.** $\mathbf{v} = \mathbf{i} - 2\mathbf{j} + 2\mathbf{k}$

7. $\mathbf{v} = \mathbf{i} + 5\mathbf{j} - 3\mathbf{k}$ **8.** $\mathbf{v} = 3\mathbf{i} - 4\mathbf{k}$

B *In Problems 9–14, find the end points of the representative $\overrightarrow{AB}$ of* **v** *from the given information.*

9. $\mathbf{v} = 2\mathbf{i} - \mathbf{j} + 3\mathbf{k}$, $A = (2, 1, 5)$ **10.** $\mathbf{v} = 3\mathbf{i} + \mathbf{j} - 4\mathbf{k}$, $A = (1, 4, 3)$

11. $\mathbf{v} = -\mathbf{i} + 2\mathbf{j} + 5\mathbf{k}$, $B = (2, 3, 8)$ **12.** $\mathbf{v} = 2\mathbf{i} + 5\mathbf{j} - 3\mathbf{k}$, $B = (4, -2, 6)$

13. $\mathbf{v} = 4\mathbf{i} - 2\mathbf{j} + \mathbf{k}$, $(2, 5, -1)$ is the midpoint of AB

14. $\mathbf{v} = 6\mathbf{i} + \mathbf{j} - 4\mathbf{k}$, $(3, 2, -5)$ is the midpoint of AB

In Problems 15–18, find the angle θ between the given vectors.

15. $\mathbf{u} = \mathbf{i} + \mathbf{j} + 2\mathbf{k}$, $\mathbf{v} = 2\mathbf{i} - \mathbf{j} + \mathbf{k}$ **16.** $\mathbf{u} = 2\mathbf{i} - 2\mathbf{j} - \mathbf{k}$, $\mathbf{v} = -\mathbf{i} + 4\mathbf{j} + 2\mathbf{k}$

17. $\mathbf{u} = 5\mathbf{i} - \mathbf{j} + 3\mathbf{k}$, $\mathbf{v} = 4\mathbf{i} + 5\mathbf{j} - 2\mathbf{k}$ **18.** $\mathbf{u} = 2\mathbf{i} + 4\mathbf{j} + 4\mathbf{k}$, $\mathbf{v} = 4\mathbf{i} - 3\mathbf{k}$

In Problems 19–22, find $\mathbf{u} + \mathbf{v}, \mathbf{u} - \mathbf{v},$ *and* $\mathbf{u} \cdot \mathbf{v}$. *Indicate whether* **u** *and* **v** *are orthogonal.*

19. $\mathbf{u} = \mathbf{i} - 2\mathbf{j} + 5\mathbf{k}$, $\mathbf{v} = 2\mathbf{i} + 4\mathbf{j} + \mathbf{k}$ **20.** $\mathbf{u} = 3\mathbf{i} + \mathbf{j} - 4\mathbf{k}$, $\mathbf{v} = 2\mathbf{i} + 6\mathbf{j} + 3\mathbf{k}$

21. $\mathbf{u} = 2\mathbf{i} - \mathbf{j} + 6\mathbf{k}$, $\mathbf{v} = \mathbf{i} - 4\mathbf{j} - \mathbf{k}$ **22.** $\mathbf{u} = 4\mathbf{i} + 3\mathbf{j} - \mathbf{k}$, $\mathbf{v} = \mathbf{i} + 2\mathbf{j} + 3\mathbf{k}$

In Problems 23–26, find the projection of **u** *upon* **v**.

23. $\mathbf{u} = 4\mathbf{i} + \mathbf{j} - \mathbf{k}$, $\mathbf{v} = \mathbf{i} + \mathbf{j} + \mathbf{k}$ **24.** $\mathbf{u} = \mathbf{i} - 2\mathbf{j} + 4\mathbf{k}$, $\mathbf{v} = 2\mathbf{j} + 3\mathbf{k}$

25. $\mathbf{u} = 4\mathbf{i} - 2\mathbf{j} - \mathbf{k}$, $\mathbf{v} = \mathbf{i} - 2\mathbf{j} + \mathbf{k}$ **26.** $\mathbf{u} = 2\mathbf{i} + \mathbf{j}$, $\mathbf{v} = \mathbf{j} - 2\mathbf{k}$

C **27.** Prove Theorem 10.4.   **28.** Prove Theorem 10.5.   **29.** Prove Theorem 10.6.

30. Prove that $(\mathbf{u} + \mathbf{v}) \cdot (\mathbf{u} + \mathbf{v}) = |\mathbf{u}|^2 + 2\mathbf{u} \cdot \mathbf{v} + |\mathbf{v}|^2$.

31. Prove the triangle inequality $|\mathbf{u} + \mathbf{v}| \leq |\mathbf{u}| + |\mathbf{v}|$.

32. Prove $|\mathbf{u} - \mathbf{v}| \geq \|\mathbf{u}| - |\mathbf{v}\|$.

10.3

DIRECTION ANGLES, COSINES, AND NUMBERS

In plane geometry, we used the inclination and slope to give the direction of a line. The corresponding terms in solid analytic geometry are direction angles and direction cosines. They are most easily defined for vectors.

Definition   *If* **v** *is a vector, then the ordered set* $\{\alpha, \beta, \gamma\}$ *is the set of* **direction angles** *for* **v** *if* α *is the angle between* **v** *and* **i**, β *is the angle between* **v** *and* **j**, *and* γ *is the angle between* **v** *and* **k**.

The direction angles for a vector **v** are illustrated in Figure 10.10. Note that the direction angles are not necessarily in the coordinate planes. The angle α is in the plane determined by **v** and **i**; β and γ are in planes determined by **v** and **j** and by **v** and **k**, respectively.

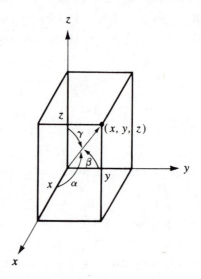

Figure 10.10

The angle between two vectors was defined in the previous section. Note again that this is not a directed angle; in fact, we have given no convention for positive and negative angles in space. Thus direction angles are never negative and never greater than 180°.

Just as the inclination of a line is less convenient than the slope, so the direction angles for a vector **v** are relatively inconvenient. More convenient to use are the cosines of the direction angles.

Definition   *If* **v** *is a vector, then the ordered set* $\{l, m, n\}$ *is the set of* **direction cosines** *for* **v** *if* $l = \cos \alpha$, $m = \cos \beta$, *and* $n = \cos \gamma$, *where* $\{\alpha, \beta, \gamma\}$ *is the set of direction angles for* **v**.

If $\overrightarrow{OP}$ is a representative of **v** with O the origin and P the point (x, y, z) (see Figure 10.10), then

$$l = \cos \alpha = \frac{x}{\rho}, \quad m = \cos \beta = \frac{y}{\rho}, \quad n = \cos \gamma = \frac{z}{\rho},$$

where $\rho = \sqrt{x^2 + y^2 + z^2}$. It is this relation to the coordinates of a point on the line which allows us to prove the next theorem.

Theorem 10.12   *If* $\{l, m, n,\}$ *is a set of direction cosines for a vector, then*

$$l^2 + m^2 + n^2 = 1.$$

Proof   From the relations above, we have

$$l^2 + m^2 + n^2 = \frac{x^2}{\rho^2} + \frac{y^2}{\rho^2} + \frac{z^2}{\rho^2}$$

$$= \frac{x^2 + y^2 + z^2}{\rho^2} = 1.$$

For most purposes, an even more convenient set of numbers is a set of direction numbers for a line.

Definition $\{a, b, c\}$ *is a set of* **direction numbers** *for the vector* **v** *if there is a nonzero constant* k *such that* $a = kl, b = km,$ *and* $c = kn,$ *where* $\{l, m, n\}$ *is the set of direction cosines for* **v**.

Theorem 10.12 may be used to find direction cosines of a vector if its direction numbers are known. Of course, additional information is needed to get the proper signs.

Example 1   Given that a vector **v** has direction numbers $\{4, 1, -2\}$ and is directed upward, find its direction cosines.

Solution
$$a^2 + b^2 + c^2 = k^2 l^2 + k^2 m^2 + k^2 n^2$$
$$= k^2(l^2 + m^2 + n^2) = k^2$$

Thus,
$$k^2 = 16 + 1 + 4 = 21 \quad \text{and} \quad k = \pm\sqrt{21}.$$

Since **v** is directed upward, $\gamma < 90°$ and $n > 0$. Thus $k = -\sqrt{21}$, and

$$l = -\frac{4}{\sqrt{21}}, \quad m = -\frac{1}{\sqrt{21}}, \quad \text{and} \quad n = \frac{2}{\sqrt{21}}.$$

The following theorem is a direct consequence of the foregoing definitions.

Theorem 10.13   *The vector* **v** $= a\mathbf{i} + b\mathbf{j} + c\mathbf{k}$ *has direction numbers* $\{a, b, c\}$.

Example 2   Give a set of direction numbers for the vector **v** represented by $\overrightarrow{AB}$, where $A = (3, -1, 2)$ and $B = (5, 2, -1)$.

Solution   By Theorem 10.5,

$$\mathbf{v} = (5 - 3)\mathbf{i} + (2 + 1)\mathbf{j} + (-1 - 2)\mathbf{k} = 2\mathbf{i} + 3\mathbf{j} - 3\mathbf{k}.$$

By Theorem 10.13, one set of direction numbers for **v** is $\{2, 3, -3\}$.

Direction angles, cosines, and numbers of vectors carry over directly to lines.

Definition   *The statement, "A vector **v** is directed along a line l," means that a representative of **v** is on l.*

Definition   *A set of direction angles, cosines, or numbers for a line l is any set of direction angles, cosines, or numbers, respectively, for any vector **v** directed along l.*

Note that every line has two sets of direction angles and two sets of direction cosines, corresponding to the two possible directions on the line. It is easily seen that if $\{l, m, n\}$ is one set of direction cosines for a line, then the other is $\{-l, -m, -n\}$.

Example 3   Find the two sets of direction cosines and direction angles for a line if $\{1, 2, 2\}$ is a set of direction numbers for it.

Solution   Since the three numbers given are a number k times the direction cosines, we have the following.

$$kl = 1, \quad km = 2, \quad \text{and} \quad kn = 2$$
$$k^2 l^2 + k^2 m^2 + k^2 n^2 = 1 + 4 + 4 = 9$$
$$k^2(l^2 + m^2 + n^2) = 9$$
$$k^2 = 9 \quad \text{(by Theorem 10.12)}$$
$$k = \pm 3$$

Thus, the two possible sets of direction cosines are $\{1/3, 2/3, 2/3\}$ and $\{-1/3, -2/3, -2/3\}$, and they give approximate direction angles $\{71°, 48°, 48°\}$ and $\{109°, 132°, 132°\}$, respectively.

Example 4   Suppose a line has direction numbers $\{2, -4, 1\}$ and contains the point $(1, 3, 4)$. Find another point on the line.

Solution   By Theorems 10.5 and 10.13, we have

$$a = x_2 - x_1 \quad b = y_2 - y_1, \quad \text{and} \quad c = z_2 - z_1,$$

or

$$x_2 = x_1 + a, \qquad y_2 = y_1 + b, \qquad \text{and} \qquad z_2 = z_1 + c.$$

Thus,

$$x_2 = 1 + 2 = 3, \qquad y_2 = 3 - 4 = -1, \qquad \text{and} \qquad z_2 = 4 + 1 = 5,$$

which give the point $(3, -1, 5)$ (see Figure 10.11). Of course, any nonzero multiple of the given direction numbers gives another set of direction numbers; these may be used to find other points on the line.

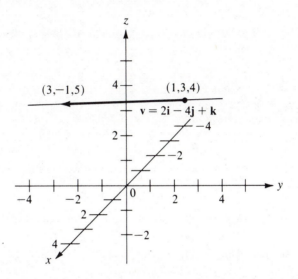

Figure 10.11

Perhaps you feel a bit uneasy about this last example. How do we know that, of all possible sets of direction numbers, the one we were given is *that* one—that is, the set of direction numbers determined by the method of Example 2? The answer is simple—all of them are. Here is the reason. No matter what pair of points we choose on a line, we must get one of the two possible sets of direction cosines. Taking one of those two sets, we can get from it any set of direction numbers by multiplying by the proper number k. In particular, if we take a set of direction cosines for the line of Example 4 and multiply it by the proper number k, we get the direction numbers $\{2, -4, 1\}$. Now there are two points on the line at a distance $|k|$ from the point $(1, 3, 4)$. By choosing the proper one of those two points, we see that that point, together with $(1, 3, 4)$ would give the direction numbers $\{2, -4, 1\}$. Of course, this argument could be repeated for any point and set of direction numbers.

It might be noted that we could, say, double the direction numbers of Example 4 to give another set of direction numbers, $\{4, -8, 2\}$. This, together with the point

$(1, 3, 4)$, gives the point $(5, -5, 6)$, which is also on the line. This could be repeated indefinitely to get as many points on the line as we choose.

It is clear that if two lines are parallel and directed the same way, they must have the same set of direction angles and, thus, the same set of direction cosines. If they are parallel and have opposite directions, their direction angles are supplementary and one set of direction cosines is the negative of the other. Thus, any set of direction numbers for one line is proportional to a set of direction numbers for the other. Furthermore, this chain of reasoning can be reversed to show that if two lines have proportional sets of direction numbers, they are parallel.

Theorem 10.14   *Two distinct lines are parallel if and only if sets of direction numbers for the two lines are proportional.*

Suppose that lines l_1 and l_2 have direction numbers $\{a_1, b_1, c_1\}$ and $\{a_2, b_2, c_2\}$, respectively. Then vectors $\mathbf{v}_1$ and $\mathbf{v}_2$ directed along lines l_1 and l_2, respectively, may be represented by

$$\mathbf{v}_1 = a_1\mathbf{i} + b_1\mathbf{j} + c_1\mathbf{k} \qquad \text{and} \qquad \mathbf{v}_2 = a_2\mathbf{i} + b_2\mathbf{j} + c_2\mathbf{k}.$$

By Theorem 10.8, $\mathbf{v}_1$ and $\mathbf{v}_2$ are orthogonal if and only if

$$\mathbf{v}_1 \cdot \mathbf{v}_2 = a_1a_2 + b_1b_2 + c_1c_2 = 0.$$

This gives the following theorem for perpendicularity of lines.

Theorem 10.15   *Two lines with direction numbers $\{a_1, b_1, c_1\}$ and $\{a_2, b_2, c_2\}$ are perpendicular if and only if*

$$a_1a_2 + b_1b_2 + c_1c_2 = 0.$$

PROBLEMS

A   *In Problems 1–6, find the set of direction angles for the vector described.*

1. Direction numbers $\{1, 4, 8\}$; directed to the right of the xz plane
2. Direction numbers $\{4, -4, 2\}$; directed to the right of the xz plane
3. Direction numbers $\{1, 2, -4\}$; directed behind the yz plane
4. Direction numbers $\{2, -1, -3\}$; directed above the xy plane
5. Direction numbers $\{1, 1, 1\}$; directed behind the yz plane
6. Direction numbers $\{1, -1, 0\}$; directed to the right of the xz plane

In Problems 7–12, find a set of direction numbers for the lines containing the two given points.

7. $(1, 4, 3)$ and $(5, 2, -1)$ **8.** $(2, 0, -4)$ and $(-1, 2, 3)$

9. $(2, 2, 1)$ and $(0, 0, 3)$ **10.** $(3, 5, -2)$ and $(-1, 4, 4)$

11. $(0, 0, 0)$ and $(5, 1, -2)$ **12.** $(-1, 4, 5)$ and $(3, -4, 0)$

B   *In Problems 13–18, find two more points on the line.*

13. Direction numbers $\{1, 5, 2\}$; containing $(2, 3, -1)$

14. Direction numbers $\{1, 4, 0\}$; containing $(-2, 1, 1)$

15. Direction numbers $\{2, 1, 2\}$; containing $(1, 3, 3)$

16. Direction numbers $\{1, 1, 1\}$; containing $(2, 4, -1)$

17. Direction numbers $\{4, 0, -1\}$; containing $(1, 3, -1)$

18. Direction numbers $\{4, 4, 3\}$; containing $(-4, -4, -3)$

19. Give the direction angles and direction cosines for the coordinate axes with their usual directions.

20. Give a set of formulas for finding all points on the line described in Example 4. [*Hint:* Consider the two paragraphs following Example 4.]

In Problems 21–30, two lines are described by a pair of points on each. Indicate whether the lines are parallel, perpendicular, coincident or none of these.

21. $(3, 4, 1), (4, 8, -1); (2, 3, -5), (0, -5, -1)$

22. $(2, 1, 5), (3, 3, -1); (4, 2, 10), (1, -4, 5)$

23. $(4, 1, -4), (3, 2, 1); (4, 1, -4), (11, 3, -3)$

24. $(4, 2, -1), (7, 6, 2); (5, 10, 3), (-4, -2, -6)$

25. $(2, 1, 4), (4, -3, 12); (1, 3, 0), (6, -7, 20)$

26. $(4, 5, 1), (3, 2, -4); (4, 1, 2), (5, -1, 3)$

27. $(2, 3, 1), (4, -2, 2); (1, 0, 3), (3, -3, 1)$

28. $(3, 1, 4), (4, 3, 3); (5, 5, 2), (0, -5, 7)$

29. $(2, 1, 3), (5, -1, 1); (3, 4, -1), (5, 3, 3)$

30. $(4, 4, -3), (1, 3, -1); (2, 1, 5), (8, 3, 1)$

10.4

THE LINE

Recall that in Chapter 9 (pages 249–250) we derived a parametric representation for a line containing two given points. The result was basically the point-of-division formulas with r as the parameter. We used a vector argument there. The same argument holds in three dimensions; but instead of using two points, let us consider the line l with direction numbers $\{a, b, c\}$ and containing the point

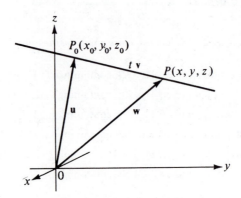

Figure 10.12

$P_0 = (x_0, y_0, z_0)$. Let $P = (x, y, z)$ be any point on l. If $\mathbf{u}$ and $\mathbf{w}$ are the vectors represented by $\overrightarrow{OP_0}$ and $\overrightarrow{OP}$, respectively (see Figure 10.12), it is seen that

$$\mathbf{u} = x_0 \mathbf{i} + y_0 \mathbf{j} + z_0 \mathbf{k} \qquad \text{and} \qquad \mathbf{w} = x \mathbf{i} + y \mathbf{j} + z \mathbf{k}.$$

Since $\{a, b, c\}$ is a set of direction numbers for l,

$$\mathbf{v} = a \mathbf{i} + b \mathbf{j} + c \mathbf{k}$$

is a vector lying along l. But $\overrightarrow{P_0 P}$ represents another vector lying along l; it must be a scalar multiple, $t\mathbf{v}$, of $\mathbf{v}$. Since

$$\mathbf{w} = \mathbf{u} + t\mathbf{v},$$

it follows that

$$x = x_0 + ta,$$
$$y = y_0 + tb,$$

and

$$z = z_0 + tc.$$

Theorem 10.16 *A parametric representation of the line containing (x_0, y_0, z_0) and having direction numbers $\{a, b, c\}$ is*

$$x = x_0 + at, \qquad y = y_0 + bt, \qquad z = z_0 + ct.$$

The following nonvector argument may also be used to establish this result. Given a set of direction numbers for a line and one point on that line, we can find another simply by adding. Furthermore, we can find other sets of direction numbers by taking a multiple of the original set. Thus if the point given is (x_0, y_0, z_0) and the set of direction numbers given is $\{a, b, c\}$, then any point (x, y, z) such that

$$x = x_0 + at,$$
$$y = y_0 + bt,$$
$$z = z_0 + ct,$$

where t is a real number, is on the given line. Furthermore, if (x, y, z) is a point on the line different from (x_0, y_0, z_0), then a set of direction numbers for the line is $\{x - x_0, y - y_0, z - z_0\}$. These must be a multiple of the given set of direction numbers $\{a, b, c\}$; that is, for some t,

$$x - x_0 = at,$$
$$y - y_0 = bt,$$
$$z - z_0 = ct.$$

These equations hold not only for every point on the line different from (x_0, y_0, z_0) but also for (x_0, y_0, z_0). Thus a point is on the given line if and only if it satisfies the set of equations given above.

Example 1   Find a parametric representation for the line containing $(1, 3, -2)$ and having direction numbers $\{3, 2, -1\}$.

Solution $x = 1 + 3t, \qquad y = 3 + 2t, \qquad z = -2 - t$

Example 2   Find a parametric representation of the line containing $(4, 2, -1)$ and $(0, 2, 3)$.

Solution   A set of direction numbers is $\{4 - 0, 2 - 2, -1 - 3\} = \{4, 0, -4\}$. Thus the line is

$$x = 4 + 4t, \qquad y = 2, \qquad z = -1 - 4t.$$

Once we have the direction numbers, we may use them with either of the two given points. Thus, another representation of the line in Example 2 is

$$x = 4s, \qquad y = 2, \qquad z = 3 - 4s.$$

Although this does not look much like the first representation, it is easily seen that they are the same. For instance, $t = 0$ gives the point $(4, 2, -1)$, as does $s = 1$; $t = -1$ gives the point $(0, 2, 3)$, as does $s = 0$, and so forth.

In fact,

$$x = 4 + 4t \qquad y = 2, \qquad z = -1 - 4t$$
$$= 4(t + 1) \qquad\qquad\qquad = 3 - 4 - 4t$$
$$= 4s, \qquad\qquad\qquad\qquad = 3 - 4(t + 1)$$
$$\qquad\qquad\qquad\qquad\qquad\qquad = 3 - 4s,$$

where $s = t + 1$. Thus, whatever point we get using a value of t can be found by choosing $s = t + 1$.

A simpler set of direction numbers can also be found. Since the ones we have are all multiples of 4, we can multiply through by $1/4$ to get another set of direction numbers, $\{1, 0, -1\}$. Using these with the first point gives

$$x = 4 + u, \qquad y = 2, \qquad z = -1 - u.$$

Again, we see that $4t = u$, so the two representations are equivalent.

Perhaps you wonder what is needed to be able to say that two parametric representations are equivalent. If a value of t and another of s both give the same point, then, for those values of t and s, the three coordinates must be equal. Eliminating x, y, and z between the two parametric representations gives three equations in t and s (in some of these, the parameters may both be absent, as they are in the representation of y here). If all give the same result when they are solved for one parameter in terms of the other, and if the domain and range are the same, then the representations are equivalent.

Suppose we eliminate the parameter in the representation given by Theorem 10.16. If none of the direction numbers is zero, we can solve each equation for t and set them equal to each other. This gives

$$\frac{x - x_0}{a} = \frac{y - y_0}{b} = \frac{z - z_0}{c}.$$

Actually, this is just a shorter way of writing the three equations

$$\frac{x - x_0}{a} = \frac{y - y_0}{b},$$

$$\frac{y - y_0}{b} = \frac{z - z_0}{c},$$

and

$$\frac{x - x_0}{a} = \frac{z - z_0}{c}.$$

But these three equations are not independent—the last can be found from the first two, Let us discard it and consider only the first two, which, as we shall see in the next section, represent planes. Any point that satisfies both equations is on both planes and therefore on the intersection of the two planes, which is a line. Thus this representation of a line gives it as the intersection of two planes. It might be noted that the equation we discarded is also a plane containing the same line. These three planes can be seen in Figure 10.13. The line PQ is projected upon each of the three coordinate planes. In each case, this projection—together with the original line—determines a plane. The plane

$$\frac{x - x_0}{a} = \frac{y - y_0}{b}$$

is the plane determined by PQ and the projection, $P_1 Q_1$, of PQ on the xy plane.

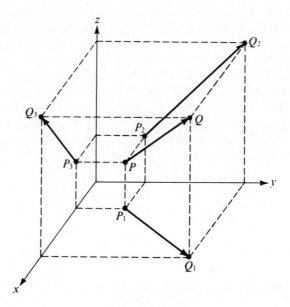

Figure 10.13

The z axis is either on or parallel to this plane. Similarly,

$$\frac{y - y_0}{b} = \frac{z - z_0}{c}$$

is the plane determined by PQ and the projection, $P_2 Q_2$, of PQ on the yz plane; and

$$\frac{x - x_0}{a} = \frac{z - z_0}{c}$$

is the plane determined by PQ and the projection, $P_3 Q_3$, of PQ on the xy plane.

What, now, if one of the direction numbers is zero? Let us suppose that $a = 0$. Then the line in parametric form is

$$x = x_0, \qquad y = y_0 + bt, \qquad z = z_0 + ct.$$

We do not have to eliminate the parameter from the first equation—it is already gone. By eliminating t between the last two equations as before, we have

$$\frac{y - y_0}{b} = \frac{z - z_0}{c}.$$

This together with $x = x_0$ (or $x - x_0 = 0$) gives the line as the intersection of two planes. In this case, plane PQQ_1P_1 and plane PQQ_3P_3 of Figure 10.13 are identical; they are both the plane $x = x_0$, which is parallel to (or on) the yz plane. Thus the line is parallel to the yz plane.

If two of the direction numbers are zero, we have two equations in which the parameter is missing. The parameter in the third equation cannot be eliminated, because there is no other equation with which to combine it. But it is not necessary to eliminate it! The two equations without the parameter already give us the necessary two planes. In this case the line is parallel to (or on) one of the coordinate axes. To illustrate this, suppose that the line PQ of Figure 10.13 is parallel to the y axis. Then $P_3 = Q_3$, and this point lies on the line PQ; hence no plane is determined by this projection. On the other hand, PQQ_1P_1 is a plane perpendicular to the xy plane and parallel to the y axis, while PQQ_2P_2 is a plane perpendicular to the yz plane and parallel to the y axis.

Theorem 10.17   *If a line contains the point (x_0, y_0, z_0) and has direction numbers $\{a, b, c\}$, then it can be represented by*

(a)
$$\frac{x - x_0}{a} = \frac{y - y_0}{b} = \frac{z - z_0}{c}$$

if none of the direction numbers is zero;

(b)
$$x - x_0 = 0 \quad and \quad \frac{y - y_0}{b} = \frac{z - z_0}{c}$$

if $a = 0$ and neither b nor c is zero (similar results follow if $b = 0$ or $c = 0$);

(c)
$$x - x_0 = 0 \quad and \quad y - y_0 = 0$$

if $a = 0$ and $b = 0$ (again, similar results follow some other pair of direction numbers equaling zero). These are called **symmetric equations** *of the line.*

Example 3   Find symmetric equations of the line containing $(4, 1, -2)$ and having direction numbers $\{1, 3, -2\}$.

Solution
$$\frac{x - 4}{1} = \frac{y - 1}{3} = \frac{z + 2}{-2}$$

Example 4   Find symmetric equations for the line containing $(4, 1, 3)$ and $(2, 1, -2)$.

Solution A set of direction numbers is $\{4 - 2, 1 - 1, 3 + 2\} = \{2, 0, 5\}$. Since $b = 0$, we have (using the first point)

$$\frac{x - 4}{2} = \frac{z - 3}{5} \quad \text{and} \quad y - 1 = 0.$$

Example 5 Find the point of intersection (if any) of the lines

$$x = 3 + 2t, \quad y = 2 - t, \quad z = 5 + t$$

and

$$x = -3 - s, \quad y = 7 + s, \quad z = 16 + 3s.$$

Solution Let us assume that there is a point of intersection. Then there is a value of t and a value of s which yield the same values of x, y, and z. For these particular values of t and s, we have

$$x = 3 + 2t = -3 - s$$
$$y = 2 - t = 7 + s$$
$$z = 5 + t = 16 + 3s$$

or

$$2t + s = -6, \quad t + s = -5, \quad t - 3s = 11.$$

If we solve the first pair simultaneously, we get

$$t = -1 \quad \text{and} \quad s = -4.$$

We see that they also satisfy the third equation. Thus there is a point of intersection which corresponds to $t = -1$ (or $s = -4$). It is $(1, 3, 4)$.

Note that this method requires that the lines be given in parametric form. If they are given as symmetric equations, they must first be changed to a parametric representation. A comparison of Theorems 10.16 and 10.17 makes this easy.

It might be noted that there are three possibilities. One is the situation in which there is a value of t and a value of s satisfying all three of the equations in t and s, as above. This results in a single point of intersection. In a second possibility, there is no value for t or s satisfying all three of the equations; that is, the values of t and s that satisfy the first two equations fail to satisfy the third. Thus, there is no point of intersection. The third possibility is that any two of the three equations in t and s are dependent; that is, any pair of values for t and s that satisfies one of them, satisfies all three. In this case we have two different representations for the same line (see the discussion following Example 2).

PROBLEMS

A   *In Problems* 1–16, *represent the given line in parametric form and in symmetric form.*

1. Containing $(5, 1, 3)$; direction numbers $\{3, -2, 4\}$
2. Containing $(2, -4, 2)$; direction numbers $\{2, 3, 1\}$
3. Containing $(5, -2, 1)$; direction numbers $\{4, 1, -2\}$
4. Containing $(2, 0, 3)$; direction numbers $\{4, -1, 3\}$
5. Containing $(1, 1, 1)$; direction numbers $\{2, 0, 1\}$
6. Containing $(1, 0, 5)$; direction numbers $\{3, 1, 0\}$
7. Containing $(4, 4, 1)$; direction numbers $\{0, 0, 1\}$
8. Containing $(3, 1, 2)$; direction numbers $\{1, 0, 0\}$
9. Containing $(4, 0, 5)$ and $(2, 3, 1)$
10. Containing $(3, 3, 1)$ and $(4, 0, 2)$
11. Containing $(8, 4, 1)$ and $(-2, 0, 3)$
12. Containing $(-4, 2, 0)$ and $(3, 1, 2)$
13. Containing $(5, 1, 3)$ and $(5, 2, 4)$
14. Containing $(2, 2, 4)$ and $(1, 2, 7)$
15. Containing $(1, -2, 3)$ and $(1, 4, 3)$
16. Containing $(2, 4, -5)$ and $(5, 4, -5)$

B   *In Problems* 17–24, *find the point of intersection (if any) of the given lines.*

17. $x = 4 + t, y = -8 - 2t, z = 12t;\quad x = 3 + 2s, y = -1 + s, z = -3 - 3s$
18. $x = 2 - t, y = 3 + 2t, z = 4 + t; x = 1 + s, y = -2 + s, z = 5 - 4s$
19. $x = 3 + t, y = 4 - 2t, z = 1 + 5t; x = 5 - s, y = 3 + 2s, z = 8 + 4s$
20. $x = 3 - t, y = 5 + 3t, z = -1 - 4t;\quad x = 8 + 2s, y = -6 - 4s, z = 5 + s$
21. $\dfrac{x - 2}{1} = \dfrac{y - 3}{-2} = \dfrac{z + 1}{1};\ \dfrac{x - 3}{2} = \dfrac{y - 1}{-4} = \dfrac{z}{2}$
22. $\dfrac{x - 5}{1} = \dfrac{y + 2}{-2} = \dfrac{z - 3}{5};\ \dfrac{x - 4}{-2} = \dfrac{y - 2}{1} = \dfrac{z - 4}{3}$
23. $\dfrac{x - 3}{1} = \dfrac{y + 3}{-4}, z + 1 = 0;\ \dfrac{x}{-2} = \dfrac{y - 2}{1} = \dfrac{z - 3}{4}$
24. $\dfrac{x - 2}{1} = \dfrac{y - 3}{-2} = \dfrac{z}{4};\quad x - 4 = 0, \dfrac{y - 2}{1} = \dfrac{z - 3}{-1}$

In Problems 25–30, *indicate whether the two given lines are parallel, perpendicular, co-incident, or none of these.*

25. $x = 3 + 5t, y = -1 - 2t, z = 4 + t;$ $x = 3, y = 4 + 2s, z = -2 + 4s$

26. $x = 4 - t, y = 3 + 2t, z = 1 + t;$ $x = 1 + 2t, y = 4 - 4t, z = 3 - 2t$

27. $x = 2 + t, y = 5 - 3t, z = 1 + 4t;$ $x = 4 - t, y = 2 + 2t, z = 3t$

28. $\dfrac{x - 2}{1} = \dfrac{y - 5}{-3} = \dfrac{z + 1}{2};$ $\dfrac{x - 4}{-3} = \dfrac{y + 1}{9} = \dfrac{z - 3}{-6}$

29. $\dfrac{x + 3}{1} = \dfrac{y - 4}{3} = \dfrac{z + 2}{-2};$ $\dfrac{x - 5}{-3} = \dfrac{y + 3}{-9} = \dfrac{z - 1}{6}$

30. $\dfrac{x - 1}{2} = \dfrac{z + 3}{4}, y - 5 = 0;$ $\dfrac{x + 2}{6} = \dfrac{y - 5}{3} = \dfrac{z}{2}$

31. Give equations for each of the coordinates axes.

10.5

THE CROSS PRODUCT

Let us now look at the other product of two vectors—the cross product.

Definition   *If* $\mathbf{u} = a_1\mathbf{i} + b_1\mathbf{j} + c_1\mathbf{k}$ *and* $\mathbf{v} = a_2\mathbf{i} + b_2\mathbf{j} + c_2\mathbf{k}$, *then the* **cross product** *(vector product, outer product) of* $\mathbf{u}$ *and* $\mathbf{v}$ *is*

$$\mathbf{u} \times \mathbf{v} = (b_1c_2 - c_1b_2)\mathbf{i} + (c_1a_2 - a_1c_2)\mathbf{j} + (a_1b_2 - b_1a_2)\mathbf{k}.$$

Some obvious questions arise. Why do we want to define a cross product this way? What is it good for? What are its properties? In some ways, all answers are the same. We define the cross product in this way to establish some interesting properties that are useful for certain applications. In a way, this is approaching the problem backward. It would be more logical to define the cross product of two vectors as that one having the desired properties and then show that such a vector must take the form given. The reason for our way of doing it is that it is by far the simpler approach. Before looking at some properties, let us consider a simpler form for the cross product.

Theorem 10.18   *If* $\mathbf{u} = a_1\mathbf{i} + b_1\mathbf{j} + c_1\mathbf{k}$ *and* $\mathbf{v} = a_2\mathbf{i} + b_2\mathbf{j} + c_2\mathbf{k}$, *then*

$$\mathbf{u} \times \mathbf{v} = \begin{vmatrix} \mathbf{i} & \mathbf{j} & \mathbf{k} \\ a_1 & b_1 & c_1 \\ a_2 & b_2 & c_2 \end{vmatrix}.$$

This theorem follows directly from the definition if we expand the above determinant by minors along the first row.

Example 1   If $u = 3i + j - 2k$ and $v = i + 2j + k$, find $u \times v$ and $v \times u$.

Solution

$$u \times v = \begin{vmatrix} i & j & k \\ 3 & 1 & -2 \\ 1 & 2 & 1 \end{vmatrix} = 5i - 5j + 5k$$

$$v \times u = \begin{vmatrix} i & j & k \\ 1 & 2 & 1 \\ 3 & 1 & -2 \end{vmatrix} = -5i + 5j - 5k$$

Note that $u \times v \neq v \times u$!

Again we are not multiplying numbers; there is no reason to assume that the cross product of two vectors has the same properties as the product of two numbers. We have already seen one difference in Example 1. The cross product has the following properties.

Theorem 10.19   *If u, v and w are vectors and a is a scalar, then the following properties hold.*

(a) $u \times v = -(v \times u)$

(b) $u \times (v + w) = u \times v + u \times w$

(c) $u \times 0 = 0 \times u = 0$

(d)   *If $u = av$, then $u \times v = 0$ (that is, the cross product of parallel vectors is 0)*

(e) $(u \times v) \cdot w = u \cdot (v \times w)$

Proof   (a)   Suppose $u = a_1 i + b_1 j + c_1 k$, $v = a_2 i + b_2 j + c_2 k$, and $w = a_3 i + b_3 j + c_3 k$. Since, by Theorem 10.18, $u \times v$ and $v \times u$ are given by determinants that are identical except for the reversal of the second and third rows, it follows that

$$u \times v = -(v \times u).$$

(b)   Since $v + w = (a_2 + a_3)i + (b_2 + b_3)j + (c_2 + c_3)k$,

$$u \times (v + w) = \begin{vmatrix} i & j & k \\ a_1 & b_1 & c_1 \\ a_2 + a_3 & b_2 + b_3 & c_2 + c_3 \end{vmatrix}$$

$$= \begin{vmatrix} i & j & k \\ a_1 & b_1 & c_1 \\ a_2 & b_2 & c_2 \end{vmatrix} + \begin{vmatrix} i & j & k \\ a_1 & b_1 & c_1 \\ a_3 & b_3 & c_3 \end{vmatrix}$$

$$= (u \times v) + (u \times w).$$

The proofs of the remaining three parts are left to the student (see Problems 30 and 31).

It might be noted that the definition of the cross product was stated in terms of three-dimensional vectors. In fact, we must have a three-dimensional vector space, for $\mathbf{u} \times \mathbf{v}$ is not in the plane determined by $\mathbf{u}$ and $\mathbf{v}$, as shown in the next theorem.

Theorem 10.20 *If* $\mathbf{u}$ *and* $\mathbf{v}$ *are nonzero vectors, then* $\mathbf{u} \times \mathbf{v}$ *is perpendicular to both* $\mathbf{u}$ *and* $\mathbf{v}$.

Proof

$$\mathbf{u} \cdot (\mathbf{u} \times \mathbf{v}) = (\mathbf{u} \times \mathbf{u}) \cdot \mathbf{v} \qquad \text{(why?)}$$
$$= \mathbf{0} \cdot \mathbf{v} \qquad \text{(why?)}$$
$$= 0 \qquad \text{(why?)}$$

Thus $\mathbf{u}$ and $\mathbf{u} \times \mathbf{v}$ are perpendicular. A similar argument shows that $\mathbf{u} \times \mathbf{v}$ and $\mathbf{v}$ are perpendicular.

This property of the cross product gives us its principal use. Certain problems in three-dimensional analytic geometry that were relatively difficult without the use of the cross product are easier now.

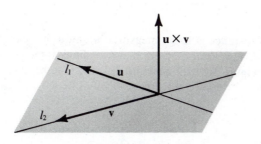

Figure 10.14

Example 2 Find a set of direction numbers for a line perpendicular to the plane containing

$$x = 1, \qquad y = 3 + 2t, \qquad z = 4 + t$$

and

$$x = 1 + 4s, \qquad y = 3 + 2s, \qquad z = 4 + 2s.$$

Solution Any line perpendicular to a given plane is perpendicular to any line in that plane. This suggests the use of the cross product. Vectors directed along the given lines are

$$\mathbf{u} = 2\mathbf{j} + \mathbf{k} \qquad \text{and} \qquad \mathbf{v} = 4\mathbf{i} + 2\mathbf{j} + 2\mathbf{k}.$$

Since $\mathbf{u} \times \mathbf{v} = 2\mathbf{i} + 4\mathbf{j} - 8\mathbf{k}$ (see Figure 10.14), we have $\{2, 4, -8\}$ as one set of direction numbers for the desired line; $\{1, 2, -4\}$ is a simpler set.

Example 3   Find equations for the line containing $(1, 4, 3)$ and perpendicular to

$$\frac{x - 1}{2} = \frac{y + 3}{1} = \frac{z - 2}{4} \quad \text{and} \quad \frac{x + 2}{3} = \frac{y - 4}{2} = \frac{z + 1}{-2}.$$

Solution   Again, vectors along the two given lines are

$$\mathbf{u} = 2\mathbf{i} + \mathbf{j} + 4\mathbf{k} \quad \text{and} \quad \mathbf{v} = 3\mathbf{i} + 2\mathbf{j} - 2\mathbf{k};$$

and $\mathbf{u} \times \mathbf{v} = -10\mathbf{i} + 16\mathbf{j} + \mathbf{k}$ is perpendicular to both of them. The desired line is, therefore,

$$\frac{x - 1}{-10} = \frac{y - 4}{16} = \frac{z - 3}{1}.$$

Example 4   Find the distance between the lines

$$x = 1 - 4t, \quad y = 2 + t, \quad z = 3 + 2t$$

and

$$x = 1 + s, \quad y = 4 - 2s, \quad z = -1 + s.$$

Solution   The desired distance is to be measured along a line perpendicular to both of the given lines. Again, vectors along the given lines are $\mathbf{u} = -4\mathbf{i} + \mathbf{j} + 2\mathbf{k}$ and $\mathbf{v} = \mathbf{i} - 2\mathbf{j} + \mathbf{k}$ (see Figure 10.15). Thus the distance is to be measured along

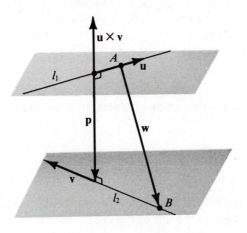

Figure 10.15

$$\mathbf{u} \times \mathbf{v} = 5\mathbf{i} + 6\mathbf{j} + 7\mathbf{k}.$$

The point $A = (1, 2, 3)$ is on the first line, and $B = (1, 4, -1)$ is on the second. The vector represented by $\overrightarrow{AB}$ is

$$\mathbf{w} = 2\mathbf{j} - 4\mathbf{k}.$$

We want a vector whose representatives are all perpendicular to both of the given lines and with one representative having its head on one line and its tail on the other. All of the representatives of $\mathbf{u} \times \mathbf{v}$ are perpendicular to both lines and one of the representatives of $\mathbf{w}$ has its end points on the given lines. Thus, the projection $\mathbf{p}$ of $\mathbf{w}$ on $\mathbf{u} \times \mathbf{v}$ has the desired properties and its length is the distance between the given lines.

$$|\mathbf{p}| = \frac{|\mathbf{w} \cdot (\mathbf{u} \times \mathbf{v})|}{|\mathbf{u} \times \mathbf{v}|} = \frac{|0 \cdot 5 + 2 \cdot 6 - 4 \cdot 7|}{\sqrt{25 + 36 + 49}} = \frac{16}{\sqrt{110}}$$

Up to this point we have been dealing exclusively with the direction of $\mathbf{u} \times \mathbf{v}$. Its length also has some interesting properties.

Theorem 10.21 *If $\mathbf{u}$ and $\mathbf{v}$ are vectors and θ is the angle between them, then*

$$|\mathbf{u} \times \mathbf{v}| = |\mathbf{u}| |\mathbf{v}| \sin \theta.$$

Proof Since $\cos \theta = \mathbf{u} \cdot \mathbf{v}/(|\mathbf{u}| |\mathbf{v}|)$ by Theorem 10.10,

$$|\mathbf{u}| |\mathbf{v}| \sin \theta = |\mathbf{u}| |\mathbf{v}| \sqrt{1 - \cos^2 \theta} \qquad \text{(see Note 1)}$$

$$= |\mathbf{u}| |\mathbf{v}| \sqrt{1 - \frac{(\mathbf{u} \cdot \mathbf{v})^2}{|\mathbf{u}|^2 |\mathbf{v}|^2}}$$

$$= \sqrt{|\mathbf{u}|^2 |\mathbf{v}|^2 - (\mathbf{u} \cdot \mathbf{v})^2}.$$

If we let $\mathbf{u} = a_1\mathbf{i} + b_1\mathbf{j} + c_1\mathbf{k}$ and $\mathbf{v} = a_2\mathbf{i} + b_2\mathbf{j} + c_2\mathbf{k}$, then

$$|\mathbf{u}| |\mathbf{v}| \sin \theta = \sqrt{(a_1^2 + b_1^2 + c_1^2)(a_2^2 + b_2^2 + c_2^2) - (a_1 a_2 + b_1 b_2 + c_1 c_2)^2}$$

$$= \sqrt{(b_1 c_2 - c_1 b_2)^2 + (c_1 a_2 - a_1 c_2)^2 + (a_1 b_2 - b_1 a_2)^2}$$

$$= |\mathbf{u} \times \mathbf{v}|. \qquad \text{(see Note 2)}$$

Note 1: By the definition of the angle between two vectors, $0° \le \theta \le 180°$; and $\sin \theta \ge 0$.

Note 2: The algebra here is routine but tedious. It is left to the student.

Note the similarity between this theorem and the first part of Theorem 10.10. One consequence of this theorem is given in Problem 25.

It appears that Theorem 10.20 and 10.21 give a geometric description of $\mathbf{u} \times \mathbf{v}$, the first giving its direction and the second, its length. Actually, this is not quite true. There are two vectors of a given length which are perpendicular to both $\mathbf{u}$ and $\mathbf{v}$; they have opposite orientations (that is, one is the negative of the other). It can be shown that $\mathbf{u} \times \mathbf{v}$ is the one that gives the system $\{\mathbf{u}, \mathbf{v}, \mathbf{u} \times \mathbf{v}\}$ a right-hand orientation; that is, if the index and second fingers of the right hand point in the directions of $\mathbf{u}$ and $\mathbf{v}$, respectively, then the thumb points in the direction of $\mathbf{u} \times \mathbf{v}$ (see Figure 10.16). This is summarized in the next theorem.

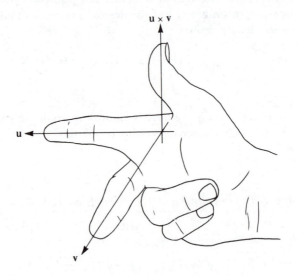

Figure 10.16

Theorem 10.22   *If $\mathbf{u}$ and $\mathbf{v}$ are vectors, θ is the angle between them, and $\mathbf{n}$ is the unit vector perpendicular to both $\mathbf{u}$ and $\mathbf{v}$ such that $\{\mathbf{u}, \mathbf{v}, \mathbf{n}\}$ forms a right-hand system, then*

$$\mathbf{u} \times \mathbf{v} = |\mathbf{u}| |\mathbf{v}| \sin \theta \mathbf{n}.$$

Some authors take this as the definition of $\mathbf{u} \times \mathbf{v}$. Another direct result from Theorem 10.21 follows.

Theorem 10.23   *If $\mathbf{u}$ and $\mathbf{v}$ are two nonzero vectors, then $\mathbf{u} \times \mathbf{v} = \mathbf{0}$ if and only if $\mathbf{u} = k\mathbf{v}$ for some scalar k.*

This extends Theorem 10.19d, which gives part of this theorem. The proof is left to the student (see Problem 35).

It might be noted that Examples 2 and 3 can also be solved without the use of vectors. We simply use direction numbers instead of vectors.

Example 2   Find a set of direction numbers for a line perpendicular to the plane containing

$$x = 1, \qquad y = 3 + 2t, \qquad z = 4 + t$$

and

$$x = 1 + 4s, \qquad y = 3 + 2s, \qquad z = 4 + 2s.$$

Alternate   A line perpendicular to the given plane must be perpendicular to any line in the
Solution   plane. Thus the desired line with direction numbers $\{a, b, c\}$ is perpendicular to the two given lines with direction numbers $\{0, 2, 1\}$ and $\{4, 2, 2\}$. This gives the equations

$$2b + c = 0$$
$$4a + 2b + 2c = 0.$$

Since we have only two equations in three unknowns, we cannot solve for all of them; however, we can solve for two of them in terms of the third. By subtracting the first equation from the second, we get

$$4a + c = 0$$

or $c = -4a$. Doubling the first equation and subtracting from the second, we have

$$4a - 2b = 0$$

or $b = 2a$. Thus $\{a, b, c\} = \{a, 2a, -4a\}$ are direction numbers for the desired line. A second set of direction numbers can be found by multiplying by $1/a$, giving $\{1, 2, -4\}$.

Example 3   Find equations for the line containing $(1, 4, 3)$ and perpendicular to

$$\frac{x - 1}{2} = \frac{y + 3}{1} = \frac{z - 2}{4} \qquad \text{and} \qquad \frac{x + 2}{3} = \frac{y - 4}{2} = \frac{z + 1}{-2}.$$

Alternate   Again, direction numbers for the two given lines are $\{2, 1, 4\}$ and $\{3, 2, -2\}$; for
Solution   the desired line, $\{a, b, c\}$. Since the desired line is perpendicular to both of the given lines, we have

$$2a + b + 4c = 0$$
$$3a + 2b - 2c = 0.$$

Doubling the second equation and adding to the first gives

$$8a + 5b = 0$$

or $b = -8a/5$. Doubling the first equation and substracting the second, we have

$$a + 10c = 0$$

or $c = -a/10$. Thus the direction numbers for the desired line are $\{a, -8a/5, -a/10\}$ or, multiplying by $-10/a$, $\{-10, 16, 1\}$. Using these with the given point, we see that symmetric equations for the line are

$$\frac{x - 1}{-10} = \frac{y - 4}{16} = \frac{z - 3}{1}.$$

Example 4 can also be solved without the use of vectors; however, the solution involves planes, which we have not yet considered.

PROBLEMS

In Problems 1–6, find **u** × **v**. *Use the dot product to verify that your result is perpendicular to both* **u** *and* **v**.

A   **1.** $u = 3i - j + 4k, v = 2i + j + k$ **2.** $u = i + j + k, v = 2i - j - 4k$

3. $u = 2i + 3j - k, v = -i + 2j$ **4.** $u = 4i + 2j, v = 3i - j$

5. $u = 3i + k, v = -i + j$ **6.** $u = 2i + j - k, v = -i - j + 3k$

B   *In Problems 7–12, find direction numbers for the line described.*

7. Perpendicular to the plane containing $(4, 1, 2)$, $(2, -1, 1)$, and $(3, 0, 4)$

8. Perpendicular to the plane containing $(2, 2, 3)$, $(-1, 4, 1)$, and $(0, 1, 2)$

9. Perpendicular to the plane containing $x = 2 + t, y = 3 - 2t, z = -t$ and $x = 2 - 2s$, $y = 3 + s, z = -s$

10. Perpendicular to the plane containing $x = 3 + 4t, y = 1 - t, z = 3$ and $x = 3 - 2s$, $y = 1 + 2s, z = 3 - s$

11. Perpendicular to the plane containing $x = 4 + t, y = -1 + 2t, z = 2t$ and $x = 2 + s$, $y = 4 + 2s, z = 1 + 2s$

12. Perpendicular to the plane containing $x = 2 + 2t, y = 3 - t, z = -1 + t$ and $x = 4 + 2s, y = 2 - s, z = 4 + s$

In Problems 13–18, find equations for the line described.

13. Containing $(3, 2, 1)$ and perpendicular to $x = 1 - 2t, y = 3 + t, z = 4 - t$ and $x = 2 + s, y = -1 + 2s, z = 3 - s$

14. Containing $(4, -1, 0)$ and perpendicular to $x = 3 + t, y = 2 - t, z = 2t$ and $x = 4$, $y = 2 + s, z = -1 + s$

15. Containing $(2, 3, 1)$ and perpendicular to the plane determined by $(2, 3, 1)$ and the line $x = 0, y = 2t, z = t$

16. Containing $(0, 4, -2)$ and perpendicular to the plane determined by $(0, 4, -2)$ and the line $x = -2 + 2t, y = 8t, z = -1 + t$

17. Containing $(2, 0, 5)$ and perpendicular to and containing a point of $x = 4 + t, y = 3 - 2t, z = 1 + t$

18. Containing $(1, 1, 2)$ and perpendicular to and containing a point of $x = 1 - t, y = 2 + 2t, z = 4t$

In Problems 19–24, find the distance between the given lines.

19. $x = 1 + t, y = -2 + 3t, z = 4 + t$ and $x = 2 - s, y = 3 + 2s, z = 1 + s$

20. $x = 2 + t, y = 1 - t, z = 4t$ and $x = 2 + s, y = 4 - 2s, z = 1 + 3s$

21. $x = 1 + t, y = 1 - 5t, z = 2 + t$ and $x = 4 + s, y = 5 + 2s, z = -3 + 4s$

22. $x = 2 + t, y = -4 + t, z = 1 - 3t$ and $x = 3 - s, y = 4 + 2s, z = 2 + s$

23. $x = 2 + 3t, y = 5 + t, z = -1 - 2t$ and $x = 2 + 3s, y = 3 + s, z = 5 - 2s$

24. $x = 4t, y = 1 + t, z = -2 - t$ and $x = 9 + 4s, y = 1 + s, z = -2 - s$

25. Suppose the vectors **u** and **v** are represented by $\overrightarrow{AB}$ and $\overrightarrow{AC}$, respectively. Show that the area of $\triangle ABC$ is $|\mathbf{u} \times \mathbf{v}|/2$. (Equivalently, the parallelogram determined by AB and AC has area $|\mathbf{u} \times \mathbf{v}|$.) [*Hint:* Use Theorem 10.21.]

In Problems 26–29, use the result of Problem 25 to find the area of the triangles with the given vertices.

26. $(1, 0, 4), (2, -1, 2), (4, 4, 1)$

27. $(3, -2, 1), (-1, 2, 0), (4, 4, 2)$

28. $(2, 4, 3), (1, 0, 1), (-2, 2, 4)$

29. $(4, 2), (3, -1), (-1, 0)$

C   30. Prove parts c and d of Theorem 10.19.

31. Show that if $\mathbf{u} = a_1\mathbf{i} + b_1\mathbf{j} + c_1\mathbf{k}$, $\mathbf{v} = a_2\mathbf{i} + b_2\mathbf{j} + c_2\mathbf{k}$, and $\mathbf{w} = a_3\mathbf{i} + b_3\mathbf{j} + c_3\mathbf{k}$, then

$$\mathbf{u} \cdot (\mathbf{v} \times \mathbf{w}) = \begin{vmatrix} a_1 & b_1 & c_1 \\ a_2 & b_2 & c_2 \\ a_3 & b_3 & c_3 \end{vmatrix}.$$

Use this result to prove Theorem 10.19e.

32. Given that $\overrightarrow{AB}$, $\overrightarrow{AC}$, and $\overrightarrow{AD}$ represent the vectors **u**, **v**, and **w**, respectively, show that the parallelepiped determined by AB, AC, and AD (see Figure 10.17) has volume $|\mathbf{u} \cdot (\mathbf{v} \times \mathbf{w})|$. [*Hint:* By Problem 25, the area of the base is $|\mathbf{v} \times \mathbf{w}|$.]

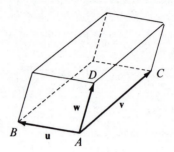

Figure 10.17

In Problems 33 and 34, use the result of Problem 32 to find the volume of the parallelepiped determined by AB, AC, and AD.

33. $A = (0, 0, 0)$, $B = (2, 1, 3)$, $C = (5, 3, 1)$, $D = (2, -1, 4)$

34. $A = (1, 3, 2)$, $B = (4, 1, 5)$, $C = (1, 5, 2)$, $D = (0, 5, -1)$

35. Prove Theorem 10.23.

10.6

THE PLANE

Let us now consider the plane. Perhaps the simplest way of determining a plane is by three noncollinear points. But, for the purpose of determining its equation, it is better to describe it by a single point and a line perpendicular to it.

Let p be a plane in space, containing the point $P_0 = (x_0, y_0, z_0)$ (see Figure 10.18); and let l, with direction numbers $\{a, b, c\}$, be a line perpendicular to p. In order to determine an equation of p, we consider any point $P = (x, y, z)$ lying in the plane. The directed line segment $\overrightarrow{P_0 P}$ represents a vector

$$\mathbf{v} = (x - x_0)\mathbf{i} + (y - y_0)\mathbf{j} + (z - z_0)\mathbf{k}$$

in the plane p. Since l is perpendicular to p, it is perpendicular to any line in this plane; in particular, it is perpendicular to $P_0 P$. Since

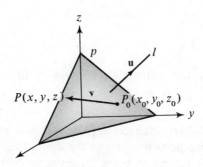

Figure 10.18

$$\mathbf{u} = a\mathbf{i} + b\mathbf{j} + c\mathbf{k}$$

is a vector lying along l, we have the following.

$$\mathbf{u} \cdot \mathbf{v} = 0$$

$$[a\mathbf{i} + b\mathbf{j} + c\mathbf{k}] \cdot [(x - x_0)\mathbf{i} + (y - y_0)\mathbf{j} + (z - z_0)\mathbf{k}] = 0$$

$$a(x - x_0) + b(y - y_0) + c(z - z_0) = 0$$

Furthermore, the argument may be traced backward to show that any point (x, y, z) that satisfies the last equation must lie in the plane p. Thus we have proved the following theorem.

Theorem 10.24 *A point is on a plane containing (x_1, y_1, z_1) and perpendicular to a line with direction numbers $\{A, B, C\}$ if and only if it satisfies the equation*

$$A(x - x_1) + B(y - y_1) + C(z - z_1) = 0.$$

This theorem can also be proved using a nonvector argument. The vectors are simply replaced by direction numbers. Suppose p is the plane in space, containing the point $P_0 = (x_0, y_0, x_0)$; and l, with direction numbers $\{a, b, c\}$, is a line perpendicular to p. Then if $P = (x, y, z)$ is any other point of p, a set of direction numbers for the line PP_0 is $\{x - x_0, y - y_0, z - z_0\}$. Since this line in the plane p must be perpendicular to l, we have

$$a(x - x_0) + b(y - y_0) + c(z - z_0) = 0.$$

Example 1 Find the equation of the plane containing $(1, 3, -2)$ and perpendicular to the line through $(2, 5, 1)$ and $(0, 1, -3)$.

Solution A set of direction numbers for the given line is $\{2, 4, 4\}$ or $\{1, 2, 2\}$. Thus the desired plane is

$$1(x - 1) + 2(y - 3) + 2(z + 2) = 0$$
$$x + 2y + 2z - 3 = 0.$$

Example 2 Find an equation of the plane containing the two lines

$$x = 1, y = 3 + 2t, z = 4 + t \qquad \text{and} \qquad x = 1 + 4s, y = 3 + 2s, x = 4 + 2s.$$

Solution These lines clearly intersect at $(1, 3, 4)$. All we need, then, is a set of direction numbers for a line perpendicular to the desired plane. This was done in Example 2 of the previous section by using the cross product of two vectors or (as shown in the alternate solution) by using the direction numbers. One such set is $\{1, 2, -4\}$. By Theorem 10.24, the corresponding plane is

$$1(x - 1) + 2(y - 3) - 4(z - 4) = 0$$
$$x + 2y - 4z + 9 = 0.$$

Example 3 Find an equation of the plane containing the two lines

$$x = 1 + t, \ y = 3 - 2t, \ z = -2 + 2t$$

and

$$x = 4 + s, \ y = 2 - 2s, \ z = -1 + 2s.$$

Solution The two lines are parallel since they both have direction numbers $\{1, -2, 2\}$. Since $(1, 3, -2)$ is on the first line and $(4, 2, -1)$ is on the second, the line through these two points intersects both of the given lines and lies in the desired plane (see Figure 10.19). Its direction numbers are $\{4 - 1, 2 - 3, -1 + 2\}$ or $\{3, -1, 1\}$. We now have

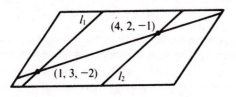

Figure 10.19

two intersecting lines with direction numbers $\{1, -2, 2\}$ and $\{3, -1, 1\}$ and lying in the desired plane. By the methods of the previous section, a line perpendicular to both of these lines (and therefore perpendicular to the plane containing them) has direction numbers $\{0, 1, 1\}$. Thus, by Theorem 10.24, the desired plane is

$$0(x - 1) + 1(y - 3) + 1(z + 2) = 0$$
$$y + z - 1 = 0.$$

The following theorem is a direct consequence of Theorem 10.24.

Theorem 10.25   *Any plane can be represented by an equation of the form*

$$Ax + By + Cz + D = 0,$$

where $\{A, B, C\}$ is a set of direction numbers for a line normal to (that is, perpendicular to) the plane. Conversely, an equation of the above form (where A, B, and C are not all zero) represents a plane with $\{A, B, C\}$ a set of direction numbers for a normal line.

Example 4   Find an equation of the plane containing the points $P_1 = (1, 0, 1)$, $P_2 = (-1, -4, 1)$, and $P_3 = (-2, -2, 2)$.

Solution   This problem may be solved using either Theorem 10.25 or Theorem 10.24. Let us do it both ways.

   By Theorem 10.25, the equation we seek is, for the proper choices of A, B, C, and D,

$$Ax + By + Cz + D = 0.$$

We get an equation in A, B, C, and D from each of the three given points.

$$P_1 = (1, 0, 1): \qquad A \qquad\quad + \ C + D = 0,$$
$$P_2 = (-1, -4, 1): \quad -A - 4B + \ C + D = 0,$$
$$P_3 = (-2, -2, 2): \quad -2A - 2B + 2C + D = 0.$$

Although we cannot solve for A, B, C, and D directly, since we have only three equations in four unknowns, we can solve for three of them in terms of the other one. If we take A to be fixed and solve for the other three, we have $B = -A/2$, $C = 2A$, and $D = -3A$. We may give A any nonzero value we want; let us choose $A = 2$. Then $B = -1, C = 4$, and $D = 6$; the resulting equation is

$$2x - y + 4z - 6 = 0.$$

We now solve the same problem using Theorem 10.24. We let $\overrightarrow{P_1 P_2}$ and $\overrightarrow{P_1 P_3}$ represent the vectors **u** and **v**, respectively. Then

$$\mathbf{u} = (-1 - 1)\mathbf{i} + (-4 - 0)\mathbf{j} + (1 - 1)\mathbf{k} = -2\mathbf{i} - 4\mathbf{j}$$
$$\mathbf{v} = (-2 - 1)\mathbf{i} + (-2 - 0)\mathbf{j} + (2 - 1)\mathbf{k} = -3\mathbf{i} - 2\mathbf{j} + \mathbf{k}.$$

Since **u** and **v** lie in the desired plane, their cross product is perpendicular to it.

$$\mathbf{u} \times \mathbf{v} = \begin{vmatrix} \mathbf{i} & \mathbf{j} & \mathbf{k} \\ -2 & -4 & 0 \\ -3 & -2 & 1 \end{vmatrix} = -4\mathbf{i} + 2\mathbf{j} - 8\mathbf{k}$$

Thus $\{-4, 2, -8\}$ is a set of direction numbers for a line perpendicular to the desired plane. A simpler set is $\{2, -1, 4\}$. Using this, together with the point $(1, 0, 1)$ in Theorem 10.24, we have

$$2(x - 1) - y + 4(z - 1) = 0$$

$$2x - y + 4z - 6 = 0.$$

Example 5 Sketch $x + 2y + 3z = 6$.

Solution By Theorem 11.25, we know that this equation represents a plane. Knowing this, we merely need to find three points to determine the plane. The simplest points to find are the intercepts (the points where the plane crosses the coordinate axes), which are $(6, 0, 0)$, $(0, 3, 0)$, and $(0, 0, 2)$. Thus we have the plane shown in Figure 10.20.

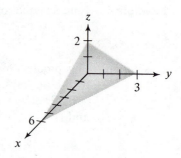

Example 6 Sketch $x + 2y = 4$.

Solution This equation represents a line if we are considering only the xy plane (for which $z = 0$). But we get the same line when $z = 1$ or $z = 2$, and so forth. Thus the result is a plane that is parallel to the z axis (see Figure 10.21).

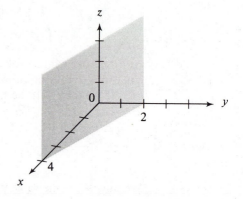

Figure 10.21

Of course the two planes

$$A_1 x + B_1 y + C_1 z + D_1 = 0$$

and

$$A_2 x + B_2 y + C_2 z + D_2 = 0$$

are parallel if and only if their normal lines are parallel. Similarly, the planes are perpendicular if and only if their normal lines are perpendicular. Thus from Theorems 10.14 and 10.15, we have the following theorem for planes.

Theorem 10.26 *The planes*

$$A_1 x + B_1 y + C_1 z + D_1 = 0$$

and

$$A_2 x + B_2 y + C_2 z + D_2 = 0$$

are parallel (or coincident) if and only if there is a number k such that

$$A_2 = kA_1, \qquad B_2 = kB_1, \qquad C_2 = kC_1;$$

they are perpendicular if and only if

$$A_1 A_2 + B_1 B_2 + C_1 C_2 = 0.$$

Example 7 Show that $3x + y - 4z = 2$ and $6x + 2y - 8z = 3$ are parallel planes and that $3x - y + 2z = 5$ is perpendicular to both of them.

Solution The coefficients of $x, y,$ and z in the three equations are

$$\begin{aligned}
A_1 &= 3, & A_2 &= 6, & A_3 &= 3, \\
B_1 &= 1, & B_2 &= 2, & B_3 &= -1, \\
C_1 &= -4, & C_2 &= -8, & C_3 &= 2.
\end{aligned}$$

Since $A_2 = 2A_1$, $B_2 = 2B_1$, and $C_2 = 2C_1$, the first and second planes are either parallel or coincident. But since $D_2 = -3 \neq 2D_1 = 2(-2)$, they are not equivalent equations—the planes are parallel. Since

$$A_1 A_3 + B_1 B_3 + C_1 C_3 = 3 \cdot 3 + 1(-1) - 4 \cdot 2 = 0,$$

the first and third planes are perpendicular. Of course, the third plane must then be perpendicular to the second as well; but this may also be checked using Theorem 10.26.

$$A_2 A_3 + B_2 B_3 + C_2 C_3 = 6 \cdot 3 + 2(-1) - 8 \cdot 2 = 0$$

PROBLEMS

A *In Problems 1–6, sketch the plane.*

1. $2x + 3y + z = 6$ **2.** $3x - y + z = 9$ **3.** $2x + y - 4z + 4 = 0$

4. $x - y - 4z = 8$ **5.** $x + 2y = 3$ **6.** $y - 5 = 0$

In Problems 7–24, find an equation(s) of the plane(s) satisfying the given conditions.

7. Containing $(3, 2, -5)$ and perpendicular to a line with direction numbers $\{3, -4, 1\}$

8. Containing $(4, 2, 3)$ and perpendicular to a line with direction numbers $\{-2, 5, 1\}$

9. Containing $(4, 1, -3)$ and perpendicular to the line $x = 2 + 3t$, $y = 4 - t$, $z = 3 - 2t$

10. Containing $(3, 2, 5)$ and perpendicular to the line $x = 1 + t$, $y = 3t$, $z = 4 + t$

11. Containing $(3, 5, 1)$ and parallel to $3x - 4y + 2z = 3$

12. Containing $(4, -1, 2)$ and parallel to $x + y - 2z = 4$

B   13. Containing $(1, 1, 0)$, $(1, 3, 2)$, and $(2, -1, 1)$

14. Containing $(2, -2, -2)$, $(1, -3, 5)$, and $(-1, 4, 1)$

15. Containing $(1, 4, 2)$, $(2, 3, -1)$, and $(5, 0, 2)$

16. Containing $(3, 1, -4)$, $(2, 3, 1)$, and $(7, 4, -2)$

17. Containing $x = 4 + t$, $y = 2 - t$, $z = 1 + 2t$ and $x = 4 - 3s$, $y = 2 + 2s$, $z = 1 - s$

18. Containing $x = 2 + 2t$, $y = -1 + t$, $z = 4 - t$ and $x = 2 - s$, $y = -1 - 2s$, $z = 4 + 3s$

19. Containing $(4, 1, 2)$ and $x = 4 - t$, $y = 1 + 2t$, $z = 3 - t$

20. Containing $(-2, 3, -4)$ and $x = 1 + t$, $y = 3 - 2t$, $z = -2 + t$

21. Containing $x = 3 + 2t$, $y = 4 - t$, $z = 1 + t$ and $x = -1 + 2s$, $y = 3 - s$, $z = 4 + s$

22. Containing $x = 4 + t$, $y = 2t$, $z = 5$ and $x = 1 + s$, $y = 3 + 2s$, $z = -2$

23. Containing $(1, 5, -2)$ and perpendicular to $3x + 2y - z + 1 = 0$ and $x - y + 2z = 0$

24. Containing $(3, 0, -4)$ and perpendicular to $2x - 5y + z = 1$ and $x - 2y - z = 3$

In Problems 25–28, find equations of the given line.

25. Containing $(2, 5, -1)$ and perpendicular to $2x - y + 3z + 2 = 0$

26. Containing $(4, -2, 3)$ and perpendicular to $3x + 2y - z + 6 = 0$

27. Containing $(2, -4, 5)$ and parallel to $x - y + 3z = 4$ and $3x - 3y + 2z = 5$

28. Containing $(4, 0, 5)$ and parallel to $2x - 5y + z + 1 = 0$ and $x + 2y - z + 2 = 0$

In Problems 29–34, indicate whether the given planes are parallel, perpendicular, coincident, or none of these.

29. $3x + y - 5z = 2$, $x + 2y + z = 4$ 30. $4x - 2y + 2z = 6$, $2x - y + z = 3$

31. $4x + y - z = 5, x - y + 2z = 2$ **32.** $4x - 2y + z = 1, x + y - 2z = 0$

33. $2x - y + 3z = 4, 6x - 3y + 9z = 5$ **34.** $x + 3y - z = 4, 2x - y + z = 3$

In Problems 35–38, find the point of intersection of the plane and the line.

35. $3x - 2y + z = 4; x = 2 + t, y = 1 - 2t, z = 2 - 5t$

36. $x + 2y - 4z = 12; x = 1 + t, y = -2 + 3t, z = -2t$

37. $3x - y + 4z = 7; x = 2 - t, y = 5 + 2t, z = 1 + t$

38. $2x + 3y - z = 2; x = 1 + t, y = -2 + 3t, z = 6 - t$

39. Does the line $x = 1, y = 5 + 4t, z = 2 + t$ lie in the plane $2x - y + 4z = 5$?

10.7

DISTANCE BETWEEN A POINT AND A PLANE OR LINE; ANGLES BETWEEN LINES OR PLANES

In Section 10.5, we used the cross product and the projection of one vector upon another to find the distance between a pair of lines in space. Similar methods can be used to find the distance between a point and a plane or between a point and a line.

Example 1 Find the distance between $(3, -4, 1)$ and
$x - 2y + 2z + 4 = 0$.

Solution A vector perpendicular to the given plane is

$$\mathbf{v} = \mathbf{i} - 2\mathbf{j} + 2\mathbf{k}$$

(see Figure 10.22). We now choose an arbitrary point on $x - 2y + 2z + 4 = 0$, say $(0, 0, -2)$, and let $\mathbf{u}$ be the vector represented by the directed line segment from $(3, -4, 1)$ to $(0, 0, -2)$.

$$\mathbf{u} = (0 - 3)\mathbf{i} + (0 + 4)\mathbf{j} + (-2 - 1)\mathbf{k}$$
$$= -3\mathbf{i} + 4\mathbf{j} - 3\mathbf{k}$$

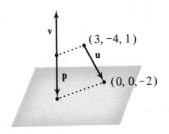

Figure 10.22

Now the distance we want is the length of the projection $\mathbf{p}$ of $\mathbf{u}$ upon $\mathbf{v}$.

$$d = \frac{|\mathbf{u} \cdot \mathbf{v}|}{|\mathbf{v}|}$$

$$= \frac{|(-3)(1) + (4)(-2) + (-3)(2)|}{\sqrt{1 + 4 + 4}} = \frac{17}{3}$$

Alternate Solution   Again there is a simple nonvector solution for this problem. Since we want the perpendicular distance, we first find equations for the line containing $(3, -4, 1)$ and perpendicular to the given plane. Knowing that this line has direction numbers $\{1, -2, 2\}$, we see that the line is

$$x = 3 + t, \ y = -4 - 2t, \ z = 1 + 2t.$$

Now we find the point of intersection of this line and the given plane. This is done by substituting the parametric equations of the line into the equation of the plane and solving for t.

$$(3 + t) - 2(-4 - 2t) + 2(1 + 2t) + 4 = 0$$
$$17 + 9t = 0$$
$$t = -\frac{17}{9}$$

Thus

$$x = 3 - \frac{17}{9} = \frac{10}{9}$$

$$y = -4 + \frac{34}{9} = -\frac{2}{9}$$

$$z = 1 - \frac{34}{9} = -\frac{25}{9}$$

The distance we seek is the distance between this point (x, y, z) and $(3, -4, 1)$.

$$d = \sqrt{\left(\frac{10}{9} - 3\right)^2 + \left(-\frac{2}{9} + 4\right)^2 + \left(-\frac{25}{9} - 1\right)^2}$$

$$= \sqrt{\left(-\frac{17}{9}\right)^2 + \left(\frac{34}{9}\right)^2 + \left(-\frac{34}{9}\right)^2}$$

$$= \sqrt{\frac{2601}{81}} = \sqrt{\frac{289}{9}} = \frac{17}{3}$$

This argument can be shortened somewhat. After we found the point (x, y, z) we used it only to find $x - 3$, $y + 4$, and $z - 1$ in the distance formula. But since the parametric equations

$$x = 3 + t, \quad y = -4 - 2t, \quad z = 1 + 2t$$

are equivalent to

$$x - 3 = t, \quad y + 4 = -2t, \quad z - 1 = 2t,$$

we can use t, $-2t$, and $2t$ in place of $x - 3$, $y + 4$, and $z - 1$. Thus after we found that $t = -17/9$, we can go directly to the distance formula.

$$d = \sqrt{t^2 + (-2t)^2 + (2t)^2}$$

$$= \sqrt{9t^2} = 3|t| = \frac{17}{3}$$

Exactly the same method can be used to find the distance between the point (x_1, y_1, z_1) and the plane $Ax + By + Cz + D = 0$. We obtain the following.

Theorem 10.27   *The distance between the point* (x_1, y_1, z_1) *and the plane* $Ax + By + Cz + D = 0$ *is*

$$d = \frac{|Ax_1 + By_1 + Cz_1 + D|}{\sqrt{A^2 + B^2 + C^2}}.$$

The proof is left to the student (see Problem 33). Notice that this formula is similar to the one on page 74 for the distance between a point and a line in two dimensions. With this formula, the distance of Example 1 is

$$d = \frac{|Ax_1 + By_1 + Cz_1 + D|}{\sqrt{A^2 + B^2 + C^2}}$$

$$= \frac{|1 \cdot 3 - 2(-4) + 2 \cdot 1 + 4|}{\sqrt{1^2 + (-2)^2 + 2^2}}$$

$$= \frac{17}{3}.$$

Example 2   Find the distance between $P = (5, 1, 3)$ and the line $x = 3$, $y = 7 + t$, $z = 1 + t$.

Solution   A vector directed along the given line is

$$\mathbf{u} = \mathbf{j} + \mathbf{k}$$

(see Figure 10.23), and the point $Q = (3, 7, 1)$ is on the line. Letting $\mathbf{v}$ be the vector represented by $\overrightarrow{QP}$,

$$\mathbf{v} = (5 - 3)\mathbf{i} + (1 - 7)\mathbf{j} + (3 - 1)\mathbf{k}$$

$$= 2\mathbf{i} - 6\mathbf{j} + 2\mathbf{k}.$$

From Figure 10.23, the distance is

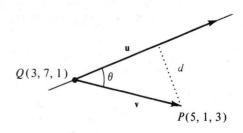

Figure 10.23

$$d = |\mathbf{v}| \sin \theta.$$

But by Theorem 10.21,

$$|\mathbf{u} \times \mathbf{v}| = |\mathbf{u}||\mathbf{v}| \sin \theta.$$

Therefore

$$d = |\mathbf{v}| \sin \theta = \frac{|\mathbf{u} \times \mathbf{v}|}{|\mathbf{u}|} = \frac{|8\mathbf{i} + 2\mathbf{j} - 2\mathbf{k}|}{|\mathbf{j} + \mathbf{k}|} = \frac{\sqrt{64 + 4 + 4}}{\sqrt{2}}$$

$$= \frac{\sqrt{72}}{\sqrt{2}} = \sqrt{36} = 6.$$

Alternate Solution   A nonvector solution for this problem is almost identical to the alternate solution of Example 1. The only difference is that we find the plane containing (5, 1, 3) and perpendicular to the given line, rather than finding a line perpendicular to a given plane. Since the line has direction numbers {0, 1, 1}, the plane containing (5, 1, 3) and perpendicular to the line is

$$0(x - 5) + 1(y - 1) + 1(z - 3) = 0$$
$$y + z = 4.$$

This plane and the given line give

$$7 + t + 1 + t = 4$$
$$2t = -4$$
$$t = -2.$$
$$x = 3, \quad y = 7 - 2 = 5, \quad z = 1 - 2 = -1$$

Thus the distance we want is the distance between (5, 1, 3) and (3, 5, −1), which is

$$d = \sqrt{(5 - 3)^2 + (1 - 5)^2 + (3 + 1)^2}$$
$$= \sqrt{4 + 16 + 16} = \sqrt{36} = 6.$$

Note that we cannot shorten the solution here as we did in Example 1, because the (3, 7, 1) of the line is not one of the two points giving the distance.

We have already considered the angle between two vectors in Section 10.2 (see page 268). The relationship between the two vectors and the angle between them was given in Theorem 10.7. This is restated here for convenience.

If $\mathbf{u} = a_1\mathbf{i} + b_1\mathbf{j} + c_1\mathbf{k}$ *and* $\mathbf{v} = a_2\mathbf{i} + b_2\mathbf{j} + c_2\mathbf{k}$ ($\mathbf{u} \neq \mathbf{0}$ *and* $\mathbf{v} \neq \mathbf{0}$) *and if* θ *is the angle between them, then*

$$\cos \theta = \frac{a_1 a_2 + b_1 b_2 + c_1 c_2}{|\mathbf{u}||\mathbf{v}|}.$$

An angle between two lines can be found in much the same way, since the direction numbers of a line are the components of some vector directed along that line.

However, an angle between two lines is the angle between two vectors directed along these lines. Since vectors directed along a line can be oriented in one of two directions, there are two angles θ_1 and θ_2 between any pair of lines. Since

$$\theta_1 + \theta_2 = 180°$$

and

$$\cos\theta_2 = -\cos\theta_1,$$

we can determine the absolute value of $\cos\theta$, where θ is an angle between two lines; but we need additional information to determine the sign of $\cos\theta$.

Theorem 10.28   *If θ is an angle between two lines with direction numbers $\{a_1, b_1, c_1\}$ and $\{a_2, b_2, c_2\}$, then*

$$|\cos\theta| = \frac{|a_1a_2 + b_1b_2 + c_1c_2|}{\sqrt{a_1^2 + b_1^2 + c_1^2}\,\sqrt{a_2^2 + b_2^2 + c_2^2}}.$$

Example 3   Find the acute angle between the lines

$$x = 2 - 3t, \qquad y = 4 + t, \qquad z = t$$

and

$$x = 3 + s, \qquad y = 1 - s, \qquad z = 2 + 2s.$$

Solution   Direction numbers for the lines are $\{-3, 1, 1\}$ and $\{1, -1, 2\}$. Thus

$$
\begin{aligned}
|\cos\theta| &= \frac{|a_1a_2 + b_1b_2 + c_1c_2|}{\sqrt{a_1^2 + b_1^2 + c_1^2}\,\sqrt{a_2^2 + b_2^2 + c_2^2}}\\[2mm]
&= \frac{|-3\cdot 1 + 1(-1) + 1\cdot 2|}{\sqrt{9 + 1 + 1}\,\sqrt{1 + 1 + 4}}\\[2mm]
&= \frac{|-2|}{\sqrt{11}\,\sqrt{6}} = \frac{2}{\sqrt{66}} \approx 0.2462.
\end{aligned}
$$

Since the angle is acute, $\cos\theta$ is positive.

$$\cos\theta \approx 0.2462$$
$$\theta \approx 76°$$

If two lines are perpendicular, then $\cos\theta = 0$. This gives us the well-known test for perpendicularity

$$a_1a_2 + b_1b_2 + c_1c_2 = 0,$$

where $\{a_1, b_1, c_1\}$ and $\{a_2, b_2, c_2\}$ are direction numbers for the lines.

Let us now consider an angle of intersection of two planes. This is defined to be an angle between their normal lines. With this definition, the problem is reduced to a familiar one. Note that we have again made no attempt to define *the* angle between two planes but only *an* angle. The particular one desired must be specified. The situation is illustrated in Figure 10.24.

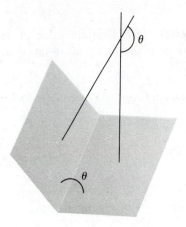

Figure 10.24

Theorem 10.29   *If θ is an angle between the planes*

$$A_1 x + B_1 y + C_1 z + D_1 = 0 \quad and \quad A_2 x + B_2 y + C_2 z + D_2 = 0,$$

then

$$|\cos \theta| = \frac{|A_1 A_2 + B_1 B_2 + C_1 C_2|}{\sqrt{A_1^2 + B_1^2 + C_1^2}\sqrt{A_2^2 + B_2^2 + C_2^2}}.$$

Example 4   Find the acute angle between the planes $2x + y - z + 3 = 0$ and $4x - y + z + 1 = 0$.

Solution

$$|\cos \theta| = \frac{|A_1 A_2 + B_1 B_2 + C_1 C_2|}{\sqrt{A_1^2 + B_1^2 + C_1^2}\sqrt{A_2^2 + B_2^2 + C_2^2}}$$

$$= \frac{|2 \cdot 4 + 1(-1) - 1 \cdot 1|}{\sqrt{4 + 1 + 1}\sqrt{16 + 1 + 1}}$$

$$= \frac{|6|}{\sqrt{6}\sqrt{18}} = \frac{6}{6\sqrt{3}} = \frac{\sqrt{3}}{3} \approx 0.5774$$

Since we want the acute angle, $\cos \theta$ is positive.

$$\cos \theta \approx 0.5774$$

$$\theta \approx 55°$$

PROBLEMS

A   *In Problems* 1–8, *find the distance between the plane and point given.*

1. $2x - 4y + 4z + 3 = 0$; $(1, 3, -2)$ **2.** $4x + y - 8z + 1 = 0$; $(2, 0, 3)$

3. $x + y - 2z - 4 = 0$; $(3, 3, 1)$ **4.** $2x - y + z + 5 = 0$; $(1, 0, 2)$

5. $x + 3y + z - 2 = 0$; $(2, 1, -3)$ **6.** $x - 2y + 4 = 0$; $(2, 2, 4)$

7. $x + z - 5 = 0$; $(3, 3, 1)$ **8.** $y + 7 = 0$; $(1, 3, 1)$

B   **9.** If the distance between $(1, 4, z)$ and $8x - y + 4z - 3 = 0$ is 1, find z.

10. If the distance between $(2, y, 3)$ and $4x - 4y + 2z - 5 = 0$ is $3/2$, find y.

In Problems 11–18, *find the distance between the point and line given.*

11. $(1, 3, -2)$; $x = 4$, $y = -3 + 4t$, $z = 11 + 5t$

12. $(4, 3, 3)$; $x = 2 + 2t$, $y = 5 - 5t$, $z = -1 - t$

13. $(2, 4, -1)$; $x = 5 + t$, $y = -2 + 3t$, $z = 3 + t$

14. $(-1, 0, 5)$; $\dfrac{x - 2}{2} = \dfrac{y - 1}{1} = \dfrac{z + 2}{3}$

15. $(4, 1, -2)$; $x = 2 - t$, $y = 3 + 2t$, $z = 1 + t$

16. $(2, 3, -1)$; $x = 4 + t$, $y = 1 - t$, $z = 3 + 2t$

17. $(1, 4, 2)$; $\dfrac{x - 1}{3} = \dfrac{y + 2}{1} = \dfrac{z - 4}{-2}$

18. $(3, -1, 4)$; $\dfrac{x + 2}{1} = \dfrac{y}{-2} = \dfrac{z + 4}{-2}$

In Problems 19–24, *find the distance between the parallel planes.*

19. $2x - y + 2z = 9$, $2x - y + 2z = -12$

20. $x - 4y - 2z = 5$, $x - 4y - 2z = 10$

21. $3x + y - 4z = 3$, $6x + 2y - 8z = -5$

22. $x + y + 4z = 6$, $2x + 2y + 8z = 9$

23. $x + 2y = 6$, $x + 2y = 1$

24. $x - y - z = 4$, $2x - 2y - 2z = -3$

In Problems 25–32, find the angle described.

25. The angle between $\mathbf{u} = \mathbf{i} + 3\mathbf{j} + 4\mathbf{k}$ and $\mathbf{v} = 3\mathbf{i} - \mathbf{j} + 4\mathbf{k}$

26. The angle between $\mathbf{u} = \mathbf{i} - 2\mathbf{j} + 6\mathbf{k}$ and $\mathbf{v} = 4\mathbf{i} + 5\mathbf{j}$

27. The acute angle between $x = 2t, y = 3t, z = -t$ and $x = 4t, y = -t, z = 2t$

28. The obtuse angle between

$$x = 1 - 2t, y = 3 + t, z = 4t, \quad \text{and} \quad x = 2 + t, y = 3 - 2t, z = 1 + 3t$$

29. The obtuse angle between

$$x = 2 + t, y = 3 - 2t, z = 2 - t \quad \text{and} \quad x = 4 - t, y = 2 + t, z = 2t$$

30. The acute angle between

$$x = 3 - t, y = 4 - 2t, z = -1 - t \quad \text{and} \quad x = t, y = -4 + t, z = 2 + t$$

31. The acute angle between $2x + y - z - 1 = 0$ and $x + y - 3z + 4 = 0$

32. The obtuse angle between $x - y + z - 4 = 0$ and $2x + y + z = 0$

C   **33.** Prove Theorem 10.27.

10.8

CYLINDERS AND SPHERES

We now turn our attention to more complex surfaces, beginning with the cylinder. A cylinder is formed by a line (**generatrix**) moving along a curve (**directrix**) while remaining parallel to a fixed line. If the generatrix is parallel to one of the coordinate axes, the equation of the cylinder is quite simple.

Theorem 10.30   *A nonlinear equation of the form*

$$f(x, y) = 0$$

is a cylinder with generatrix parallel to the z axis and directrix $f(x, y) = 0$ in the xy plane. Similar statements hold when one of the other variables is absent.

It is a simple matter to see why this is so. If $x = x_0$ and $y = y_0$ satisfies the equation $f(x, y) = 0$, then any point of the form (x_0, y_0, z), for *any* choice of z, is on the surface. But the set of all such points is a line parallel to the z axis. Thus any point on the curve $f(x, y) = 0$ in the xy plane determines a line parallel to the z axis in space. The result is then a cylinder.

Example 1   Sketch $x^2 + y^2 = 4$.

Solution   The surface is a cylinder with generatrix parallel to
the z axis and directrix a circle in the xy plane.
A portion of the cylinder is given in Figure 10.25.

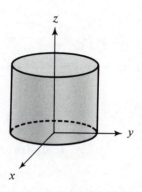

Figure 10.25

Example 2   Sketch $y = z^2$.

Solution   The surface is a cylinder with generatrix
parallel to (or on) the x axis and directrix
a parabola in the yz plane. A portion of
the cylinder is given in Figure 10.26.

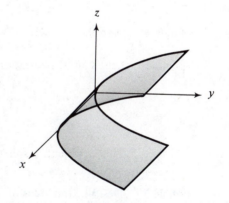

Figure 10.26

Another relatively simple surface is the sphere. The following theorems concerning the sphere are analogous to those for a circle and are proved in much the
same way.

Theorem 10.31   *A point (x, y, z) is on the sphere of radius r and center at (h, k, l) if and only if it
satisfies the equation*

$$(x - h)^2 + (y - k)^2 + (z - l)^2 = r^2.$$

Theorem 10.32   *Any sphere can be represented by an equation of the form*

$$Ax^2 + Ay^2 + Az^2 + Gx + Hy + Iz + J = 0,$$

where $A \neq 0$.

Theorem 10.33   *An equation of the form*

$$Ax^2 + Ay^2 + Az^2 + Gx + Hy + Iz + J = 0,$$

where $A \neq 0$, *represents a sphere, a point, or no locus.*

The proofs are left to the student.

Example 3   Give an equation for the sphere with center $(1, 3, -2)$ and radius 3.

Solution   By Theorem 10.31, the equation is

$$(x - 1)^2 + (y - 3)^2 + (z + 2)^2 = 9$$

or

$$x^2 + y^2 + z^2 - 2x - 6y + 4z + 5 = 0.$$

Example 4   Describe the locus of $x^2 + y^2 + z^2 + 2x - 4y - 8z + 5 = 0$.

Solution   Let us put the equation into the form of Theorem 11.31 by completing squares.

$$x^2 + 2x \qquad + y^2 - 4y \qquad + z^2 - 8z \qquad = -5$$
$$x^2 + 2x + 1 + y^2 - 4y + 4 + z^2 - 8z + 16 = -5 + 1 + 4 + 16$$
$$(x + 1)^2 + (y - 2)^2 + (z - 4)^2 = 16$$

This represents a sphere with center $(-1, 2, 4)$ and radius 4.

Example 5   Describe the locus of

$$2x^2 + 2y^2 + 2z^2 - 2x + 6y - 4z + 7 = 0.$$

Solution

$$x^2 + y^2 + z^2 - x + 3y - 2z + \frac{7}{2} = 0$$

$$x^2 - x \qquad + y^2 + 3y \qquad + z^2 - 2z \qquad = -\frac{7}{2}$$

$$x^2 - x + \frac{1}{4} + y^2 + 3y + \frac{9}{4} + z^2 - 2z + 1 = -\frac{7}{2} + \frac{1}{4} + \frac{9}{4} + 1$$

$$\left(x - \frac{1}{2}\right)^2 + \left(y + \frac{3}{2}\right)^2 + (z - 1)^2 = 0.$$

The equation represents the point $(1/2, -3/2, 1)$.

PROBLEMS

A   *In Problems 1–10, sketch the given surface.*

1. $y^2 + z^2 = 1$ 2. $x^2 + z^2 = 4$

3. $y = x^2$ 4. $x^2 - z^2 = 1$

5. $xy = 4$ 6. $x^2 + z^2 + 2x = 0$

7. $z = 4 - y^2$ 8. $x = \sin z$

9. $y = \ln x$ 10. $z = e^y$

In Problems 11–20, identify the equation as representing a sphere, a point, or no locus. If it is a sphere, give its center and radius. If it is a point give its coordinates.

11. $x^2 + y^2 + z^2 - 2x + 4z - 4 = 0$

12. $x^2 + y^2 + z^2 + 6x - 10y + 2z + 19 = 0$

13. $x^2 + y^2 + z^2 - 8x + 4y - 10z + 46 = 0$

14. $x^2 + y^2 + z^2 + 6x - 8y - 2z + 22 = 0$

15. $2x^2 + 2y^2 + 2z^2 + 2x - 6y + 4z - 1 = 0$

16. $2x^2 + 2y^2 + 2z^2 - 2x + 2y - 10z + 13 = 0$

17. $9x^2 + 9y^2 + 9z^2 - 6x + 6y + 12z - 2 = 0$

18. $3x^2 + 3y^2 + 3z^2 + 4x - 2y - 8z + 7 = 0$

19. $4x^2 + 4y^2 + 4z^2 - 8x - 4y + 16z + 21 = 0$

20. $6x^2 + 6y^2 + 6z^2 - 6x - 4y - 3z = 0$

B   *In Problems 21–28, find an equation(s) in the general form of the sphere(s) described.*

21. Center $(4, 1, -2)$ and radius 3

22. Center $(3, 1, 1)$ and containing the origin

23. Center $(2, 4, 7)$ and tangent to $4x - 8y + z = 1$

24. Center $(4, 1, -3)$ and tangent to $2x - y - 2z = 4$

25. Tangent to $x - 3y + 4z + 23 = 0$ at $(1, 4, -3)$ with radius $\sqrt{26}$

26. Tangent to $x + 2y + 2z - 17 = 0$ at $(1, 4, 4)$ with radius 3

27. Containing $(3, 1, -1)$, $(2, 5, 2)$, $(-3, 0, 1)$, and $(-1, 0, 0)$.

28. Containing $(4, 1, 0)$, $(-2, -1, 0)$, $(0, 2, 1)$, and $(1, 1, 1)$

C   29. Prove Theorem 10.31.

   30. Prove Theorem 10.32.

   31. Prove Theorem 10.33.

10.9

QUADRIC SURFACES

In the plane, a second-degree equation represents a parabola, ellipse, hyperbola, or a degenerate case of one of these. There are far more variations in space, where we have already seen that certain cylinders and the sphere are represented by second-degree equations. The **traces** in the coordinate planes of a given surface are simply the intersections of the surface with the coordinate planes. The traces in the coordinate planes of quadric surfaces, represented by second-degree equations, are conics or degenerate conics. We say that a quadric surface is in the standard position if its traces in the coordinate planes are in the standard position—that is, they have their center or vertex at the origin and axes along the coordinate axes. In the following discussion, all surfaces are assumed to be in the standard position.

The **ellipsoid** (see Figure 10.27) is represented by an equation of the form

$$\frac{x^2}{a^2} + \frac{y^2}{b^2} + \frac{z^2}{c^2} = 1.$$

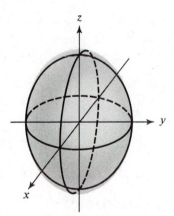

Figure 10.27

Its traces in the coordinate planes are ellipses (or circles). We have already seen a special case of this, in which $a = b = c$. In that case, we have a sphere. There are two other special cases. One is the **prolate spheroid**. Here, two of the denominators are equal and both are less than the third. It has the shape of a football and may be generated by rotating an ellipse about its major axis. The other case is the **oblate spheroid**, in which two of the denominators are equal and both greater than the third. It has the shape of a doorknob and may be generated by rotating an ellipse about its minor axis.

The **hyperboloid of one sheet** (see Figure 10.28) is represented by an equation of the form

$$\frac{x^2}{a^2} + \frac{y^2}{b^2} - \frac{z^2}{c^2} = 1.$$

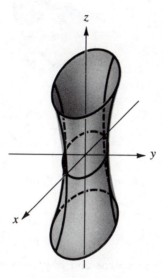

Figure 10.28

Its traces in the xz plane and yz plane are hyperbolas; in the xy plane, it is an ellipse. If $a = b$, it may be generated by rotating a hyperbola about its conjugate axis.

The **hyperboloid of two sheets** (see Figure 10.29) is represented by an equation of the form

$$\frac{x^2}{a^2} - \frac{y^2}{b^2} - \frac{z^2}{c^2} = 1.$$

Its traces in the xy plane and xz plane are hyperbolas. It has no trace in the yz plane; however, if $|x| > a$, its intersection with a plane parallel to the yz plane is an ellipse. If $b = c$, it may be generated by rotating a hyperbola about its transverse axis.

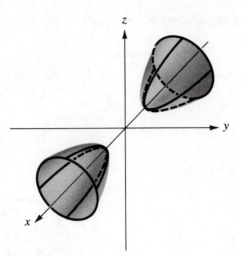

Figure 10.29

The **elliptic paraboloid** (see Figure 10.30) is represented by an equation of the form

$$\frac{x^2}{a^2} + \frac{y^2}{b^2} = \frac{z}{c}.$$

Its traces in the xz plane and yz plane are parabolas. Its trace in the xy plane is a single point. If $c > 0$, then its intersection with a plane parallel to and above the

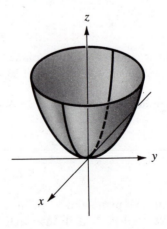

Figure 10.30

xy plane is an ellipse; below the xy plane there is no intersection. This situation is reversed if $c < 0$. In Figure 10.30, $c > 0$. If $a = b$, the elliptic paraboloid is generated by rotating a parabola about its axis.

The **hyperbolic paraboloid** (see Figure 10.31) is represented by an equation of the form

$$\frac{x^2}{a^2} - \frac{y^2}{b^2} = \frac{z}{c}.$$

Its traces in the xz plane and yz plane are parabolas, one opening upward and the other down. Its trace in the xy plane is a pair of lines intersecting at the origin (a degenerate hyperbola). Its intersection with a plane parallel to the xy plane is a hyperbola. If $c > 0$, those hyperbolas above the xy plane have the transverse axis parallel to the x axis, while those below have it parallel to the y axis. If $c < 0$, this situation is reversed. In Figure 10.31, $c < 0$.

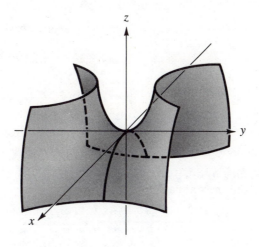

Figure 10.31

The **elliptic cone** (see Figure 10.32) is represented by an equation of the form

$$\frac{x^2}{a^2} + \frac{y^2}{b^2} - \frac{z^2}{c^2} = 0.$$

Its trace in the xz plane is a pair of lines intersecting at the origin. Its trace in the yz plane is also a pair of lines intersecting at the origin. Its trace in the xy plane is a single point at the origin. Its intersection with a plane parallel to the xy plane is an ellipse. If $a = b$, it is a circular cone.

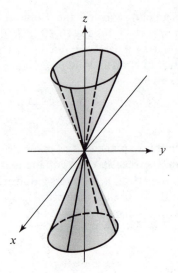

Figure 10.32

Example 1   Describe and sketch

$$9x^2 + 9y^2 - 4z^2 = 36.$$

Solution   Dividing by 36, we have a hyperboloid of one sheet in the standard form.

$$\frac{x^2}{4} + \frac{y^2}{4} - \frac{z^2}{9} = 1$$

Since the denominators of the x^2 and y^2 terms are equal, the trace in the xy plane, as well as in any plane parallel to it, is a circle. The surface is shown in Figure 10.33.

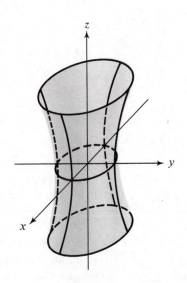

Figure 10.33

Example 2 Describe and sketch

$$9x^2 + 9y^2 - 4z^2 = 0.$$

Solution Again dividing by 36, we have

$$\frac{x^2}{4} + \frac{y^2}{4} - \frac{z^2}{9} = 0,$$

which is a circular cone with its axis the z axis. The cone is given in Figure 10.34.

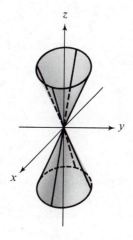

Figure 10.34

If the given equation does not fit any of these forms, determine its traces in the coordinate planes and planes parallel to them in order to have some idea of the shape.

PROBLEMS

B *Describe and sketch the quadric surfaces in Problems* 1–25.

1. $36x^2 + 9y^2 + 4z^2 = 36$
2. $4x^2 + 9y^2 + 9z^2 = 36$
3. $x - y^2 - z^2 = 0$
4. $4x^2 + 4y^2 - z^2 + 16 = 0$
5. $x^2 - y^2 + z^2 = 0$
6. $x^2 - y - z^2 = 0$
7. $x^2 + 2y + z^2 = 0$
8. $x^2 - 4y^2 - 4z^2 = 0$
9. $25x^2 - 4y^2 + 25z^2 + 100 = 0$
10. $16x^2 - 9y^2 - 9z^2 + 144 = 0$
11. $x^2 + 4y - z^2 = 0$
12. $25x^2 + 16y^2 + 25z^2 = 400$
13. $36x^2 - 9y^2 + 16z^2 + 144 = 0$
14. $16x^2 - 9y + 16z^2 = 0$
15. $16x^2 - 9y^2 - 9z^2 = 0$
16. $36x^2 - 4y - 9z^2 = 0$
17. $9x^2 - 36y^2 + 16z^2 + 144 = 0$
18. $x^2 + y^2 - 4z = 0$
19. $x^2 + y^2 - 4z^2 = 4$
20. $x^2 + y^2 + 4z^2 = 4$

21. $x^2 - y^2 - 9z = 0$ **22.** $9x^2 - y^2 - z^2 = 9$

23. $9x^2 - y^2 - z^2 + 0$ **24.** $x^2 + y^2 + 2z = 0$

25. $25x^2 - 4y^2 + 25z^2 = 100$

C   *If the coordinate axes are translated with the origin moved to* (h, k, l), *the equations of translation are*

$$x' = x - h, \qquad y' = y - k, \qquad z' = z - l.$$

In many cases, translations may be carried out by completing the square. In Problems 26–30, translate and sketch.

26. $z = 4 - x^2 - 2y^2$

27. $z = 1 + x^2 + y^2$

28. $z = x^2 + y^2 + 2x + 4y + 7$

29. $x^2 + 4y^2 + 9z^2 + 2x + 16y - 18z - 10 = 0$

30. $x^2 + 4y^2 - z^2 - 2x - 24y - 8z + 17 = 0$

10.10

CYLINDRICAL AND SPHERICAL COORDINATES

Here we look at two other coordinate systems in space that are useful. The first of these is called a **cylindrical coordinate system**, which is especially convenient when the z axis is a line of symmetry. In this system, a point P with projection Q on the xy plane (see Figure 10.35) is represented by (r, θ, z), where (r, θ) is a polar representation of Q and z is the (directed) distance of P from the xy plane.

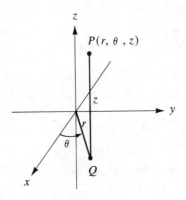

Figure 10.35

The relations between rectangular and cylindrical coordinates are the same as the relations between rectangular and polar coordinates as given by the following theorem.

Theorem 10.34 *If (x, y, z) and (r, θ, z) represent the same point in rectangular and cylindrical coordinates, then*

$$x = r \cos \theta, \qquad y = r \sin \theta, \qquad z = z$$

or

$$r = \pm \sqrt{x^2 + y^2}, \qquad \theta = \arctan \frac{y}{x}, \qquad z = z.$$

Example 1 Express in rectangular coordinates the point having cylindrical coordinates $(4, 30°, -2)$.

Solution

$$x = r \cos \theta = 4 \cos 30° = 4 \cdot \frac{\sqrt{3}}{2} = 2\sqrt{3}$$

$$y = r \sin \theta = 4 \sin 30° = 4 \cdot \frac{1}{2} = 2$$

$$z = -2$$

Thus, the point is $(2\sqrt{3}, 2, -2)$.

Example 2 Express in cylindrical coordinates the point having rectangular coordinates $(2, 2, 4)$.

Solution

$$r = \pm \sqrt{x^2 + y^2} = \pm \sqrt{8} = \pm 2\sqrt{2}$$

$$\theta = \arctan \frac{y}{x} = \arctan 1 = \frac{\pi}{4} + \pi n$$

$$z = 4$$

There are two choices for r and infinitely many for θ. The choices we make are not independent of each other—the choice of one of them puts restrictions on the other. If we choose θ to be $\pi/4$ (or any first-quadrant angle), we must choose $r = 2\sqrt{2}$; if we choose θ to be $5\pi/4$ (or any third-quadrant angle), we must choose $r = -2\sqrt{2}$. Thus, two possible representations are

$$(2\sqrt{2}, \pi/4, 4) = (-2\sqrt{2}, 5\pi/4, 4).$$

Example 3 Express in cylindrical coordinates the rectangular coordinate equation $x + y + z = 1$.

Solution Substituting $x = r \cos \theta$ and $y = r \sin \theta$, we have

$$r \cos \theta + r \sin \theta + z = 1.$$

Example 4 Express in rectangular coordinates the cylindrical coordinate equation $r = z \sin \theta$.

Solution Multiplying both sides by r, we have

$$r^2 = z \cdot r \sin \theta \qquad \text{or} \qquad x^2 + y^2 = yz.$$

Another useful system for representing points in space uses **spherical coordinates**. In this system, a point is represented by (ρ, θ, φ), where ρ is the distance of the point from the origin, θ has the same meaning as in cylindrical coordinates, and φ is the angle between the positive end of the z axis and the segment joining

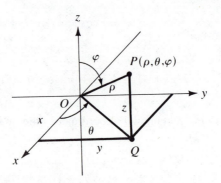

Figure 10.36

the origin to the given point (see Figure 10.36). Since φ is undirected, it is never negative. Furthermore

$$0° \leq \varphi \leq 180°.$$

Likewise, we restrict $\rho : \rho \geq 0$.

From Fig. 10.36, we see that $\overline{OQ} = \rho \sin \varphi$. Thus

$$x = \overline{OQ} \cos \theta = \rho \sin \varphi \cos \theta,$$
$$y = \overline{OQ} \sin \theta = \rho \sin \varphi \sin \theta,$$

and

$$z = \rho \cos \varphi.$$

We can easily see from these that

$$\rho^2 = x^2 + y^2 + z^2.$$

Thus we have the following theorem.

Theorem 10.35   *If (x, y, z) and (ρ, θ, φ) represent the same point in rectangular and spherical coordinates, respectively, then*

$$x = \rho \sin \varphi \cos \theta,$$
$$y = \rho \sin \varphi \sin \theta,$$
$$z = \rho \cos \varphi,$$

and

$$\rho^2 = x^2 + y^2 + z^2.$$

Example 5 Express in rectangular coordinates the point having spherical coordinates $(2, \pi/6, 2\pi/3)$.

Solution

$$x = \rho \sin \varphi \cos \theta$$

$$= 2 \sin \frac{2\pi}{3} \cdot \cos \frac{\pi}{6} = 2 \cdot \frac{\sqrt{3}}{2} \cdot \frac{\sqrt{3}}{2} = \frac{3}{2}$$

$$y = \rho \sin \varphi \sin \theta$$

$$= 2 \sin \frac{2\pi}{3} \sin \frac{\pi}{6} = 2 \cdot \frac{\sqrt{3}}{2} \cdot \frac{1}{2} = \frac{\sqrt{3}}{2}$$

$$z = \rho \cos \varphi$$

$$= 2 \cos \frac{2\pi}{3} = 2 \left(-\frac{1}{2} \right) = -1$$

Thus the point is $(3/2, \sqrt{3}/2, -1)$.

Example 6 Express in spherical coordinates the point having rectangular coordinates $(\sqrt{3}, \sqrt{3}, -\sqrt{2})$.

Solution

$$\rho^2 = x^2 + y^2 + z^2$$

$$= 3 + 3 + 2 = 8$$

$$\rho = 2\sqrt{2}$$

$$z = \rho \cos \varphi$$

$$-\sqrt{2} = 2\sqrt{2} \cos \varphi$$

$$\cos \varphi = -\frac{1}{2}$$

$$\varphi = 120°$$

$$x = \rho \sin \varphi \cos \theta$$

$$\sqrt{3} = 2\sqrt{2} \cdot \frac{\sqrt{3}}{2} \cos \theta$$

$$\cos \theta = \frac{1}{\sqrt{2}}$$

$$\theta = 45°$$

The point is $(2\sqrt{2}, 45°, 120°)$.

Example 7 Express in spherical coordinates the rectangular coordinate equation $x^2 + y^2 - z^2 = 0$.

Solution
$$\rho^2 \sin^2 \varphi \cos^2 \theta + \rho^2 \sin^2 \varphi \sin^2 \theta - \rho^2 \cos^2 \varphi = 0$$
$$\rho^2 [\sin^2 \varphi (\cos^2 \theta + \sin^2 \theta) - \cos^2 \varphi] = 0$$
$$\rho^2 (\sin^2 \varphi - \cos^2 \varphi) = 0$$

$$\sin^2 \varphi - \cos^2 \varphi = 0 \qquad \text{or} \qquad \rho = 0$$
$$\cos 2\varphi = 0$$
$$2\varphi = \frac{\pi}{2}, \frac{3\pi}{2}$$
$$\varphi = \frac{\pi}{4}, \frac{3\pi}{4}$$

Thus, we have $\varphi = \pi/4$, $\varphi = 3\pi/4$, and $\rho = 0$. $\varphi = \pi/4$ gives the top half of the cone, $\varphi = 3\pi/4$ gives the bottom half, and $\rho = 0$ gives a single point—the origin. Since $\rho = 0$ is included in both of the others, we may drop it. The final result is

$$\varphi = \frac{\pi}{4} \qquad \text{and} \qquad \varphi = \frac{3\pi}{4}.$$

Example 8 Express in rectangular coordinates the spherical coordinate equation $\rho^2 \sin \varphi \cos \varphi \cos \theta = 1$.

Solution
$$\rho^2 \sin \varphi \cos \varphi \cos \theta = 1$$
$$(\rho \sin \varphi \cos \theta)(\rho \cos \varphi) = 1$$
$$xz = 1$$

PROBLEMS

A **1.** The following points are given in cylindrical coordinates. Express them in rectangular coordinates.

    a. $(2, 45°, 1)$ **b.** $(3, 2\pi/3, -2)$
    c. $(1, 0°, 2)$ **d.** $(0, \pi/4, -3)$

2. The following points are given in rectangular coordinates. Express them in cylindrical coordinates.

    a. $(1, 1, 3)$ **b.** $(0, 2, -2)$
    c. $(-1, \sqrt{3}, 4)$ **d.** $(-2\sqrt{3}, -2, 3)$

3. The following points are given in spherical coordinates. Express them in rectangular coordinates.

    a. $(3, 45°, 30°)$ **b.** $(1, \pi/6, 0)$
    c. $(1, 90°, 45°)$ **d.** $(2, 5\pi/6, 3\pi/4)$

4. The following points are given in rectangular coordinates. Express them in spherical coordinates.

    a. $(2, 2, 0)$ **b.** $(2, 1, -2)$
    c. $(2, -\sqrt{3}, 4)$ **d.** $(1, 1, \sqrt{2})$

B

5. The following points are given in cylindrical coordinates. Express them in spherical coordinates.

    a. $(3, 30°, 4)$ **b.** $(2, \pi/4, -2)$
    c. $(0, 45°, 3)$ **d.** $(2, \pi/2, -4)$

6. The following points are given in spherical coordinates. Express them in cylindrical coordinates.

    a. $(4, 45°, 30°)$ **b.** $(2, 2\pi/3, \pi/2)$
    c. $(2, 210°, 135°)$ **d.** $(3, \pi/6, 2\pi/3)$

In Problems 7–14, express the given rectangular coordinate equations in cylindrical and spherical coordinates.

7. $x^2 + y^2 = 4$ **8.** $x^2 + y^2 + z^2 = 4$
9. $x^2 + y^2 = z$ **10.** $x^2 - y^2 = z$
11. $x^2 - y^2 - z^2 = 1$ **12.** $x^2 - y^2 + z^2 = 1$
13. $x^2 + y^2 - z^2 = 1$ **14.** $4x^2 + 9y^2 + 9z^2 = 1$

In Problems 15–22, express the given equations in rectangular coordinates.

15. $z = r^2 \sin 2\theta$ **16.** $z = r^2 \cos 2\theta$
17. $z = r^2$ **18.** $z = 1 + \sin \theta$
19. $\rho \sin \varphi \tan \varphi \sin 2\theta = 2$ **20.** $\rho \sin \varphi = 1$
21. $\rho = \sin \varphi \cos \theta$ **22.** $\rho^2 \sin \varphi \cos \varphi = 1$

REVIEW PROBLEMS

A

1. Find the point 1/3 of the way from $(-2, 1, 5)$ to $(1, 4, -4)$.

2. Find the projection of $\mathbf{u} = \mathbf{i} - 3\mathbf{j} + \mathbf{k}$ upon $\mathbf{v} = 4\mathbf{i} + 2\mathbf{j} + 3\mathbf{k}$.

3. Find the set of direction angles for the vector $\mathbf{v}$ with direction numbers $\{1, -1, \sqrt{2}\}$ if $\mathbf{v}$ is directed upward.

4. Suppose the line l_1 contains $(2, -2, 4)$ and $(5, 3, 0)$, while l_2 contains $(4, -3, 1)$ and $(3, -4, -1)$. Are l_1 and l_2 parallel, perpendicular, coincident, or none of these?

5. Suppose the line l_1 contains $(5, 1, -2)$ and $(2, -3, 1)$, while l_2 contains $(3, 8, 1)$ and $(-3, 0, 7)$. Are l_1 and l_2 parallel, perpendicular, coincident, or none of these?

In Problems 6 and 7, represent the given line in parametric form and in symmetric form.

6. Containing $(4, 0, 3)$ and $(5, 1, 3)$

7. Containing $(2, -3, 5)$; direction numbers $\{3, -2, 1\}$

In Problems 8 and 9, find the point of intersection (if any) of the given lines.

8. $\dfrac{x}{2} = \dfrac{y - 6}{-4} = \dfrac{z - 3}{1}$; $\dfrac{x - 2}{-1} = \dfrac{y - 4}{3} = \dfrac{z - 7}{1}$

9. $x = 2 + t$, $y = 3 - 2t$, $z = 3 + 5t$; $x = -3 + 2s$, $y = -1 + 3s$, $z = -s$

10. Find the distance between the lines
$$x = 4 - t,\ y = 3 + 2t,\ z = 4 + t \quad \text{and} \quad x = 3 + 2s,\ y = 5 - s,\ z = -1 + 4s.$$

In Problems 11 and 12, find an equation(s) of the plane(s) satisfying the given conditions.

11. Containing $(4, 0, -2)$ and perpendicular to $x = 4 - t$, $y = 3 + 2t$, $z = 1$

12. Containing $(2, 5, 1)$, $(3, -2, 4)$, and $(1, 0, 2)$

13. Find the distance between $(4, 3, 1)$ and $2x - 2y + z = 4$.

14. Find the distance between $x - 4y + z = 3$ and $x - 4y + z = 8$.

15. Find the distance between $(2, 5, -2)$ and the line $x = 3 + t$, $y = 4 - t$, $z = 2 + 2t$.

16. Find the distance between the lines
$$x = 1 + 2t,\ y = 4 + 3t,\ z = 2 - t \quad \text{and} \quad x = 4 + 2s,\ y = 2 + 3s,\ z = 5 - s.$$

17. Find the acute angle between the planes $2x - 5y + z = 3$ and $x + y - 2z = 1$.

In Problems 18 and 19, describe the locus of the given equation.

18. $x^2 + y^2 + z^2 - 2x - 4y + 8z + 5 = 0$

19. $2x^2 + 2y^2 + 2z^2 - 2x - 4y + 6z + 7 = 0$

In Problems 20–23, sketch and describe.

20. $9x^2 + 9y^2 - 4z^2 = 36$ 21. $z^2 - x^2 = 4$

22. $9x^2 - 36y^2 + 4z^2 = 0$ 23. $x^2 + y - z^2 = 0$

24. The following points are given in rectangular coordinates. Express them in cylindrical and spherical coordinates.
    (a) $(2, 2, 1)$ (b) $(-1, 1, -2)$

25. Express $z = r \cos \theta$ in rectangular coordinates.

26. Express $z = 4x^2 + 4y^2$ in cylindrical and spherical coordinates.

27. Express $\rho = \sin \varphi\ (2 \cos \theta - \sin \theta)$ in rectangular coordinates.

B 28. The point $(5, 3, -2)$ is a distance 3 from the midpoint of the segment joining $(5, 7, 2)$ and $(1, 1, z)$. Find z.

29. Find the end points of the representative $\overrightarrow{AB}$ of $\mathbf{v}$ if $\mathbf{v} = 3\mathbf{i} - \mathbf{j} + 4\mathbf{k}$ and the midpoint of AB is $(2, 4, -3)$.

30. Suppose $\mathbf{u} = 2\mathbf{i} - \mathbf{j} + 3\mathbf{k}$ and $\mathbf{v} = \mathbf{i} + 4\mathbf{j} - \mathbf{k}$. Express $\mathbf{u}$ in the form $\mathbf{u} = \mathbf{p} + \mathbf{q}$, where $\mathbf{p} = k\mathbf{v}$ and $\mathbf{q} \cdot \mathbf{v} = 0$. Interpret geometrically.

In Problems 31 and 32, find equations for the line described.

31. Containing $(3, 5, -2)$ and perpendicular to the plane containing
$$x = 5 - t,\ y = 2 + 3t,\ z = 4 + t \quad \text{and} \quad x = 3 - s,\ y = 5 + 3s,\ z = -1 + s$$

32. Perpendicular at $(4, 2, 3)$ to the plane containing $(4, 2, 3)$, $(-1, 3, 1)$, and $(2, 5, -3)$.

33. Find an equation(s) of the plane(s) containing $(4, 2, -3)$ and perpendicular to
$$2x - y + 4z = 5 \quad \text{and} \quad x + 3y - z = 2.$$

34. Find, in parametric form, the line of intersection of
$$2x - y + 3z = 2 \quad \text{and} \quad x + 3y + 2z = 4.$$

35. Sketch and describe $x^2 + y^2 + 4z^2 - 2x + 4y + 1 = 0$.

36. Sketch the graph of $\mathbf{f}(t) = (t + 1)\mathbf{i} + (t^2 - 1)\mathbf{j} + (t + 1)\mathbf{k}$.

C 37. Prove or show to be false: $\mathbf{u} \times (\mathbf{v} \times \mathbf{w}) = (\mathbf{u} \times \mathbf{v}) \times \mathbf{w}$.

TABLES

TABLE 1

COMMON LOGARITHMS

x	0	1	2	3	4	5	6	7	8	9
1.0	.0000	.0043	.0086	.0128	.0170	.0212	.0253	.0294	.0334	.0374
1.1	.0414	.0453	.0492	.0531	.0569	.0607	.0645	.0682	.0719	.0755
1.2	.0792	.0828	.0864	.0899	.0934	.0969	.1004	.1038	.1072	.1106
1.3	.1139	.1173	.1206	.1239	.1271	.1303	.1335	.1367	.1399	.1430
1.4	.1461	.1492	.1523	.1553	.1584	.1614	.1644	.1673	.1703	.1732
1.5	.1761	.1790	.1818	.1847	.1875	.1903	.1931	.1959	.1987	.2014
1.6	.2041	.2068	.2095	.2122	.2148	.2175	.2201	.2227	.2253	.2279
1.7	.2304	.2330	.2355	.2380	.2405	.2430	.2455	.2480	.2504	.2529
1.8	.2553	.2577	.2601	.2625	.2648	.2672	.2695	.2718	.2742	.2765
1.9	.2788	.2810	.2833	.2856	.2878	.2900	.2923	.2945	.2967	.2989
2.0	.3010	.3032	.3054	.3075	.3096	.3118	.3139	.3160	.3181	.3201
2.1	.3222	.3243	.3263	.3284	.3304	.3324	.3345	.3365	.3385	.3404
2.2	.3424	.3444	.3464	.3483	.3502	.3522	.3541	.3560	.3579	.3598
2.3	.3617	.3636	.3655	.3674	.3692	.3711	.3729	.3747	.3766	.3784
2.4	.3802	.3820	.3838	.3856	.3874	.3892	.3909	.3927	.3945	.3962
2.5	.3979	.3997	.4014	.4031	.4048	.4065	.4082	.4099	.4116	.4133
2.6	.4150	.4166	.4183	.4200	.4216	.4232	.4249	.4265	.4281	.4298
2.7	.4314	.4330	.4346	.4362	.4378	.4393	.4409	.4425	.4440	.4456
2.8	.4472	.4487	.4502	.4518	.4533	.4548	.4564	.4579	.4594	.4609
2.9	.4624	.4639	.4654	.4669	.4683	.4698	.4713	.4728	.4742	.4757
3.0	.4771	.4786	.4800	.4814	.4829	.4843	.4857	.4871	.4886	.4900
3.1	.4914	.4928	.4942	.4955	.4969	.4983	.4997	.5011	.5024	.5038
3.2	.5051	.5065	.5079	.5092	.5105	.5119	.5132	.5145	.5159	.5172
3.3	.5185	.5198	.5211	.5224	.5237	.5250	.5263	.5276	.5289	.5302
3.4	.5315	.5328	.5340	.5353	.5366	.5378	.5391	.5403	.5416	.5428
3.5	.5441	.5453	.5465	.5478	.5490	.5502	.5514	.5527	.5539	.5551
3.6	.5563	.5575	.5587	.5599	.5611	.5623	.5635	.5647	.5658	.5670
3.7	.5682	.5694	.5705	.5717	.5729	.5740	.5752	.5763	.5775	.5786
3.8	.5798	.5809	.5821	.5832	.5843	.5855	.5866	.5877	.5888	.5899
3.9	.5911	.5922	.5933	.5944	.5955	.5966	.5977	.5988	.5999	.6010
4.0	.6021	.6031	.6042	.6053	.6064	.6075	.6085	.6096	.6107	.6117
4.1	.6128	.6138	.6149	.6160	.6170	.6180	.6191	.6201	.6212	.6222
4.2	.6232	.6243	.6253	.6263	.6274	.6284	.6294	.6304	.6314	.6325
4.3	.6335	.6345	.6355	.6365	.6375	.6385	.6395	.6405	.6415	.6425
4.4	.6435	.6444	.6454	.6464	.6474	.6484	.6493	.6503	.6513	.6522
4.5	.6532	.6542	.6551	.6561	.6571	.6580	.6590	.6599	.6609	.6618
4.6	.6628	.6637	.6646	.6656	.6665	.6675	.6684	.6693	.6702	.6712
4.7	.6721	.6730	.6739	.6749	.6758	.6767	.6776	.6785	.6794	.6803
4.8	.6812	.6821	.6830	.6839	.6848	.6857	.6866	.6875	.6884	.6893
4.9	.6902	.6911	.6920	.6928	.6937	.6946	.6955	.6964	.6972	.6981
5.0	.6990	.6998	.7007	.7016	.7024	.7033	.7042	.7050	.7059	.7067
5.1	.7076	.7084	.7093	.7101	.7110	.7118	.7126	.7135	.7143	.7152
5.2	.7160	.7168	.7177	.7185	.7193	.7202	.7210	.7218	.7226	.7235
5.3	.7243	.7251	.7259	.7267	.7275	.7284	.7292	.7300	.7308	.7316
5.4	.7324	.7332	.7340	.7348	.7356	.7364	.7372	.7380	.7388	.7396
x	0	1	2	3	4	5	6	7	8	9

TABLE 1

(*continued*)

x	0	1	2	3	4	5	6	7	8	9
5.5	.7404	.7412	.7419	.7427	.7435	.7443	.7451	.7459	.7466	.7474
5.6	.7482	.7490	.7497	.7505	.7513	.7520	.7528	.7536	.7543	.7551
5.7	.7559	.7566	.7574	.7582	.7589	.7597	.7604	.7612	.7619	.7627
5.8	.7634	.7642	.7649	.7657	.7664	.7672	.7679	.7686	.7694	.7701
5.9	.7709	.7716	.7723	.7731	.7738	.7745	.7752	.7760	.7767	.7774
6.0	.7782	.7789	.7796	.7803	.7810	.7818	.7825	.7832	.7839	.7846
6.1	.7853	.7860	.7868	.7875	.7882	.7889	.7896	.7903	.7910	.7917
6.2	.7924	.7931	.7938	.7945	.7952	.7959	.7966	.7973	.7980	.7987
6.3	.7993	.8000	.8007	.8014	.8021	.8028	.8035	.8041	.8048	.8055
6.4	.8062	.8069	.8075	.8082	.8089	.8096	.8102	.8109	.8116	.8122
6.5	.8129	.8136	.8142	.8149	.8156	.8162	.8169	.8176	.8182	.8189
6.6	.8195	.8202	.8209	.8215	.8222	.8228	.8235	.8241	.8248	.8254
6.7	.8261	.8267	.8274	.8280	.8287	.8293	.8299	.8306	.8312	.8319
6.8	.8325	.8331	.8338	.8344	.8351	.8357	.8363	.8370	.8376	.8382
6.9	.8388	.8395	.8401	.8407	.8414	.8420	.8426	.8432	.8439	.8445
7.0	.8451	.8457	.8463	.8470	.8476	.8482	.8488	.8494	.8500	.8506
7.1	.8513	.8519	.8525	.8531	.8537	.8543	.8549	.8555	.8561	.8567
7.2	.8573	.8579	.8585	.8591	.8597	.8603	.8609	.8615	.8621	.8627
7.3	.8633	.8639	.8645	.8651	.8657	.8663	.8669	.8675	.8681	.8686
7.4	.8692	.8698	.8704	.8710	.8716	.8722	.8727	.8733	.8739	.8745
7.5	.8751	.8756	.8762	.8768	.8774	.8779	.8785	.8791	.8797	.8802
7.6	.8808	.8814	.8820	.8825	.8831	.8837	.8842	.8848	.8854	.8859
7.7	.8865	.8871	.8876	.8882	.8887	.8893	.8899	.8904	.8910	.8915
7.8	.8921	.8927	.8932	.8938	.8943	.8949	.8954	.8960	.8965	.8971
7.9	.8976	.8982	.8987	.8993	.8998	.9004	.9009	.9015	.9020	.9025
8.0	.9031	.9036	.9042	.9047	.9053	.9058	.9063	.9069	.9074	.9079
8.1	.9085	.9090	.9096	.9101	.9106	.9112	.9117	.9122	.9128	.9133
8.2	.9138	.9143	.9149	.9154	.9159	.9165	.9170	.9175	.9180	.9186
8.3	.9191	.9196	.9201	.9206	.9212	.9217	.9222	.9227	.9232	.9238
8.4	.9243	.9248	.9253	.9258	.9263	.9269	.9274	.9279	.9284	.9289
8.5	.9294	.9299	.9304	.9309	.9315	.9320	.9325	.9330	.9335	.9340
8.6	.9345	.9350	.9355	.9360	.9365	.9370	.9375	.9380	.9385	.9390
8.7	.9395	.9400	.9405	.9410	.9415	.9420	.9425	.9430	.9435	.9440
8.8	.9445	.9450	.9455	.9460	.9465	.9469	.9474	.9479	.9484	.9489
8.9	.9494	.9499	.9504	.9509	.9513	.9518	.9523	.9528	.9533	.9538
9.0	.9542	.9547	.9552	.9557	.9562	.9566	.9571	.9576	.9581	.9586
9.1	.9590	.9595	.9600	.9605	.9609	.9614	.9619	.9624	.9628	.9633
9.2	.9638	.9643	.9647	.9652	.9657	.9661	.9666	.9671	.9675	.9680
9.3	.9685	.9689	.9694	.9699	.9703	.9708	.9713	.9717	.9722	.9727
9.4	.9731	.9736	.9741	.9745	.9750	.9754	.9759	.9763	.9768	.9773
9.5	.9777	.9782	.9786	.9791	.9795	.9800	.9805	.9809	.9814	.9818
9.6	.9823	.9827	.9832	.9836	.9841	.9845	.9850	.9854	.9859	.9863
9.7	.9868	.9872	.9877	.9881	.9886	.9890	.9894	.9899	.9903	.9908
9.8	.9912	.9917	.9921	.9926	.9930	.9934	.9939	.9943	.9948	.9952
9.9	.9956	.9961	.9965	.9969	.9974	.9978	.9983	.9987	.9991	.9996
x	0	1	2	3	4	5	6	7	8	9

TABLE 2

TRIGONOMETRIC FUNCTIONS

Degrees	Radians	sin	cos	tan	cot		
0	0.0000	0.0000	1.0000	0.0000		1.5708	90
1	0.0175	0.0175	0.9998	0.0175	57.290	1.5533	89
2	0.0349	0.0349	0.9994	0.0349	28.636	1.5359	88
3	0.0524	0.0523	0.9986	0.0524	19.081	1.5184	87
4	0.0698	0.0698	0.9976	0.0699	14.301	1.5010	86
5	0.0873	0.0872	0.9962	0.0875	11.430	1.4835	85
6	0.1047	0.1045	0.9945	0.1051	9.5144	1.4661	84
7	0.1222	0.1219	0.9925	0.1228	8.1443	1.4486	83
8	0.1396	0.1392	0.9903	0.1405	7.1154	1.4312	82
9	0.1571	0.1564	0.9877	0.1584	6.3138	1.4137	81
10	0.1745	0.1736	0.9848	0.1763	5.6713	1.3963	80
11	0.1920	0.1908	0.9816	0.1944	5.1446	1.3788	79
12	0.2094	0.2079	0.9781	0.2126	4.7046	1.3614	78
13	0.2269	0.2250	0.9744	0.2309	4.3315	1.3439	77
14	0.2443	0.2419	0.9703	0.2493	4.0108	1.3265	76
15	0.2618	0.2588	0.9659	0.2679	3.7321	1.3090	75
16	0.2793	0.2756	0.9613	0.2867	3.4874	1.2915	74
17	0.2967	0.2924	0.9563	0.3057	3.2709	1.2741	73
18	0.3142	0.3090	0.9511	0.3249	3.0777	1.2566	72
19	0.3316	0.3256	0.9455	0.3443	2.9042	1.2392	71
20	0.3491	0.3420	0.9397	0.3640	2.7475	1.2217	70
21	0.3665	0.3584	0.9336	0.3839	2.6051	1.2043	69
22	0.3840	0.3746	0.9272	0.4040	2.4751	1.1868	68
23	0.4014	0.3907	0.9205	0.4245	2.3559	1.1694	67
24	0.4189	0.4067	0.9135	0.4452	2.2460	1.1519	66
25	0.4363	0.4226	0.9063	0.4663	2.1445	1.1345	65
26	0.4538	0.4384	0.8988	0.4877	2.0503	1.1170	64
27	0.4712	0.4540	0.8910	0.5095	1.9626	1.0996	63
28	0.4887	0.4695	0.8829	0.5317	1.8807	1.0821	62
29	0.5061	0.4848	0.8746	0.5543	1.8040	1.0647	61
30	0.5236	0.5000	0.8660	0.5774	1.7321	1.0472	60
31	0.5411	0.5150	0.8572	0.6009	1.6643	1.0297	59
32	0.5585	0.5299	0.8480	0.6249	1.6003	1.0123	58
33	0.5760	0.5446	0.8387	0.6494	1.5399	0.9948	57
34	0.5934	0.5592	0.8290	0.6745	1.4826	0.9774	56
35	0.6109	0.5736	0.8192	0.7002	1.4281	0.9599	55
36	0.6283	0.5878	0.8090	0.7265	1.3764	0.9425	54
37	0.6458	0.6018	0.7986	0.7536	1.3270	0.9250	53
38	0.6632	0.6157	0.7880	0.7813	1.2799	0.9076	52
39	0.6807	0.6293	0.7771	0.8098	1.2349	0.8901	51
40	0.6981	0.6428	0.7660	0.8391	1.1918	0.8727	50
41	0.7156	0.6561	0.7547	0.8693	1.1504	0.8552	49
42	0.7330	0.6691	0.7431	0.9004	1.1106	0.8378	48
43	0.7505	0.6820	0.7314	0.9325	1.0724	0.8203	47
44	0.7679	0.6947	0.7193	0.9657	1.0355	0.8029	46
45	0.7854	0.7071	0.7071	1.0000	1.0000	0.7854	45
		cos	sin	cot	tan	Radians	Degrees

TABLE 3

SQUARES, SQUARE ROOTS, AND PRIME FACTORS

Number	Square	Square root	Factors	Number	Square	Square root	Factors
1	1	1.000		51	2,601	7.141	$3 \cdot 17$
2	4	1.414	2	52	2,704	7.211	$2^2 \cdot 13$
3	9	1.732	3	53	2,809	7.280	53
4	16	2.000	2^2	54	2,916	7.348	$2 \cdot 3^3$
5	25	2.236	5	55	3,025	7.416	$5 \cdot 11$
6	36	2.449	$2 \cdot 3$	56	3,136	7.483	$2^3 \cdot 7$
7	49	2.646	7	57	3,249	7.550	$3 \cdot 19$
8	64	2.828	2^3	58	3,364	7.616	$2 \cdot 29$
9	81	3.000	3^2	59	3,481	7.681	59
10	100	3.162	$2 \cdot 5$	60	3,600	7.746	$2^2 \cdot 3 \cdot 5$
11	121	3.317	11	61	3,721	7.810	61
12	144	3.464	$2^2 \cdot 3$	62	3,844	7.874	$2 \cdot 31$
13	169	3.606	13	63	3,969	7.937	$3^2 \cdot 7$
14	196	3.742	$2 \cdot 7$	64	4,096	8.000	2^6
15	225	3.873	$3 \cdot 5$	65	4,225	8.062	$5 \cdot 13$
16	256	4.000	2^4	66	4,356	8.124	$2 \cdot 3 \cdot 11$
17	289	4.123	17	67	4,489	8.185	67
18	324	4.243	$2 \cdot 3^2$	68	4,624	8.246	$2^2 \cdot 17$
19	361	4.359	19	69	4,761	8.307	$3 \cdot 23$
20	400	4.472	$2^2 \cdot 5$	70	4,900	8.367	$2 \cdot 5 \cdot 7$
21	441	4.583	$3 \cdot 7$	71	5,041	8.426	71
22	484	4.690	$2 \cdot 11$	72	5,184	8.485	$2^3 \cdot 3^2$
23	529	4.796	23	73	5,329	8.544	73
24	576	4.899	$2^3 \cdot 3$	74	5,476	8.602	$2 \cdot 37$
25	625	5.000	5^2	75	5,625	8.660	$3 \cdot 5^2$
26	676	5.099	$2 \cdot 13$	76	5,776	8.718	$2^2 \cdot 19$
27	729	5.196	3^3	77	5,929	8.775	$7 \cdot 11$
28	784	5.292	$2^2 \cdot 7$	78	6,084	8.832	$2 \cdot 3 \cdot 13$
29	841	5.385	29	79	6,241	8.888	79
30	900	5.477	$2 \cdot 3 \cdot 5$	80	6,400	8.944	$2^4 \cdot 5$
31	961	5.568	31	81	6,561	9.000	3^4
32	1,024	5.657	2^5	82	6,724	9.055	$2 \cdot 41$
33	1,089	5.745	$3 \cdot 11$	83	6,889	9.110	83
34	1,156	5.831	$2 \cdot 17$	84	7,056	9.165	$2^2 \cdot 3 \cdot 7$
35	1,225	5.916	$5 \cdot 7$	85	7,225	9.220	$5 \cdot 17$
36	1,296	6.000	$2^2 \cdot 3^2$	86	7,396	9.274	$2 \cdot 43$
37	1,369	6.083	37	87	7,569	9.327	$3 \cdot 29$
38	1,444	6.164	$2 \cdot 19$	88	7,744	9.381	$2^3 \cdot 11$
39	1,521	6.245	$3 \cdot 13$	89	7,921	9.434	89
40	1,600	6.325	$2^3 \cdot 5$	90	8,100	9.487	$2 \cdot 3^2 \cdot 5$
41	1,681	6.403	41	91	8,281	9.539	$7 \cdot 13$
42	1,764	6.481	$2 \cdot 3 \cdot 7$	92	8,464	9.592	$2^2 \cdot 23$
43	1,849	6.557	43	93	8,649	9.644	$3 \cdot 31$
44	1,936	6.633	$2^2 \cdot 11$	94	8,836	9.695	$2 \cdot 47$
45	2,025	6.708	$3^2 \cdot 5$	95	9,025	9.747	$5 \cdot 19$
46	2,116	6.782	$2 \cdot 23$	96	9,216	9.798	$2^5 \cdot 3$
47	2,209	6.856	47	97	9,409	9.849	97
48	2,304	6.928	$2^4 \cdot 3$	98	9,604	9.899	$2 \cdot 7^2$
49	2,401	7.000	7^2	99	9,801	9.950	$3^2 \cdot 11$
50	2,500	7.071	$2 \cdot 5^2$	100	10,000	10.000	$2^2 \cdot 5^2$

ANSWERS TO ODD-NUMBERED PROBLEMS

CHAPTER 1

Section 1.2 [pages 8–10]

1. $\sqrt{65}$ **3.** 2 **5.** $\sqrt{37}/2$ **7.** $\sqrt{6}$ **9.** Collinear **11.** Not collinear

13. Collinear **15.** Right triangle **17.** Right triangle **19.** $-3, 5$ **21.** 2, 3

23. Since $(5, 2)$ is equidistant from A and B $(d = \sqrt{17})$, it is on the perpendicular bisector.

25. Lettering the vertices A, B, C, and D, respectively, it follows that $\overline{AB} = \overline{CD} = 3$ and $\overline{BC} = \overline{AD} = \sqrt{10}$.

27. $(0, -1)$ inside; $(1, 7)$, $(2, 0)$, $(-5, 7)$, $(-5, -1)$, $(-6, 6)$ on; $(-3, 8)$, $(4, 2)$ outside

29. $A = R - A_1 - A_2 - A_3$

$$= (x_2 - x_1)(y_3 - y_2) - \frac{1}{2}(x_2 - x_1)(y_1 - y_2) - \frac{1}{2}(x_2 - x_3)(y_3 - y_2) - \frac{1}{2}(x_3 - x_1)(y_3 - y_1)$$

$$= \frac{1}{2}(2x_2y_3 - 2x_2y_2 - 2x_1y_3 + 2x_1y_2 - x_2y_1 + x_2y_2 + x_1y_1 - x_1y_2 - x_2y_3 + x_2y_2$$

$$+ x_3y_3 - x_3y_2 - x_3y_3 + x_3y_1 + x_1y_3 - x_1y_1)$$

$$= \frac{1}{2}(x_1y_2 + x_2y_3 + x_3y_1 - x_1y_3 - x_2y_1 - x_3y_2)$$

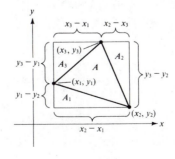

If the points are numbered in clockwise order, rather than the counterclockwise order given, the result will be the negative of the one above. Thus the absolute value is needed.

31. Placing the axes as shown, we see that $\overline{AB} = a + b$; but $\overline{AC} = \sqrt{a^2 + c^2} > a$ and $\overline{BC} = \sqrt{b^2 + c^2} > b$. Thus $\overline{AC} + \overline{BC} > \overline{AB}$.

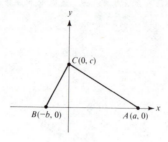

Section 1.3 [pages 15–16]

1. $(4, 3)$ **3.** $(16/5, -9/5)$ **5.** $(17, 22)$ **7.** $(2, 1)$ **9.** $(7/2, 1)$ **11.** $(12, -4)$

13. $(0, -7)$ **15.** $(3, 0)$ **17.** $(14/5, 7/5)$ **19.** $(6, -7)$ **21.** $(4/3, 5/3)$ **23.** $(2, -1/2)$

25. Placing the axes as shown, we see that the midpoints of the three sides are $D = (0, 0)$, $E = (a/2, c/2)$, and $F = (-a/2, c/2)$. Thus $\overline{CE} = \overline{ED} = \overline{DF} = \overline{FC} = \sqrt{a^2 + c^2}/2$.

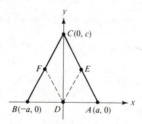

27. $7, -9$ **29.** $(13, -2), (-9, -10), (-5, 4)$

31. Placing the axes as shown, we see that $\overline{AB} = \overline{CD} = a$ and $\overline{BC} = \overline{DA} = \sqrt{b^2 + c^2}$. Thus $\overline{AB}^2 + \overline{BC}^2 + \overline{CD}^2 + \overline{DA}^2 = 2(a^2 + b^2 + c^2)$. Since $\overline{AC} = \sqrt{(a - b)^2 + c^2}$ and $\overline{BD} = \sqrt{(a + b)^2 + c^2}$, it follows that $\overline{AC}^2 + \overline{BD}^2 = (a - b)^2 + c^2 + (a + b)^2 + c^2 = 2(a^2 + b^2 + c^2)$.

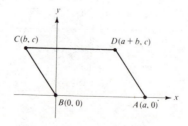

Section 1.5 [pages 22–24]

1. 5/3, 59°   **3.** 2/3, 34°   **5.** No slope, 90°   **7.** 1, 45°

9.

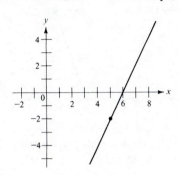

11.

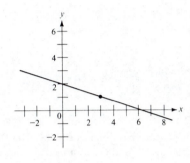

13.

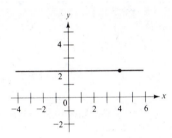

15. Parallel   **17.** None   **19.** Coincident   **21.** Perpendicular   **23.** 62.5%

25. 14/3   **27.** 10/7, 11   **29.** 9

31. Labelling the points A, B, C, and D, respectively, we have $m_{AB} = m_{CD} = -1$ and $m_{bc} = m_{ad} = 1$. Thus $AB \parallel CD$ and $BC \parallel AD$, implying that it is a parallelogram. $AB \perp BC$, implying that it is a rectangle. Since m_{AC} does not exist and $m_{BD} = 0$, $AC \perp BD$, implying that it is a square.

33. $m_{AC} = \dfrac{a - 0}{0 - a} = -1$; $m_{BD} = \dfrac{a - 0}{a - 0} = 1$; $AC \perp BD$.

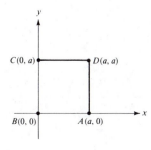

35. Since $WXYZ$ is a rhombus,
$\overline{WX} = \overline{WZ}$,
$a = \sqrt{b^2 + c^2}$,
$a^2 = b^2 + c^2$.
$m_{WY} = \dfrac{c}{a+b}$
$m_{XZ} = \dfrac{c}{b-a} = -\dfrac{c}{a-b}$
$m_{WY} \cdot m_{XZ} = -\dfrac{c^2}{a^2-b^2} = -\dfrac{c^2}{c^2} = -1$
$WY \perp XZ$

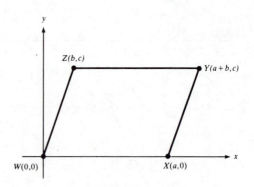

Clearly a and b must have opposite signs. Thus $a = -b$ (or $b = -a$). Now it is clear that the origin, O, is the midpoint of AB; and $\overline{OC}$ is a median. Since $\overline{OC} \perp \overline{AB}$, $\overline{OC}$ is also an altitude.

Section 1.6 [pages 29–30]

1. $135°$ **3.** $135°$ **5.** $6°$ **7.** $56°$ **9.** $12°$ **11.** $60°$ **13.** $135°$ **15.** $27°$

17. $-7 - 5\sqrt{2}$ **19.** $1 + \sqrt{2}$ **21.** $-8 + \sqrt{65}$ **23.** 2 **25.** $37°, 72°, 72°$ **27.** $1/5$

29. 2 **31.** -2 **33.** ± 1

Section 1.7 [pages 32–33]

1. Function

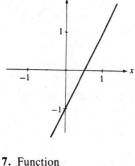

3. Function

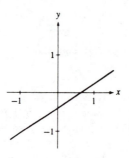

5. Function

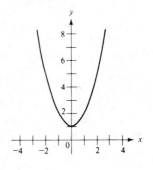

7. Function

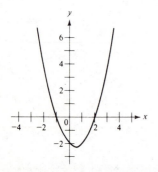

9. Not a function

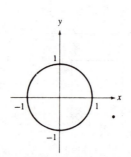

11. Function

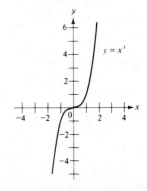

13.

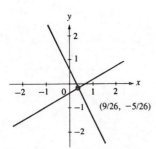

(9/26, −5/26)

15.

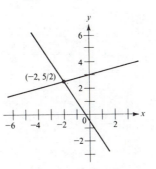

(−2, 5/2)

17. Not a function

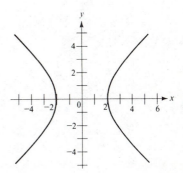

19. Not a function

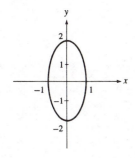

21. Function

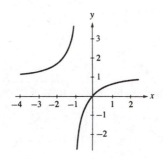

23. Not a function

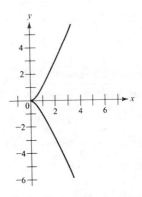

25.

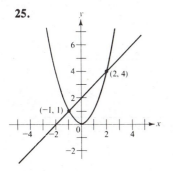

(2, 4)

(−1, 1)

27.

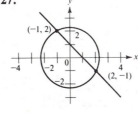

(−1, 2)

(2, −1)

29.

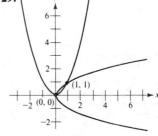

(1, 1)

(0, 0)

31.

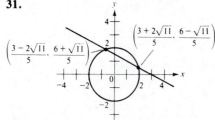

33.

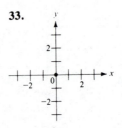

35.

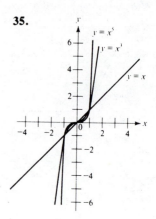

37.

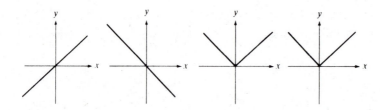

Section 1.8 [pages 35–36]

1. $14x + 8y - 69 = 0$ **3.** $x^2 + y^2 - 10x - 16y + 80 = 0$ **5.** $2x - y - 8 = 0$

7. $5x - 2y - 19 = 0$ **9.** $x^2 + y^2 - 9x + 18 = 0$ **11.** $x^2 + y^2 - x - 9y + 18 = 0$

13. $y^2 - 8x + 16 = 0$ **15.** $3x^2 - y^2 + 6x - 9 = 0$ **17.** $25x^2 + 9y^2 = 225$

19. $x^2 + y^2 = 16$ **21.** $8x^2 - y^2 = 8$ **23.** $xy = \pm 4$

Review Problems [pages 36–37]

1. $(22/7, -9/7)$ **3.** $\sqrt{221}/2, \sqrt{41}, \sqrt{185}/2$ **5.** 2   **7.** $-8/5, -1/2, 5$ **9.** $3x - 2y - 8 = 0$

11. $-1, -9$ **13.** $(2/3, 1)$ **15.** $(3, -2)$ **17.** $(1, 4), (4, 6)$ **19.** $x = y \cot y$

CHAPTER 2

Section 2.1 [pages 46–48]

1. $-6\mathbf{i} - 2\mathbf{j}$ **3.** $7\mathbf{i} + \mathbf{j}$ **5.** $-\mathbf{i} + 5\mathbf{j}$ **7.** $4\mathbf{i} - 3\mathbf{j}$ **9.** $(3/\sqrt{10})\mathbf{i} - (1/\sqrt{10})\mathbf{j}$

11. $-(1/\sqrt{5})\mathbf{i} + (2/\sqrt{5})\mathbf{j}$ **13.** $(1/\sqrt{5})\mathbf{i} + (2/\sqrt{5})\mathbf{j}$ **15.** $-(4/5)\mathbf{i} + (3/5)\mathbf{j}$

17. $B = (4, 3)$ **19.** $A = (5, 0)$ **21.** $B = (6, 0)$ **23.** $A = (1, 3)$

25. $4\mathbf{i} + \mathbf{j}, 2\mathbf{i} - 3\mathbf{j}$ **27.** $3\mathbf{i} + \mathbf{j}, -1 - 3\mathbf{j}$ **29.** $3\mathbf{i} + \mathbf{j}, -\mathbf{i} - 7\mathbf{j}$

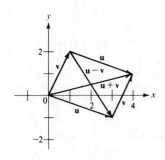   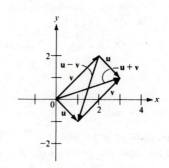   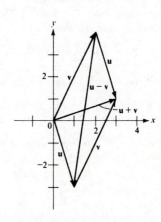

31. $5\mathbf{i} - 3\mathbf{j}, \mathbf{i} + \mathbf{j}$ **33.**

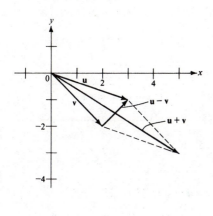   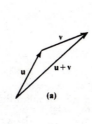

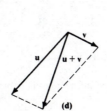   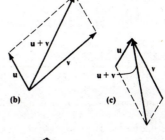

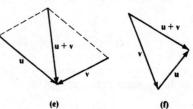

35. $A = (5/2, -3/2), B = (11/2, 7/2)$ **37.** $A = (3/2, -1/2), B = (5/2, 1/2)$ **39.** $8\mathbf{i} + 5\mathbf{j}$

41. $-5\mathbf{i} - 3\mathbf{j}$

43. **a.** $\overrightarrow{AB} \equiv \overrightarrow{AB}$ follows directly from the definition of equivalence.

   b. Since any true statement about $\overrightarrow{AB}$ and $\overrightarrow{CD}$ is also a true statement about $\overrightarrow{CD}$ and $\overrightarrow{AB}$, it follows that if $\overrightarrow{AB} \equiv \overrightarrow{CD}$, then $\overrightarrow{CD} \equiv \overrightarrow{AB}$.

c. Suppose $\overrightarrow{AB} \equiv \overrightarrow{CD}$ and $\overrightarrow{CD} \equiv \overrightarrow{EF}$. If $\overrightarrow{AB}$ has length 0, then $\overrightarrow{CD}$ has length 0; if $\overrightarrow{CD}$ has length 0, then $\overrightarrow{EF}$ has length 0. Thus $\overrightarrow{AB} \equiv \overrightarrow{EF}$. Suppose AB is not of length 0. Then $\overrightarrow{AB}$ and $\overrightarrow{CD}$ have the same length, both lie on the same or parallel lines, and they are directed in the same way. A similar statement can be made about $\overrightarrow{CD}$ and $\overrightarrow{EF}$. Thus $\overrightarrow{AB}$ and $\overrightarrow{EF}$ have the same length, both lie on the same or parallel lines, and both are directed in the same way. Thus $\overrightarrow{AB} \equiv \overrightarrow{EF}$.

45. a. $a_1\mathbf{i} + b_1\mathbf{j}$ is represented by $\overrightarrow{OP}_1$ from $(0, 0)$ to (a_1, b_1). $a_2\mathbf{i} + b_2\mathbf{j}$ is represented by $\overrightarrow{OP}_2$ from $(0,0)$ to (a_2, b_2), or by $\overrightarrow{P_1P_3}$ from (a_1, b_1) to $(a_1 + a_2, b_1 + b_2)$. Hence the sum is represented by $\overrightarrow{OP}_3$ from $(0, 0)$ to $(a_1 + a_2, b_1 + b_2)$, giving $(a_1\mathbf{i} + b_1\mathbf{j}) + (a_2\mathbf{i} + b_2\mathbf{j}) = (a_1 + a_2)\mathbf{i} + (b_1 + b_2)\mathbf{j}$.

b. Let $(a_1\mathbf{i} + b_1\mathbf{j}) - (a_2\mathbf{i} + b_2\mathbf{j}) = a\mathbf{i} + b\mathbf{j}$. Then $a_1\mathbf{i} + b_1\mathbf{j} = (a_2\mathbf{i} + b_2\mathbf{j}) + (a\mathbf{i} + b\mathbf{j}) = (a_2 + a)\mathbf{i} + (b_2 + b)\mathbf{j}$. Thus $a_1 = a_2 + a$ or $a = a_1 - a_2$, and $b_1 = b_2 + b$ or $b = b_1 - b_2$.

c. $\mathbf{v} = a\mathbf{i} + b\mathbf{j}$ is represented by $\overrightarrow{OP}$ from $(0, 0)$ to (a, b); $\mathbf{w} = da\mathbf{i} + db\mathbf{j}$ is represented by $\overrightarrow{OQ}$ from $(0, 0)$ to (da, db). Clearly these points lie on the same line; so $\mathbf{w}$ is in the same direction as or opposite direction from $\mathbf{v}$, depending upon the sign of d. Furthermore, $|\mathbf{w}| = \sqrt{d^2a^2 + d^2b^2} = \sqrt{d^2(a^2 + b^2)} = |d| \sqrt{a^2 + b^2} = |d| \, |\mathbf{v}|$.

d. $a\mathbf{i} + b\mathbf{j}$ is represented by $\overrightarrow{OP}$ from $(0, 0)$ to (a, b). Its length is $\sqrt{(a - 0)^2 + (b - 0)^2} = \sqrt{a^2 + b^2}$.

47. Clearly if $a = c$ and $b = d$, then $a\mathbf{i} + b\mathbf{j} = c\mathbf{i} + d\mathbf{j}$. Suppose $a\mathbf{i} + b\mathbf{j} = c\mathbf{i} + d\mathbf{j}$, $(a - c)\mathbf{i} + (b - d)\mathbf{j} = \mathbf{0}$. This vector has length 0. Therefore $(a - c)^2 + (b - d)^2 = 0$.
$$a - c = 0 \quad \text{and} \quad b - d = 0$$
$$a = c \quad \text{and} \quad b = d$$

Section 2.2 [pages 53–55]

1. Arccos $1/5\sqrt{2} = 82°$ **3.** $90°$ **5.** $90°$ **7.** Arccos $3/\sqrt{34} = 59°$ **9.** 1, not orthogonal

11. 4, not orthogonal   **13.** -1, not orthogonal   **15.** 3, not orthogonal   **17.** $(1/2)\mathbf{i} + (1/2)\mathbf{j}$

19. $-(6/5)\mathbf{i} + (12/5)\mathbf{j}$ **21.** $(2/5)\mathbf{i} + (1/5)\mathbf{j}$ **23.** $(6/5)\mathbf{i} - (3/5)\mathbf{j}$ **25.** $30\sqrt{2}$

27. 3   **29.** -8 **31.** $-1/2$ **33.** -1 **35.** $\pm 2\sqrt{3}$ **37.** $(8 - 5\sqrt{3})/11$

39. $-(2/17)\mathbf{i} - (8/17)\mathbf{j}, -(19/17)\mathbf{i} - (76/17)\mathbf{j}$ **41.** $-(2/5)\mathbf{i} - (4/5)\mathbf{j}, (3/5)\mathbf{i} + (6/5)\mathbf{j}$

43. Since the zero vector is orthogonal to every other vector and $\mathbf{0} \cdot \mathbf{v} = 0$, that case is trivial. Suppose now that neither $\mathbf{u}$ nor $\mathbf{v}$ is the zero vector. If $\mathbf{u}$ and $\mathbf{v}$ are orthogonal, then
$$\frac{\mathbf{u} \cdot \mathbf{v}}{|\mathbf{u}| \cdot |\mathbf{v}|} = \cos 90° = 0$$
Thus $\mathbf{u} \cdot \mathbf{v} = 0$. On the other hand, if $\mathbf{u} \cdot \mathbf{v} = 0$, $\cos \theta = 0$ and $\theta = 90°$.

45. $\cos \theta = \dfrac{a_1a_2 + b_1b_2}{|\mathbf{u}| \cdot |\mathbf{v}|} = \dfrac{\mathbf{u} \cdot \mathbf{v}}{|\mathbf{u}| \cdot |\mathbf{v}|}$
$\mathbf{u} \cdot \mathbf{v} = |\mathbf{u}| \cdot |\mathbf{v}| \cos \theta$
$\mathbf{v} \cdot \mathbf{v} = a_1^2 + b_1^2 = |\mathbf{v}|^2$

Section 2.3 [pages 58–60]

1. $6\mathbf{i} + 2\mathbf{j}$

3. $5\sqrt{5}$ lb to the right and inclined upward at an angle of $63°$ with the horizontal

5. $\sqrt{25 - 12\sqrt{2}}$ lb to the right and inclined downward at an angle of 3° with the horizontal

7. $3\sqrt{3}\mathbf{i} + 3\mathbf{j}$ **9.** $3\sqrt{2}/2\mathbf{i} - 3\sqrt{2}/2\mathbf{j}$ **11.** $1.638\mathbf{i} + 1.147\mathbf{j}$ **13.** $\mathbf{j}$

15. $\sqrt{29}$ lb to the right and inclined downward at an angle of 68° with the horizontal

17. $\sqrt{34 + 15\sqrt{2}}$ lb to the left and inclined downward at an angle of 28° with the horizontal

19. $4\sqrt{2}\mathbf{i} + (4\sqrt{2} - 9)\mathbf{j}$ **21.** $(3\sqrt{2} - 8)\mathbf{i} + 3\sqrt{2}\mathbf{j}$ **23.** $(4\sqrt{3} - 5)\mathbf{i} + (5\sqrt{3} - 24)\mathbf{j}$

25. $\mathbf{q} = 1/2\mathbf{u} + \mathbf{v} + 1/2\mathbf{w}$
$\mathbf{q} = -1/2\mathbf{u} + \mathbf{p} - 1/2\mathbf{w}$
Adding, we have
$$2\mathbf{q} = \mathbf{p} + \mathbf{v}, \quad \mathbf{q} = \frac{\mathbf{p} + \mathbf{v}}{2}$$

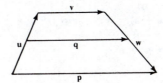

27. $\mathbf{a} = \mathbf{u} - \mathbf{v}, \quad \mathbf{b} = \mathbf{u} + \mathbf{v}$
$\mathbf{a} \cdot \mathbf{b} = (\mathbf{u} - \mathbf{v}) \cdot (\mathbf{u} + \mathbf{v})$
$\quad = \mathbf{u} \cdot \mathbf{u} + \mathbf{u} \cdot \mathbf{v} - \mathbf{v} \cdot \mathbf{u} - \mathbf{v} \cdot \mathbf{v}$
$\quad = |\mathbf{u}|^2 - |\mathbf{v}|^2 = 0$
$|\mathbf{u}| = |\mathbf{v}|$

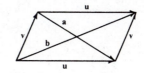

29. $\mathbf{u} = \mathbf{a} + \mathbf{d} = \mathbf{a} - \mathbf{b}, \quad \mathbf{v} = \mathbf{a} + \mathbf{b}$
$|\mathbf{u}|^2 + |\mathbf{v}|^2 = |\mathbf{a} - \mathbf{b}|^2 + |\mathbf{a} + \mathbf{b}|^2$
$\quad = |\mathbf{a}|^2 + |-\mathbf{b}|^2 - 2|\mathbf{a}||-\mathbf{b}| \cos \beta$
$\quad + |\mathbf{a}|^2 + |\mathbf{b}|^2 - 2|\mathbf{a}||\mathbf{b}| \cos \alpha$

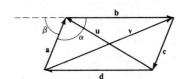

31. $|\mathbf{u}| = |\mathbf{v}|, \quad \mathbf{v} = \mathbf{u} + \mathbf{w}$
$$\cos \alpha = \frac{\mathbf{u} \cdot (-\mathbf{w})}{|\mathbf{u}||-\mathbf{w}|} = \frac{-\mathbf{u} \cdot \mathbf{w}}{|\mathbf{u}||\mathbf{w}|}$$
$$\cos \beta = \frac{\mathbf{w} \cdot \mathbf{v}}{|\mathbf{w}||\mathbf{v}|} = \frac{\mathbf{w} \cdot (\mathbf{u} + \mathbf{w})}{|\mathbf{w}||\mathbf{v}|}$$
$$= \frac{\mathbf{w} \cdot \mathbf{u} + \mathbf{w} \cdot \mathbf{w}}{|\mathbf{w}||\mathbf{u}|}$$
$|\mathbf{w}|^2 = |\mathbf{u}|^2 + |\mathbf{v}|^2 - 2|\mathbf{u}||\mathbf{v}| \cos (180° - \alpha - \beta)$
$\quad = |\mathbf{u}|^2 + |\mathbf{v}|^2 - 2\mathbf{u} \cdot \mathbf{v} = 2|\mathbf{u}|^2 - 2\mathbf{u} \cdot (\mathbf{u} + \mathbf{w})$
$\quad = 2\mathbf{u} \cdot \mathbf{u} - 2\mathbf{u} \cdot \mathbf{u} - 2\mathbf{u} \cdot \mathbf{w} = -2\mathbf{u} \cdot \mathbf{w}$
$$\cos \beta = \frac{\mathbf{u} \cdot \mathbf{w} + \mathbf{w} \cdot \mathbf{w}}{|\mathbf{w}||\mathbf{u}|} = \frac{\mathbf{u} \cdot \mathbf{w} - 2\mathbf{u} \cdot \mathbf{w}}{|\mathbf{w}||\mathbf{u}|} = -\frac{\mathbf{u} \cdot \mathbf{w}}{|\mathbf{u}||\mathbf{w}|} = \cos \alpha$$

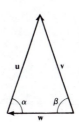

33. $\mathbf{a} + \mathbf{b} + \mathbf{c} = 0$
$\mathbf{u} = -\mathbf{a} - 1/2\mathbf{c}$
$\mathbf{v} = \mathbf{a} + 1/2\mathbf{b}$
$s\mathbf{v} = \mathbf{a} + r\mathbf{u}$
$s(\mathbf{a} + 1/2\mathbf{b}) = \mathbf{a} + r(-\mathbf{a} - 1/2\mathbf{c}), \quad s\mathbf{a} + s/2\mathbf{b} = (1 - r)\mathbf{a} - r/2\mathbf{c}$
$s/2\mathbf{b} + r/2\mathbf{c} = (1 - r - s)\mathbf{a} = (1 - r - s)(-\mathbf{b} - \mathbf{c})$
$(s/2 + 1 - r - s)\mathbf{b} + (r/2 + 1 - r - s)\mathbf{c} = 0$
$1 - r - s/2 = 1 - r/2 - s \doteq 0, \quad 2r + s = r + 2s = 2$
$2r + s = 2, \quad r + 2s = 2, \quad r = s = 2/3$

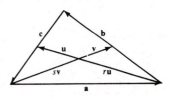

Review Problems [pages 60–61]

1. $\dfrac{1}{\sqrt{13}}\mathbf{i} - \dfrac{3}{\sqrt{13}}\mathbf{j}$ **3.** $A = (1, 3/2), B = (5, 1/2)$ **5.** 8   **7.** $3/5\mathbf{i} - 1/5\mathbf{j}$ **9.** 3/2

11. $(4\sqrt{3} - 5\sqrt{2})\mathbf{i} + (10 - 5\sqrt{2})\mathbf{j}$

13. $\mathbf{u} = \mathbf{a} + \mathbf{b},\quad \mathbf{v} = \mathbf{a} + \mathbf{d}$
$\mathbf{d} = -n\mathbf{b},\quad r\mathbf{v} = \mathbf{d} + s\mathbf{u}$
$r(\mathbf{a} + \mathbf{d}) = -n\mathbf{b} + s(\mathbf{a} + \mathbf{b})$
$r(\mathbf{a} - n\mathbf{b}) = -n\mathbf{b} + s(\mathbf{a} + \mathbf{b}),\quad (r - s)\mathbf{a} = (rn - n + s)\mathbf{b}$
$r - s = 0,\quad rn - n + s = 0$
$rn + r - n = 0,\quad r(1 + n) = n$
$r = \dfrac{n}{1 + n},\quad s = \dfrac{n}{1 + n},\quad 1 - r = 1 - s = \dfrac{1}{1 + n}$
$1 - r : r = 1 - s : s = 1 : n$

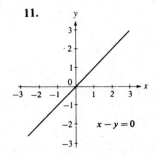

15. $15\sqrt{3}$ lb, 15 lb

CHAPTER 3

Section 3.1 [pages 66–68]

1.

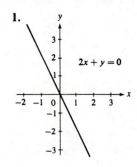

$2x + y = 0$

3.

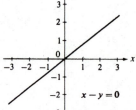

$x - y = 0$

5.

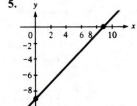

$x - y - 9 = 0$

7.

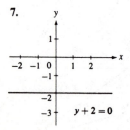

$y + 2 = 0$

9.

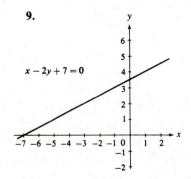

$x - 2y + 7 = 0$

11.

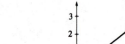

$x - y = 0$

13.

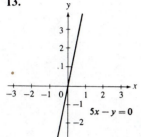

15.

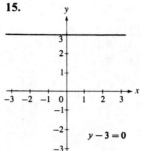

17. $2x + y - 6 = 0, x - 2y - 3 = 0,$
$3x - y + 1 = 0$

19. $x - 2y - 3 = 0, 2x + y - 6 = 0,$
$x + 3y - 3 = 0$

21. $5x - 6y + 2 = 0, 2x + 9y - 3 = 0,$
$7x + 3y - 1 = 0$

25. $3x - 2y + 5 = 0$

27. The given circles do not intersect.   **29.** $x - 3y + 3 = 0$

31. $7x - y - 20 = 0, 7x - y - 20 = 0$ **33.** $9C - 5F + 160 = 0$

23.

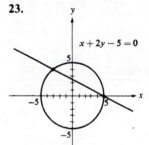

35. $(x_1, y_1),\quad (x_2, y_2),\quad m = \dfrac{y_1 - y_2}{x_1 - x_2}$

$$y - y_1 = \frac{y_1 - y_2}{x_1 - x_2}(x - x_1)$$
$$y(x_1 - x_2) - y_1(x_1 - x_2) = x(y_1 - y_2) - x_1(y_1 - y_2)$$
$$x(y_1 - y_2) - y(x_1 - x_2) + (x_1 y_2 - x_2 y_1) = 0$$

If the determinant is expanded by minors with respect to the first row, it is seen to be the same as the above result.

37. Since no two of the three lines are parallel, $A_1/B_1 \neq A_2/B_2, A_2/B_2 \neq A_3/B_3, A_3/B_3 \neq A_1/B_1$. Thus $A_1 B_2 \neq B_1 A_2$ and $A_1 B_2 - B_1 A_2 \neq 0$. Solving the first two equations simultaneously, we have

$$x = -\frac{\begin{vmatrix} C_1 & B_1 \\ C_2 & B_2 \end{vmatrix}}{\begin{vmatrix} A_1 & B_1 \\ A_2 & B_2 \end{vmatrix}}$$

$$y = -\frac{\begin{vmatrix} A_1 & C_1 \\ A_2 & C_2 \end{vmatrix}}{\begin{vmatrix} A_1 & B_1 \\ A_2 & B_2 \end{vmatrix}}$$

If the lines are concurrent, the above point of intersection is on the third line. Thus

$$-A_3 \frac{\begin{vmatrix} C_1 & B_1 \\ C_2 & B_2 \end{vmatrix}}{\begin{vmatrix} A_1 & B_1 \\ A_2 & B_2 \end{vmatrix}} - B_3 \frac{\begin{vmatrix} A_1 & C_1 \\ A_2 & C_2 \end{vmatrix}}{\begin{vmatrix} A_1 & B_1 \\ A_2 & B_2 \end{vmatrix}} + C_3 = 0$$

$$A_3 \begin{vmatrix} B_1 & C_1 \\ B_2 & C_2 \end{vmatrix} - B_3 \begin{vmatrix} A_1 & C_1 \\ A_2 & C_2 \end{vmatrix} + C_3 \begin{vmatrix} A_1 & B_1 \\ A_2 & B_2 \end{vmatrix} = 0$$

$$\begin{vmatrix} A_1 & B_1 & C_1 \\ A_2 & B_2 & C_2 \\ A_3 & B_3 & C_3 \end{vmatrix} = 0$$

The above steps are reversible.

Section 3.2 [pages 73–74]

1.

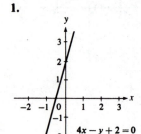

$4x - y + 2 = 0$

3.

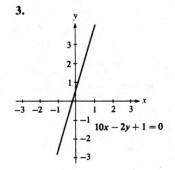

$10x - 2y + 1 = 0$

5.

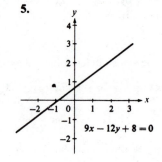

$9x - 12y + 8 = 0$

7.

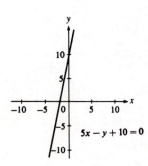

$5x - y + 10 = 0$

9.

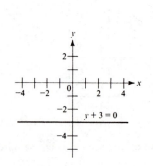

$y + 3 = 0$

11.

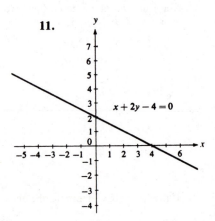

$x + 2y - 4 = 0$

13.

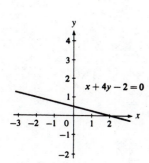

$x + 4y - 2 = 0$

15.

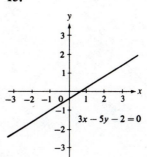

$3x - 5y - 2 = 0$

17.

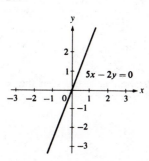

$5x - 2y = 0$

19.

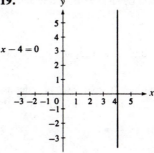

$x - 4 = 0$

21. $2x - 5y + 11 = 0$

23. $x + y - 7 = 0$

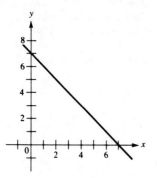

25. $x + 3y - 6 = 0; x + y - 4 = 0$

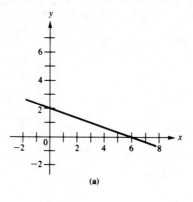

(a)

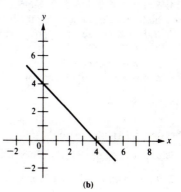

(b)

27. $4x + y - 3 = 0$ **29.** $(15/14, -5/14)$ **31.** $(-1, -7)$ **33.** $5/2$ **35.** 1

37.
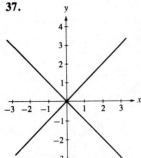

39.
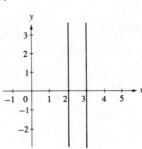

41. A representative of **v** is $\overrightarrow{OP}$, where $O = (0, 0)$ and $P = (A, B)$. The slope of OP is B/A provided $A \neq 0$. The slope of $Ax + By + C = 0$ is $-A/B$ provided $B \neq 0$. Thus they are perpendicular if neither A nor B is 0. If $A = 0$, then OP is a vertical line and $Ax + By + C = 0$ a horizontal line; if $B = 0$, OP is horizontal and $Ax + By + C = 0$ vertical.

43. $\begin{vmatrix} x & y & 1 \\ x_1 & y_1 & 1 \\ x_2 & y_2 & 1 \end{vmatrix} = 0$ is linear in x and y. By Theorem 2.5 it represents a line. The points (x_1, y_1) and (x_2, y_2)

satisfy the equation. (If two rows of a determinant are equal, the value of the determinant is 0). Thus the equation represents the line through (x_1, y_1) and (x_2, y_2).

45. $(8, 3), (3, -2), (-4, -1)$

Section 3.3 [pages 79–81]

1. $\sqrt{2}$ **3.** $9/\sqrt{41}$ **5.** $2/5$ **7.** $18/\sqrt{29}$ **9.** $10/3$ **11.** $4/\sqrt{29}$ **13.** $7/2\sqrt{5}$

15. $8/\sqrt{5}$ **17.** $13/\sqrt{10}, 26/\sqrt{29}, 26/5$ **19.** 13 **21.** $7x - 7y + 2 = 0$

23. $39x + 13y - 24 = 0$ **25.** $2(\sqrt{2} - 1)x - 2y + (4 - 3\sqrt{2}) = 0$

27. $\pm 3/\sqrt{7}$ **29.** $\pm 3/2\sqrt{2}$ **31.** $(1, 1), (-2, 2), (3, -3), (6, 6)$

33. Either $A \neq 0$ or $B \neq 0$. Suppose $B \neq 0$. If P is on the same side as the origin O, then either P and O are both above or both below the line. Thus $Ax_1 + By_1 + C$ and $A \cdot 0 + B \cdot 0 + C = C$ have the same sign. If P and O are on opposite sides of the line, then one is above and the other below the line. Thus $Ax_1 + By_1 + C$ and $A \cdot 0 + B \cdot 0 + C = C$ have opposite signs.

If $B = 0$, then $A \neq 0$ and the same argument can be used, using right and left of the line instead of above and below (see Problem 32).

35. $2(\sqrt{3} - 1)$

Section 3.4 [pages 86–88]

1. Lines through $(-1, 4)$. Does not include $x = -1$.

3. Lines with x intercept 2. Does not include $x = 2$.

5. All lines through the origin.

7. Lines having both an x intercept and a y intercept. Does not include any line through the origin.

9. Lines containing the point of intersection of $2x + 3y + 1 = 0$ and $4x + 2y - 5 = 0$. Does not include $4x + 2y - 5 = 0$.

11. Lines with y intercept twice the x intercept. Does not include any line through the origin.

13. All lines with y intercept equal to their slope. **15.** $\{3x - 5y = k \mid k \text{ real}\}$

17. $\{y - 5 = m(x - 2) \mid m \text{ real}\} \cup \{x = 2\}$

19. $\{3x - 5y + 1 + k(2x + 3y - 7) = 0 \mid k \text{ real}\} \cup \{2x + 3y - 7 = 0\}$

21. $\{Ax + By = 0 \mid A, B \text{ real}\}$ **23.** $\{Ax + By + C = 0 \mid |6A + C|/\sqrt{A^2 + B^2} = 5\}$

25. (a) $3x - 5y + 25 = 0$, (b) $5x + 3y - 49 = 0$ **27.** (a) $2x + y - 5 = 0$, (b) $x - 2y + 10 = 0$

29. $5x - y \pm 3\sqrt{26} = 0$ **31.** $15x - 8y - 43 = 0$, $3x - 4y + 1 = 0$

33. $11x - 60y - 17 = 0$, $x - 7 = 0$ **35.** $16x - 8y - 27 = 0$

37. $x - 2y - 10 = 0$, $3x + 2y - 6 = 0$

39. $x + 2y - 8 = 0$, $9x + 2y - 24 = 0$, $(11 \pm 4\sqrt{7})x - 2y - (16 \pm 8\sqrt{7}) = 0$

41. $x + y - (4 + 2\sqrt{2}) = 0$ **43.** $\sqrt{3}x - y + (6 - 4\sqrt{3}) = 0$, $\sqrt{3}x + y - (6 + 4\sqrt{3}) = 0$

Section 3.5 [pages 92–93]

1. $m = 4.17$ **3.** $k = 0.412$ **5.** $m = 2.14$, $b = 4.18$ **7.** $p = 2.40$, $q = 4.13$ **9.** $k = 11.2$

11. $\Delta H = 0.0703$ cal/mole

Review Problems [pages 93–94]

1. a. $2x - 3y + 13 = 0$, **b.** $2x - y - 6 = 0$, **c.** $3x + 3y - 1 = 0$, **d.** $x - 2 = 0$

3. a. $m = 1/4$, $a = -1$, $b = 1/4$; **b.** $m = -2/3$, $a = -5/2$, $b = -5/3$; **c.** $m = -5/2$, $a = 0$, $b = 0$; **d.** No slope, $a = -1/3$, no y intercept

5. $17/2\sqrt{10}$

7. a. $\{y + 1 = m(x - 5) \mid m \text{ real}\} \cup \{x = 5\}$ **b.** $\{2x - 3y = k \mid k \text{ real}\}$
c. $\{Ax + By + C = 0 \mid |2A + 5B + C|/\sqrt{A^2 + B^2} = 3\}$

9. $x + y - 1 = 0$ **11.** $x - 1 = 0$, $2x + 9y - 23 = 0$, $10x + 9y - 31 = 0$

13. $(0, 4)$, $(-4/3, 0)$ **15.** $y - 1 = 0$, $5x - 2y - 13 = 0$ **17.**

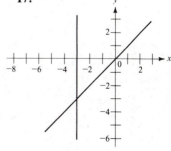

CHAPTER 4

Section 4.1 [pages 99–100]

1. $(x - 1)^2 + (y - 3)^2 = 25$
$x^2 + y^2 - 2x - 6y - 15 = 0$

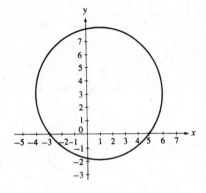

3. $(x - 5)^2 + (y + 2)^2 = 4,$
$x^2 + y^2 - 10x + 4y + 25 = 0$

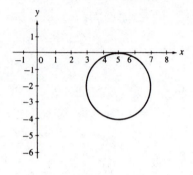

5. $(x - 1/2)^2 + (y + 3/2)^2 = 4$
$2x^2 + 2y^2 - 2x + 6y - 3 = 0$

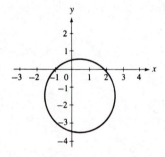

7. $(x - 4)^2 + (y + 2)^2 = 26$
$x^2 + y^2 - 8x + 4y - 6 = 0$

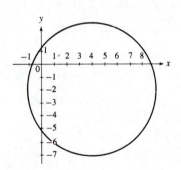

9. $x^2 + (y + 3/2)^2 = 25/4$
$x^2 + y^2 + 3y - 4 = 0$

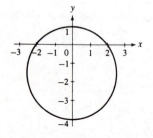

11. $(x - 3)^2 + (y - 3)^2 = 9$
$x^2 + y^2 - 6x - 6y + 9 = 0$

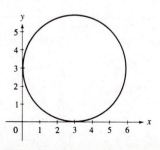

13. $(x - 4)^2 + (y - 1)^2 = 4$
$x^2 + y^2 - 8x - 2y + 13 = 0$

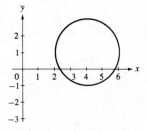

15. $(x - 4)^2 + (y + 4)^2 = 16$
$x^2 + y^2 - 8x + 8y + 16 = 0$

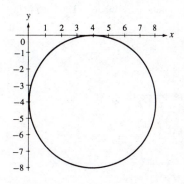

17. $(x - 1)^2 + (y - 2)^2 = 4$

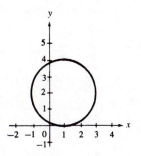

19. $(x + 3)^2 + y^2 = 25$

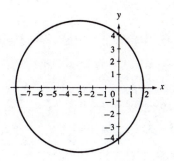

21. $(x - 1/2)^2 + (y - 3/2)^2 = 9/4$

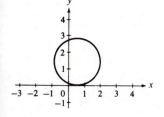

23. $(x - 4/5)^2 + (y - 2/5)^2 = 25$

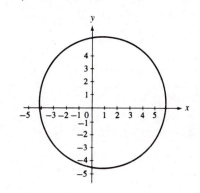

25. $(x - 1/3)^2 + (y + 1)^2 = -1/9$

27. $(x - 2/3)^2 + (y - 1/2)^2 = 0$

29. $(2, -1), (3, 3)$

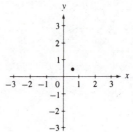

31. $(1, 4), (3, 1)$

33. No solution; the circle and line do not intersect.

35. $x + y + 1 = 0$

37. $x^2 + y^2 = a^2$

$$m_1 = \frac{y}{x + a}$$

$$m_2 = \frac{y}{x - a}$$

$$m_1 m_2 = \frac{y^2}{x^2 - a^2} = \frac{y^2}{-y^2} = -1$$

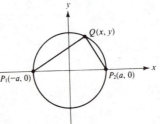

39. a. $D^2 + E^2 - 4AF > 0$ **b.** $= 0$,   **c.** < 0

Section 4.2 [pages 107–108]

1. $3x^2 + 3y^2 - 10x - 10y - 5 = 0$ **3.** $5x^2 + 5y^2 + 13x - 17y - 40 = 0$

5. $x^2 + y^2 - 13x - y = 0$ **7.** $x^2 + y^2 + 4x - 10y + 4 = 0, x^2 + y^2 - 6x + 10y + 9 = 0$

9. $x^2 + y^2 - 4x + 8y - 5 = 0$ **11.** $2x^2 + 2y^2 + 8x + 12y + 1 = 0, 2x^2 + 2y^2 - 12x - 48y + 81 = 0$

13. $x^2 + y^2 - 2x + 6y = 0$ **15.** $x^2 + y^2 + 4x + 6y - 37 = 0$

17. $13x^2 + 13y^2 + 4x + 19y - 49 = 0$

19. $x^2 + y^2 + 6x + 6y + 9 = 0, x^2 + y^2 + 6x - 6y + 9 = 0, x^2 + y^2 - 6x + 6y + 9 = 0,$
$x^2 + y^2 - 6x - 6y + 9 = 0$

21. $5x^2 + 5y^2 + 52x - 56y + 47 = 0, 5x^2 + 5y^2 - 32x + 56y - 37 = 0$

23. $x^2 + y^2 + 4x - 21 = 0$

25. Since the three points are collinear, by Problem 36, Section 3.1, the coefficient of x^2 and y^2 is 0. Now we need to show that either the coefficient of x or the coefficient of y is different from 0. Since the points are distinct, we cannot have both $x_1 = x_2 = x_3$ and $y_1 = y_2 = y_3$. Suppose y_1, y_2, and y_3 are not all the same. Suppose furthermore that $y_2 \neq y_3$.

$$\begin{vmatrix} x^2 + y^2 & y & 1 \\ x_2^2 + y_2^2 & y_2 & 1 \\ x_3^2 + y_3^2 & y_3 & 1 \end{vmatrix} = 0$$

represents a circle (since $y_2 \neq y_3$, the coefficient of $x^2 + y^2$, which is $y_2 - y_3$, is not 0) containing (x_2, y_2) and (x_3, y_3). If

$$\begin{vmatrix} x_1^2 + y_1^2 & y_1 & 1 \\ x_2^2 + y_2^2 & y_2 & 1 \\ x_3^2 + y_3^2 & y_3 & 1 \end{vmatrix} = 0,$$

then (x_1, y_1) is on this circle. But since the three points are given to be collinear, (x_1, y_1) cannot be on the circle. Thus the last determinant (which is the negative of the coefficient of x in the original determinant) is not 0.

27. $x - 3y - 7 \pm 13\sqrt{10} = 0$

Section 4.3 [pages 111–114]

1. $7x^2 + 7y^2 + 26x + 32y - 17 = 0$ **3.** $3x^2 + 3y^2 + 14x - 11y - 23 = 0$

5. $3x^2 + 3y^2 - 2x - 10y - 2 = 0$ **7.** $x^2 + y^2 - 6x + 2y + 3 = 0$

9. $13x^2 + 13y^2 + 20x - 22y - 35 = 0,\ x^2 + y^2 - 4x + 2y + 1 = 0$ **11.** $4x + 8y - 11 = 0$

13. $x^2 + y^2 + 2x + 2y - 7 + k(x^2 + y^2 - 4x - 6y + 9) = 0$
$9 + 0 + 6 + 0 - 7 + k(9 + 0 - 12 - 0 + 9) = 0$
$8 + 6k = 0$
$k = -4/3$
$x^2 + y^2 + 2x + 2y - 7 - 4/3(x^2 + y^2 - 4x - 6y + 9) = 0$
$x^2 + y^2 - 22x - 30y + 57 = 0$
$x^2 + 2x + 1 + y^2 + 2y = 7 \quad x^2 + y^2 - 4x - 6y + 9 = 0$
$(x + 1)^2 + (y + 1)^2 = 9 \quad\quad (x - 2)^2 + (y - 3)^2 = 4$

The distance between the centers is the sum of the radii. The given circles are tangent. There are many circles through this point and the point $(3, 0)$.

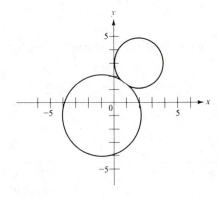

15. The family of all circles of radius 1, with center on the line $y = 1$.

17. The family of all circles in the first and third quadrants, which are tangent to both axes, together with the origin.

19. The family of all circles of radius 1 and containing the origin.

21. $\sqrt{13}$ **23.** $\sqrt{33}$

25. If P_1 and P_2 are two points on any circle, then the perpendicular bisector of $P_1 P_2$ contains the center of that circle. Thus if three circles all contain P_1 and P_2, the perpendicular bisector of $P_1 P_2$ contains the centers of all three circles.

27. $\overline{TP}^2 = \overline{CP}^2 - \overline{CT}^2$

$\qquad = (x_1 - h)^2 + (y_1 - k)^2 - r^2$

$\overline{TP} = \sqrt{(x_1 - h)^2 + (y_1 - k)^2 - r^2}$

29. Since we are dealing only with the nonnegative square roots,

$$x_1^2 + y_1^2 + Dx_1 + Ey_1 + F = x_1^2 + y_1^2 + D'x_1 + E'y_1 + F'$$

implies

$$\sqrt{x_1^2 + y_1^2 + Dx_1 + Ey_1 + F} = \sqrt{x_1^2 + y_1^2 + D'x_1 + E'y_1 + F'}.$$

Review 4 [page 114]

1. Circle with center $(5, -2)$ and radius 4.   **3.** Point $(1/3, -3/2)$ **5.** $4x + 4y - 7 = 0$

7. $25x^2 + 25y^2 - 200x - 50y + 229 = 0$ **9.** $x^2 + y^2 + 18x - 8y - 24 = 0$

11. The circle has equation $x^2 + y^2 = r^2$.

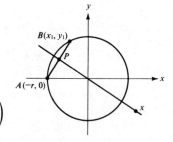

$$m_{AB} = \frac{y_1}{x_1 + r} \qquad m_{PX} = -\frac{x_1 + r}{y_1} \qquad P = \left(\frac{x_1 - r}{2}, \frac{y_1}{2}\right)$$

$$PX: y - \frac{y_1}{2} = -\frac{x_1 + r}{y_1}\left(x - \frac{x_1 - r}{2}\right)$$

This line contains the center $(0, 0)$ since substitution of $(0, 0)$ into this equation gives $x_1^2 + y_1^2 = r^2$, which is known to be true.

CHAPTER 5

Section 5.2 [page 122]

1. Axis: x axis, $V(0, 0)$, $F(4, 0)$,
    D: $x = -4$, lr $= 16$

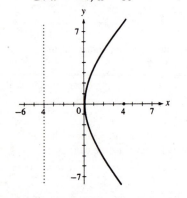

3. Axis: y axis, $V(0, 0)$,
    $F(0, 1)$, D: $y = -1$, lr $= 4$

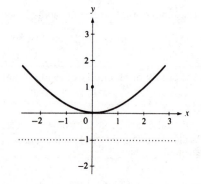

5. Axis: x axis, $V(0, 0)$, $F(5/2, 0)$,
   D: $x = -5/2$, lr = 10

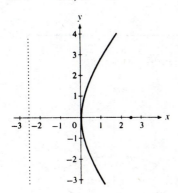

7. Axis: y axis, $V(0, 0)$, $F(0, 5/4)$,
   D: $y = -5/4$, lr = 5

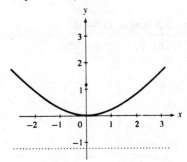

9. Axis: y axis, $V(0, 0)$, $F(0, -1/2)$,
   D: $y = 1/2$, lr = 2

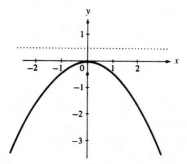

11. Axis: y axis, $V(0, 0)$, $F(0, 3/2)$,
   D: $y = -3/2$, lr = 6

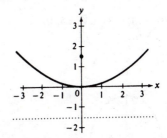

13. $y^2 = 25x$ **15.** $y^2 = 5x$, $y^2 = -5x$ **17.** $y^2 = -12x$ **19.** $x^2 = 4y/3$ **21.** $y^2 = 8x - 16$

23. $x^2 - 2xy + y^2 + 8x + 8y - 16 = 0$ **25.** $2x - y - 1 = 0$ **27.** $x + y - 4$

29. $\overline{CP} = \overline{PD}$
   $\sqrt{x^2 + (y - c)^2} = |y + c|$
   $x^2 + y^2 - 2cy + c^2 = y^2 + 2cy + c^2$
   $x^2 = 4cy$.

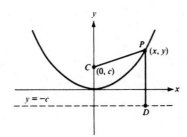

31. We assumed that $m^2x^2 + (4m^2 + 8m + 8)x + (4m^2 + 16m + 16) = 0$ is really a quadratic equation, which requires that $m \neq 0$.

Section 5.3 [pages 129–131]

1. $C(0, 0)$, $V(\pm 13, 0)$, $CV(0, \pm 5)$,
  $F(\pm 12, 0)$, lr = 50/13

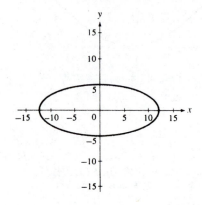

3. $C(0, 0)$, $V(\pm 5, 0)$, $CV(0, \pm 2)$
  $F(\pm \sqrt{21}, 0)$, lr = 8/5

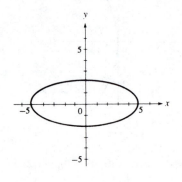

5. $C(0, 0)$, $V(0, \pm 7)$, $CV(\pm 5, 0)$,
  $F(0, \pm 2\sqrt{6})$, lr = 50/7

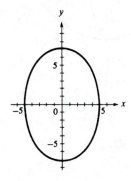

7. $C(0, 0)$, $V(0, \pm 3)$, $CV(\pm 2, 0)$,
  $F(0, \pm \sqrt{5})$, lr = 8/3

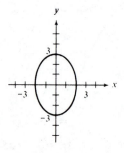

9. $C(0, 0)$, $V(0, \pm 4)$, $CV(\pm 3, 0)$,
  $F(0, \pm \sqrt{7})$, lr = 9/2

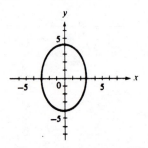

11. $x^2/144 + y^2/169 = 1$

13. $x^2/25 + y^2/10 = 1$

15. $x^2/36 + y^2/9 = 1$

17. $x^2/100 + y^2/64 = 1$

19. 0.0167

21. $4x + 3y - 11 = 0$

23. $x - 4y + 14 = 0$, $11x + 12y - 70 = 0$

25. $F(-c, 0)$,   D: $x = -a^2/c$, $e = c/a$

$$\frac{\overline{PF}}{\overline{PD}} = e$$

$$\frac{\sqrt{(x + c)^2 + y^2}}{x + a^2/c} = \frac{c}{a}$$

$$\sqrt{(x + c)^2 + y^2} = \frac{cx}{a} + a$$

$$x^2 + 2cx + c^2 + y^2 = \frac{c^2 x^2}{a^2} + 2cx + a^2$$

$$\frac{(a^2 - c^2)x^2}{a^2} + y^2 = a^2 - c^2$$

$$\frac{x^2}{a^2} + \frac{y^2}{a^2 - c^2} = 1$$

27. The ellipse has equation $b^2 x^2 + a^2 y^2 = a^2 b^2$. The tangent line has equation $y - y_0 = m(x - x_0)$. Solving the second for y and substituting into the first, we have

$$b^2 x^2 + a^2[y_0^2 + 2my_0(x - x_0) + m^2(x - x_0)^2] = a^2 b^2,$$

which simplifies to the quadratic equation

$$(b^2 + a^2 m^2)x^2 + 2a^2 m(y_0 - mx_0)x + a^2[(y_0 - mx_0)^2 - b^2] = 0.$$

Since the ellipse and its tangent line have only one point in common, this quadratic has only one solution. Thus its discriminant, $B^2 - 4AC$, is 0:

$$4a^4 m^2(y_0 - mx_0)^2 - 4(b^2 + a^2 m^2)a[(y_0 - mx_0)^2 - b^2] = 0.$$

This simplifies to

$$m^2(a^2 b^2 - b^2 x_0^2) + 2mb^2 x_0 y_0 + b^2(b^2 - y_0^2) = 0.$$

Since (x_0, y_0) is on the ellipse, it satisfies the equation of the ellipse: $b^2 x_0^2 + a^2 y_0^2 = a^2 b^2$. Thus $a^2 b^2 - b^2 x_0^2$ can be replaced by $a^2 y_0^2$, giving $a^2 m^2 y_0^2 + 2mb^2 x_0 y_0 + b^2(b^2 - y_0^2) = 0$. Similarly, we may multiply by a^2 and replace $a^2 b^2 - a^2 y_0^2$ by $b^2 x_0^2$, giving

$$a^4 m^2 y_0^2 + 2mb^2 x_0 y_0 + b^4 x_0^2 = 0.$$
$$(a^2 my_0 + b^2 x_0)^2 = 0$$
$$m = -\frac{b^2 x_0}{a^2 y_0}$$

Substituting this back into $y - y_0 = m(x - x_0)$ and simplifying, we get the desired result.

Section 5.4 [pages 137–138]

1. $C(0, 0)$, $V(\pm 4, 0)$, $F(\pm 5, 0)$,
  A: $y = \pm 3x/4$, lr $= 9/2$

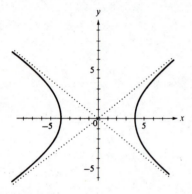

3. $C(0, 0)$, $V(0, \pm 3)$, $F(0, \pm \sqrt{13})$,
  A: $y = \pm 3x/2$, lr $= 8/3$

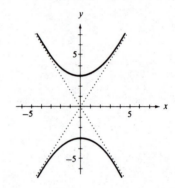

5. $C(0, 0)$, $V(\pm 12, 0)$, $F(\pm 13, 0)$,
  A: $y = \pm 5x/12$, lr $= 25/6$

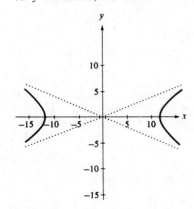

7. $C(0, 0)$, $V(0, \pm 5)$, $F(0, \pm \sqrt{34})$,
  A: $y = \pm 5x/3$, lr $= 18/5$

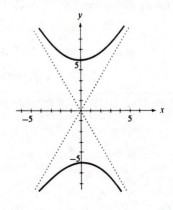

9. $C(0, 0)$, $V(\pm 1, 0)$, $F(\pm \sqrt{5}, 0)$
A: $y = \pm 2x$, lr $= 8$

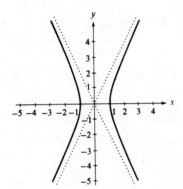

11. $C(0, 0)$, $V(\pm 3, 0)$, $F(\pm 3\sqrt{2}, 0)$,
A: $y = \pm x$, lr $= 6$

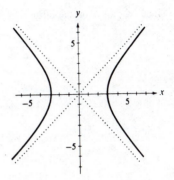

13. $C(0, 0)$, $V(0, \pm 5/2)$, $F(0, \pm \sqrt{34}/2)$
A: $y = \pm 5x/3$, lr $= 9/5$

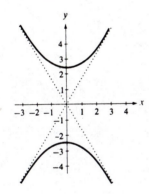

15. $x^2/4 - y^2/12 = 1$

17. $x^2/36 - y^2/16 = 1$

19. $16x^2 - 9y^2 = 144$

21. None

23. $y^2/9 - x^2/16 = 1$

25. $x^2/9 - y^2/16 = 1$

27. $52x - 15y - 144 = 0$

29. $5x - 4y - 9 = 0$

31. Using the hyperbola $x^2/a^2 - y^2/b^2 = 1$ (or $b^2x^2 - a^2y^2 = a^2b^2$), we find that its asymptotes are $bx + ay = 0$ and $bx - ay = 0$. If (x_0, y_0) is any point of the hyperbola, then the product of its distances from the asymptotes is

$$P = \frac{|bx_0 + ay_0|}{\sqrt{a^2 + b^2}} \cdot \frac{|bx_0 - ay_0|}{\sqrt{a^2 + b^2}} = \frac{|b^2x_0^2 - a^2y_0^2|}{a^2 + b^2}.$$

Since (x_0, y_0) is on the hyperbola, it satisfies the equation of the hyperbola, giving $b^2x_0^2 - a^2y_0^2 = a^2b^2$. Thus

$$P = \frac{|a^2b^2|}{a^2 + b^2} = \frac{a^2b^2}{a^2 + b^2},$$

which is seen to be independent of the point chosen.

33. $279x^2 - 121y^2 = 84,397,500$

Section 5.5 [pages 146–147]

1. See answer to Problem 27, Section 5.3.　　　**3.** See answer to Problem 27, Section 5.3.

5. Since (x_0, y_0) is on the ellipse, $x_0^2/a^2 + y_0^2/b^2 = 1$ or $b^2x_0^2 + a^2y_0^2 = a^2b^2$. In addition, $c^2 = a^2 - b^2$.

$$m_{\tan} = -\frac{b^2x_0}{a^2y_0}, \qquad m_{CP} = \frac{y_0}{x_0 - c}, \qquad m_{C'P} = \frac{y_0}{x_0 + c}$$

$$\tan \alpha = \frac{m_{CP} - m_{\tan}}{1 + m_{CP} \cdot m_{\tan}} = \frac{\dfrac{y_0}{x_0 - c} + \dfrac{b^2x_0}{a^2y_0}}{1 + \dfrac{y_0}{x_0 - c} \cdot \dfrac{-b^2x_0}{a^2y_0}}$$

$$= \frac{a^2y_0^2 + b^2x_0^2 - b^2cx_0}{a^2x_0y_0 - a^2cy_0 - b^2x_0y_0} = \frac{a^2b^2 - b^2cx_0}{c^2x_0y_0 - a^2cy_0} = -\frac{b^2}{cy_0}$$

$$\tan \beta = \frac{m_{\tan} - m_{C'P}}{1 + m_{\tan} \cdot m_{C'P}} = \frac{-\dfrac{b^2x_0}{a^2y_0} - \dfrac{y_0}{x_0 + c}}{1 - \dfrac{b^2x_0}{a^2y_0} \cdot \dfrac{y_0}{x_0 + c}}$$

$$= -\frac{b^2x_0^2 + b^2cx_0 + a^2y_0^2}{a^2x_0y_0 + a^2cy_0 - b^2x_0y_0} = -\frac{a^2b^2 + b^2cx_0}{c^2x_0y_0 + a^2cy_0} = -\frac{b^2}{cy_0}$$

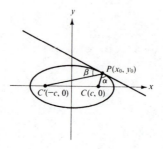

Section 5.6 [page 150]

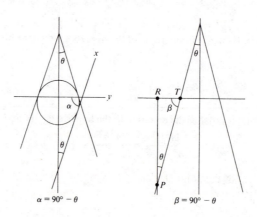

$\alpha = 90° - \theta$ 　　　　$\beta = 90° - \theta$

1. If there is only one sphere inscribed in the cone and tangent to the intersecting plane, then the intersecting plane must be parallel to a line lying on the cone and containing the apex. In relating the cone to the focus-directrix property, we see from the figures that $\alpha = \beta$. Thus $e = 1$ and the conic is a parabola.

3. The line joining the centers is the line of symmetry of the cone and contains the point of tangency. Since the intersecting plane is perpendicular to this line, the conic section must be a circle.

Review 5 [pages 150–151]

1. Axis: x axis, $V(0, 0)$, $F(-6, 0)$, D: $x = 6$, lr $= 24$

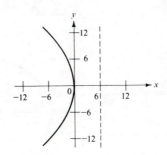

3. Axis: x axis
   $V(0, 0)$
   $F(4, 0)$
   D: $x = -4$
   lr $= 4 \cdot 4 = 16$

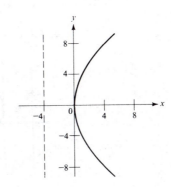

5. $C(0, 0)$
   $V(0, \pm 1)$
   $F(0, \pm \sqrt{2})$
   A: $y = \pm x$
   lr $= \dfrac{2b^2}{a} = \dfrac{2 \cdot 1}{1} = 1$

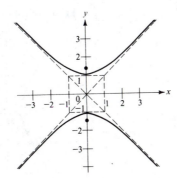

7. $y^2 = 20x$ **9.** $16x^2 + 25y^2 = 1600$ **11.** $16x^2 - 9y^2 + 576 = 0$

13. $2x + 3y - 8 = 0$

CHAPTER 6

Section 6.1 [pages 162–163]

1. $x'^2 = 8y'$

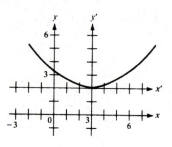

3. $(x+3)^2/25 + y^2/16 = 1,\ x'^2/25 + y'^2/16 = 1$

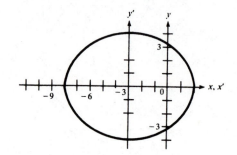

5. $(x+2)^2/4 + (y+1)^2/9 = 1,\ x'^2/4 + y'^2/9 = 1$

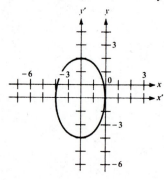

7. $(y-1)^2 = 4(x-2),\ y'^2 = 4x'$

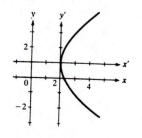

9. $(x+3)^2/4 + (y-1)^2/16 = 1,\ x'^2/4 + y'^2/16 = 1$

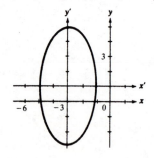

11. $(x+5)^2/4 - (y-4)^2/9 = 1,\ x'^2/4 - y'^2/9 = 1$

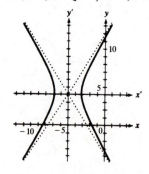

13. $9(x-4)^2+4(y+2)^2=0,\ 9x'^2+4y'^2=0$

15. $(x-\frac{1}{2})^2=y+\frac{3}{2},\ x'^2=y'$

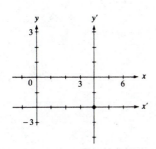

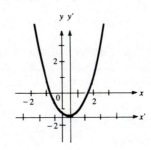

17. $(y-\frac{1}{2})^2/4-(x+\frac{3}{2})^2/16=1,\ y'^2/4-x'^2/16=1$

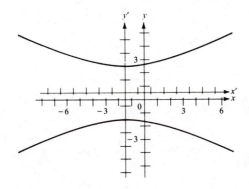

19. $25x'^2+4y'^2=-25$ **21.** $y^2-8x-10y+33=0$

23. $3x^2-y^2-12x+2y-1=0$ **25.** $9x^2-4y^2-24x-8y-184=0$

27. $x^2+36y^2-2x+216y+156=0$ **29.** $21x^2-4y^2+42x-63=0$

31. The parabola must have an equation of the form $Ax^2+Dx+Ey+F=0$ and be satisfied by the three given points. The first row of the determinant assures us of the proper form. The three points must satisfy the given equation because replacing the x and y of the determinant by the coordinates of one of the points gives us two identical rows, which implies that the determinant is zero.

Section 6.2 [pages 167–168]

1. $x'^2-2x'y'+4y'^2-5=0$ **3.** $x'^2+4x'y'-y'^2+1=0$ **5.** $x'y'+16=0$

7. $y'=x'^3-x'$ **9.** $y'=x'^3$ **11.** $y'=x'^4-4x'^3+6x'^2$ **13.** $x'^2y'-2x'^2-4=0$

15. $Ax^2 + Bxy + Cy^2 + Dx + Ey + F = 0$

$x = x' + h, \quad y = y' + k$

$A(x' + h)^2 + B(x' + h)(y' + k) + C(y' + k)^2 + D(x' + h) + E(y' + k) + F = 0$

This gives

$Ax'^2 + Bx'y' + Cy'^2 + (2Ah + Bk + D)x' + (Bh + 2Ck + E)y'$

$\qquad\qquad\qquad + (Ah^2 + Bhk + Ck^2 + Dh + Ek + F) = 0$

The first three terms above prove that A, B, and C are invariant.

Section 6.3 [page 172]

1. $\sqrt{13}x' = 6$

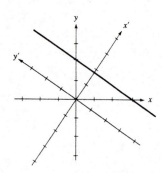

3. $x'^2 - y'^2 = 8$

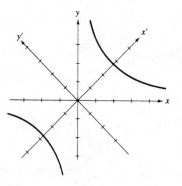

5. $\sqrt{2}y'^2 + x' = 0$

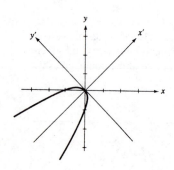

7. $x'^2 - 4y' = 0$

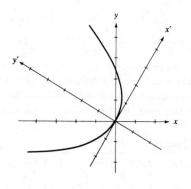

9. $17x'^2 - 9y'^2 = 8$

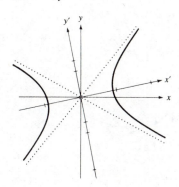

11. $17x'^2 + 7x'y' - 7y'^2 + 20 = 0$

13. $7x'^2 - 8x'y' + y'^2 - 10 = 0$

15. $3\sqrt{3}x'^2 - (6 - 8\sqrt{3})x'y'$
$\quad -(8 + 3\sqrt{3})y'^2 - 16 = 0$

17. $x = x' \cos\theta - y' \sin\theta, \quad y = x' \sin\theta + y' \cos\theta$
$\quad x'^2 \cos^2\theta - 2x'y' \sin\theta \cos\theta + y'^2 \sin^2\theta + x'^2 \sin^2\theta$
$\quad\quad + 2x'y' \sin\theta \cos\theta + y'^2 \cos^2\theta = 25$
$\quad x'^2(\cos^2\theta + \sin^2\theta) + y'^2(\sin^2\theta + \cos^2\theta) = 25$
$\quad x'^2 + y'^2 = 25$

Section 6.4 [pages 178–179]

1. $3x'^2 + y'^2 - 16y' = 0$

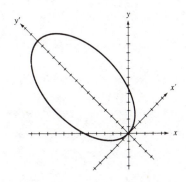

3. $4x'^2 - y'^2 - 16 = 0$

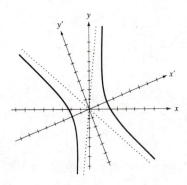

5. $5x'^2 + 20y'^2 = 20$

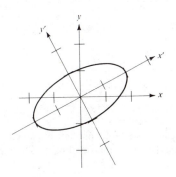

7. $4x'^2 + 9y'^2 - 36 = 0$

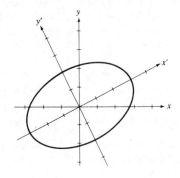

9. $x'^2 + 4x' - 5 = 0$

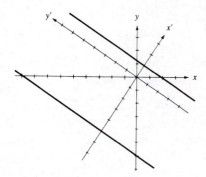

11. $y'^2 - 12x' = 0$

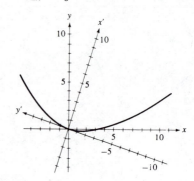

13. $x'^2 - y'^2 - 4 = 0$

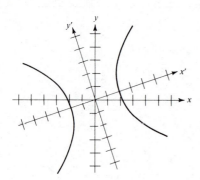

15. $x = x' \cos \theta - y' \sin \theta, \quad y = x' \sin \theta + y' \cos \theta.$

$Ax'^2 \cos^2 \theta - 2Ax'y' \sin \theta \cos \theta + Ay'^2 \sin^2 \theta + Bx'^2 \sin \theta \cos \theta$
$\quad + B(\cos^2 \theta - \sin^2 \theta)x'y' - By'^2 \sin \theta \cos \theta + Cx'^2 \sin^2 \theta$
$\quad + 2Cx'y' \sin \theta \cos \theta + Cy'^2 \cos^2 \theta + Dx' \cos \theta - Dy' \sin \theta$
$\quad + Ex' \sin \theta + Ey' \cos \theta + F = 0$

$(A \cos^2 \theta + B \sin \theta \cos \theta + C \sin^2 \theta)x'^2 + [-2A \sin \theta \cos \theta$
$\quad + B(\cos^2 \theta - \sin^2 \theta) + 2C \sin \theta \cos \theta]x'y' + (A \sin^2 \theta$
$\quad - B \sin \theta \cos \theta + C \cos^2 \theta)y'^2 + (D \cos \theta + E \sin \theta)x'$
$\quad + (-D \sin \theta + E \cos \theta)y' + F = 0$

$A' + C' = (A \cos^2 \theta + B \sin \theta \cos \theta + C \sin^2 \theta)$
$\qquad\qquad + (A \sin^2 \theta - B \sin \theta \cos \theta + C \cos^2 \theta)$
$\qquad\qquad = A(\cos^2 \theta + \sin^2 \theta) + C(\sin^2 \theta + \cos^2 \theta) = A + C$

17. $\tan \theta_1 \cdot \tan \theta_2 = \dfrac{(C - A)^2 - [(C - A)^2 + B^2]}{B^2} = -1$

Since the product is negative, $\tan \theta_1$ and $\tan \theta_2$ must have opposite signs. If we look upon $\tan \theta_1$ and $\tan \theta_2$ as the slopes of lines l_1 and l_2, respectively, then l_1 and l_2 are perpendicular. Thus their inclinations, θ_1 and θ_2, differ by an odd multiple of $90°$.

Review 6 [pages 179–180]

1. $C(1, -3)$, $V(1, -3 \pm 2)$,
$CV(1 \pm 1, -3)$, $F(1, -3 \pm \sqrt{5})$
lr = 1.

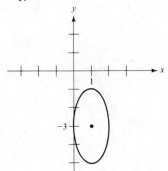

3. Axis: $y = -2$
$V(3, -2)$, $F(5, -2)$, D: $x = 1$,
lr = 8.

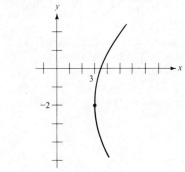

5. $C(-1, -2)$, $V(-1 \pm 5, -2)$,
$CV(-1, -2 \pm 3)$
$F(-1 \pm 4, -2)$, lr = 18/5.

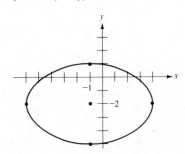

7. Axis: $x = 2$
$V(2, -1)$, $F(2, 0)$,
D: $y = -3$
lr = 4.

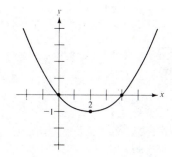

9. $2x + 3y + 5 = 0$,
$2x - 3y - 13 = 0$.

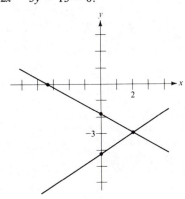

11. $y^2 - 8x - 10y + 33 = 0$

13. $x^2 - 4x - 8y - 20 = 0$, $y^2 - x + 6y + 11 = 0$

15. $16x^2 - 9y^2 - 128x - 18y + 103 = 0$

17. $y = x^4 - 16x^3 + 88x^2 - 192x + 140$

$x = x' + h, \quad y = y' + k,$

$y' + k = (x' + h)^4 - 16(x' + h)^3 + 88(x' + h)^2 - 192(x' + h) + 140$

$\qquad = x'^4 + 4hx'^3 + 6h^2x'^2 + 4h^3x' + h^4 - 16x'^3$

$\qquad\quad - 48hx'^2 - 48h^2x' - 16h^3 + 88x'^2 + 176hx'$

$\qquad\quad + 88h^2 - 192x' - 192h + 140$

$y' = x'^4 + (4h - 16)x'^3 + (6h^2 - 48h + 88)x'^2$

$\qquad + (4h^3 - 48h^2 + 176h - 192)x'$

$\qquad + (h^4 - 16h^3 + 88h^2 - 192h + 140 - k)$

$4h - 16 = 0, \quad h = 4$

$k = h^4 - 16h^3 + 88h^2 - 192h + 140$

$\quad = 256 - 1024 + 1408 - 768 + 140 = 12$

$y' = x'^4 - 8x'^2 = x'^2(x'^2 - 8).$

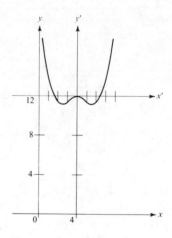

19. $\tan 2\theta = \dfrac{-\sqrt{3}}{2 - 1} = -\sqrt{3},$

$2\theta = 120°, \quad \theta = 60°$

$\sin \theta = \sqrt{3}/2, \quad \cos \theta = 1/2$

$x = \dfrac{x' - \sqrt{3}y'}{2},$

$y = \dfrac{\sqrt{3}x' + y'}{2}$

$1/4(2x'^2 - 4\sqrt{3}x'y' + 6y'^2$

$\quad - 3x'^2 + 2\sqrt{3}x'y' + 3y'^2$

$\quad + 3x'^2 + 2\sqrt{3}x'y' + y'^2) - 10 = 0$

$x'^2 + 5y'^2 - 20 = 0$

$x'^2/20 + y'^2/4 = 1.$

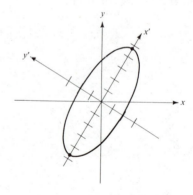

21. $B^2 - 4AC = (-2)^2 - 4 \cdot 2 \cdot 1 = -4 < 0.$

The equation is that of an ellipse (see Problem 26 below).

$y^2 - 2xy + (2x^2 - 9) = 0$

$y = \dfrac{2x \pm \sqrt{4x^2 - 4(2x^2 - 9)}}{2}$

$\quad = x \pm \sqrt{9 - x^2}$

$\quad = x, \quad x^2 + y^2 = 9.$

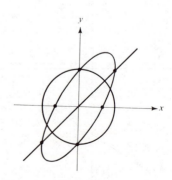

CHAPTER 7

Section 7.1 [pages 189–190]

1. $(0, 0), (-3, 0)$ **3.** $(1, 0), (-1, 0), (0, -1)$ **5.** $(-1/4, 0), (2, 0), (-3/2, 0), (0, -18)$

7. $(1, 0), (0, -1)$ **9.** $(1/3, 0), (-1/2, 0), (0, 8)$ **11.** $(0, 0)$ **13.** $(3, 0), (0, 9)$

15. None   **17.** None   **19.** $x = 1, y = 0$ **21.** $x = -3, y = 1$ **23.** $x = -1, y = 2$

25. $x = -1, x = 3, y = 0$ **27.** $y = 1$ **29.** $x = -3/2, x = -1, y = 0$ **31.** $y = 0$

33. $y = 0$ **35. a.** $y = 0$ **b.** $y = a_n/b_m$ **c.** none   **37.** 80,000; 9

Section 7.2 [pages 197–198]

1. y axis   **3.** None   **5.** x axis   **7.** Origin   **9.** Origin

11.

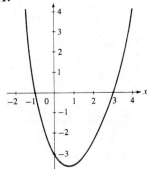

13.

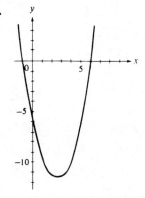

15.

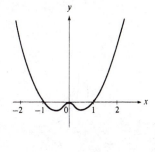

17.

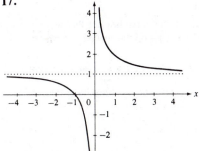

19.

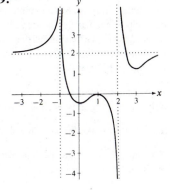

21.

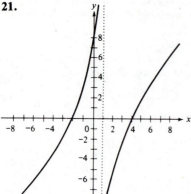

23.

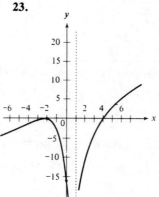

25.

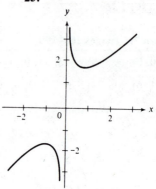

27.

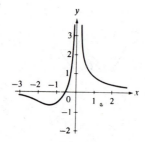

29.

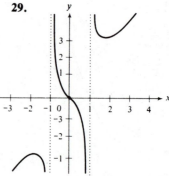

31. a. Suppose we have symmetry about both axes. Then (x, y) on the curve implies that $(-x, y)$ is on the curve by symmetry about the y axis; this, in turn, implies that $(-x, -y)$ is on the curve by symmetry about the x axis. Thus we have symmetry about the origin.

  b. Suppose we have symmetry about the x axis and the origin. Then (x, y) on the curve implies that $(x, -y)$ is on the curve by symmetry about the x axis; this, in turn, implies that $(-x, y)$ is on the curve by symmetry about the origin. Thus we have symmetry about the y axis.

  c. Symmetry about the y axis and the origin implies symmetry about the x axis by a similar argument.

33. A graph can have (at least) two points of symmetry (a line, a sine curve, etc.), but it cannot have two and only two such points because two points of symmetry imply infinitely many such points.

35. Translate to make the axes the lines of symmetry and use the argument of Problem 31.

37. No, since (x, y) on the graph implies that $(x, -y)$ is on it. Since a function must be single-valued, this cannot be the graph of a function.

Section 7.3 [pages 202–203]

1.

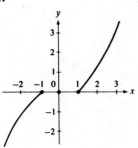

3.

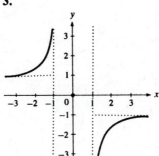

5.

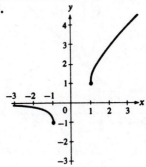

7.

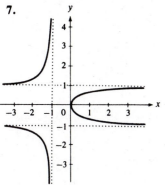

9.

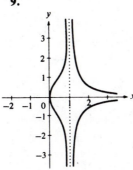

11.

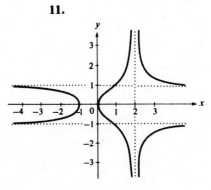

13.

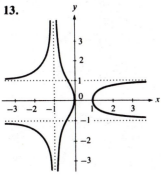

15.

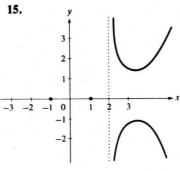

17.

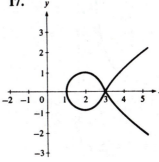

19.

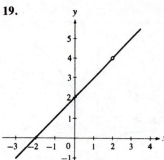

21.

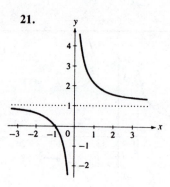

23.

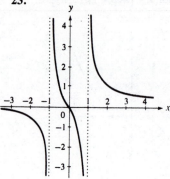

25.

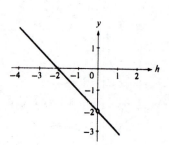

27.

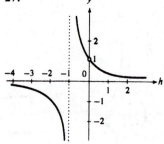

Section 7.4 [page 206]

1.

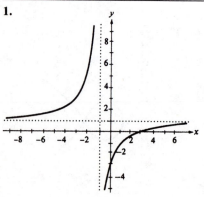

3.

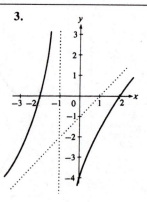

5.

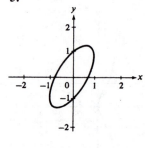

7.

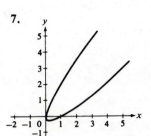

9.

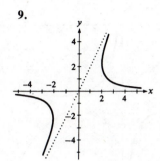

11.

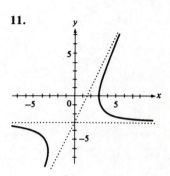

13.

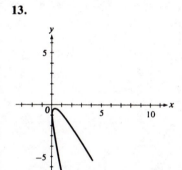

15.

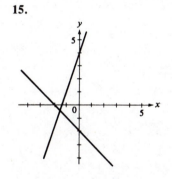

17.

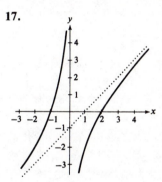

19.

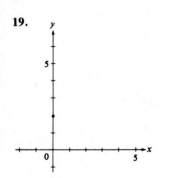

21. $\sqrt{x} + \sqrt{y} = \sqrt{a}$
$x + 2\sqrt{xy} + y = a$
$2\sqrt{xy} = a - x - y$
$4xy = a^2 + x^2 + y^2 - 2ax - 2ay + 2xy$
$x^2 - 2xy + y^2 - 2ax - 2ay + a^2 = 0$
$B^2 - 4AC = 4 - 4 \cdot 1 \cdot 1 = 0.$

Review 7 [page 207]

1.

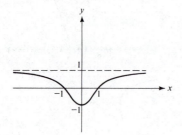

3.

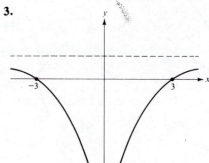

5.

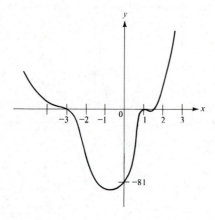

7.

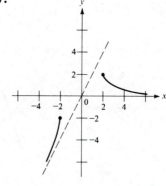

9.

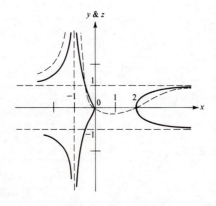

11.

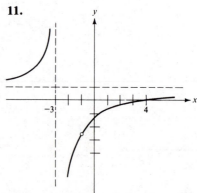

13.

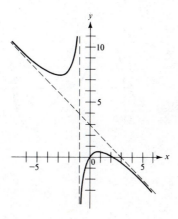

15.

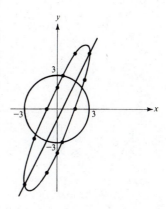

CHAPTER 8

Section 8.1 [pages 211–212]

1. $\pi/4$, $-7\pi/6$, $3\pi/2$, $\pi/6$ **2.** $60°$, $180°$, $135°$, $-90°$

3.

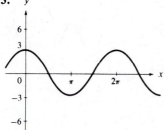

5.

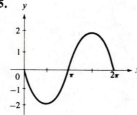

7.

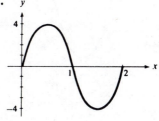

9.

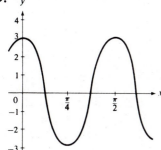

11.

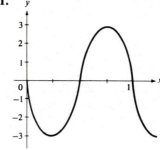

13.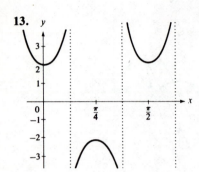

15.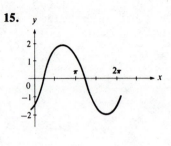

17.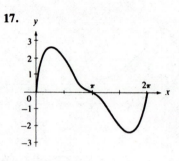

19.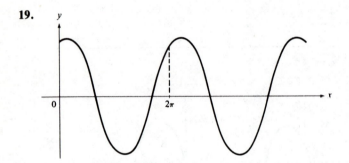

21.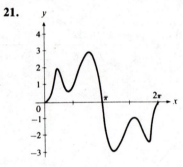

23.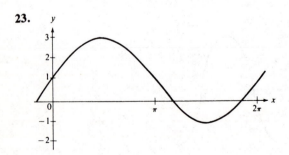

25.

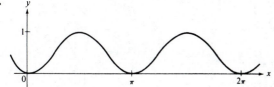

27.

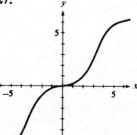

29.

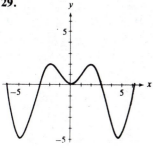

31.

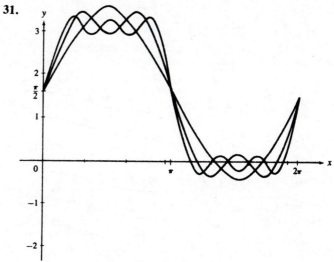

Section 8.2 [page 215]

1.

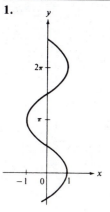

3.

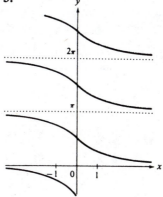

5.

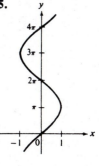

7.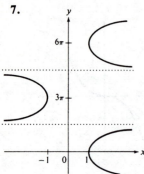

9.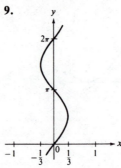

11.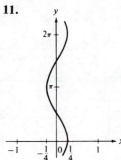

13.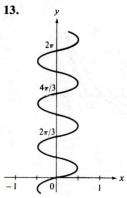

15.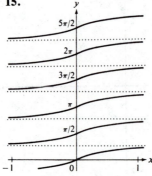

17.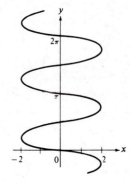

19.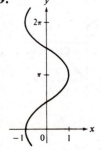

21.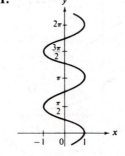

23.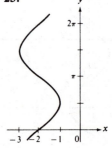

25.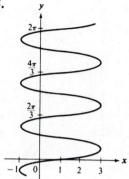

27.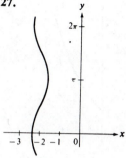

29.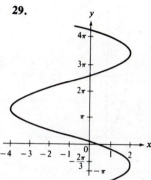

Section 8.3 [page 219]

1.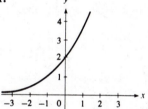

3.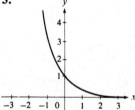

5.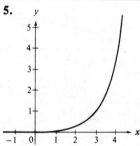

7.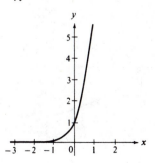

9.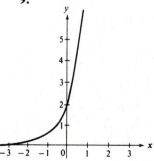

11.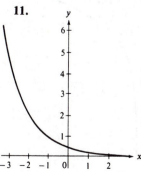

13.

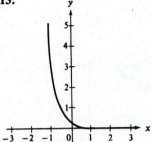

15.

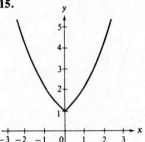

17.

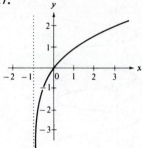

19.

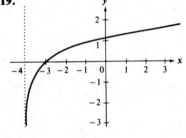

21.

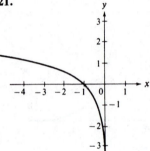

23.

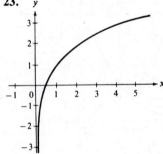

25.

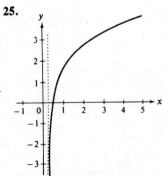

27.

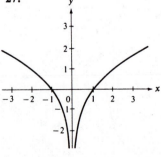

29.

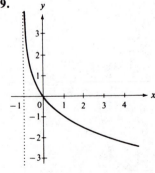

Section 8.4 [page 222]

1.

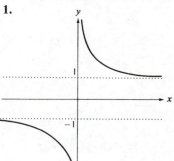

3.

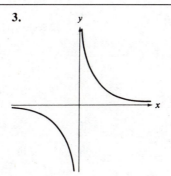

5.

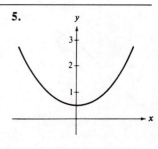

7.

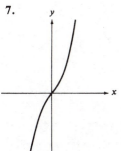

9.

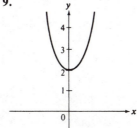

11.

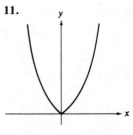

13.

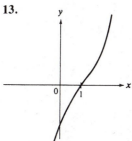

15.

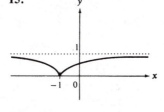

17.

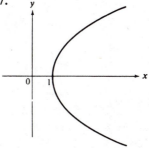

19.

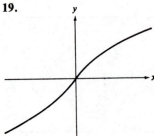

21. $\dfrac{e^x - e^{-x}}{e^x + e^{-x}}$

23. $y = \sinh^{-1} x$

$x = \sinh y = \dfrac{e^y - e^{-y}}{2}$

$2x = e^y - e^{-y}$

$e^y - 2x - e^{-y} = 0$

$e^{2y} - 2xe^y - 1 = 0$

$e^y = \dfrac{2x \pm \sqrt{4x^2 + 4}}{2} = x \pm \sqrt{x^2 + 1}$

Since $e^y > 0$, we may drop the minus.

$e^y = x + \sqrt{x^2 + 1}$

$y = \ln(x + \sqrt{x^2 + 1})$

Review 8 [page 223]

1.

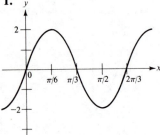

3.

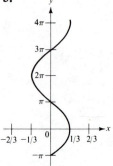

5.

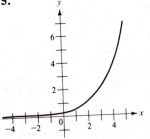

7.

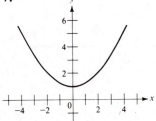

9.

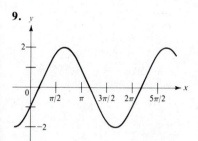

11.

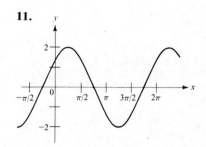

13.

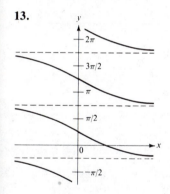

15.

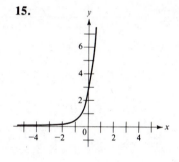

17.

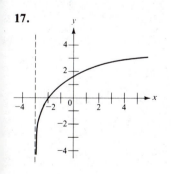

19.

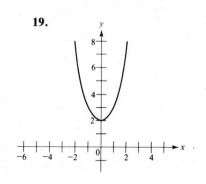

CHAPTER 9

Section 9.2 [pages 231–232]

1.

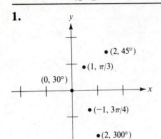

2. $(-4, 150°)$, $(-2, 60°)$, $(-1, 30°)$

3. $(4, 5\pi/3)$, $(3, 5\pi/3)$, $(0, 0)$

5.

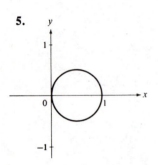

7.

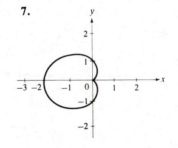

9.

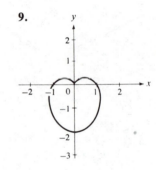

11.

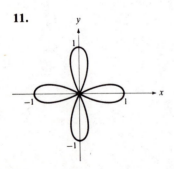

13.

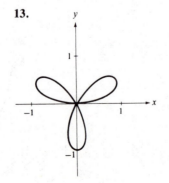

15.

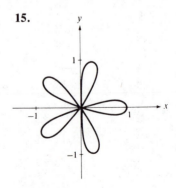

17.

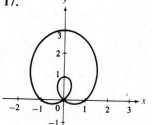

19.

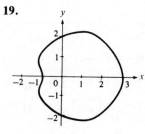

21.

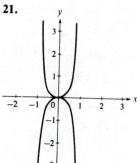

23.

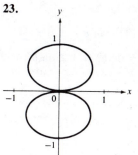

25.

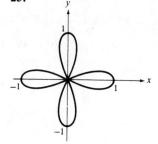

27.

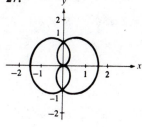

29.

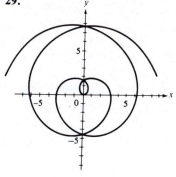

31.

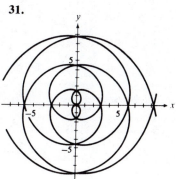

33.

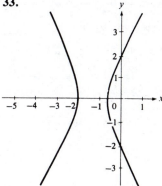

35. There are many answers. Two possible ones are:

$$\begin{cases} r \to -r \\ \theta \to -\theta, \end{cases} \qquad \begin{cases} r \to r \\ \theta \to -\pi - \theta. \end{cases}$$

Section 9.3 [pages 234–235]

1. $(\sqrt{2}, \pi/4), (\sqrt{2}, 7\pi/4)$ **3.** $(2, 0), (2, \pi)$ **5.** $(1/2, \pi/3), (1/2, 5\pi/3), (0, \pi/2) = (0, 0)$

7. $(\sqrt{3}/2, \pi/3), (-\sqrt{3}/2, 5\pi/3), (0, 0)$ **9.** $(\sqrt{2}, \pi/4)$ **11.** $(3, 109.5°), (3, 250.5°)$

13. $(2 + \sqrt{2}, \pi/4), (2 - \sqrt{2}, 3\pi/4), (2 - \sqrt{2}, 5\pi/4), (2 + \sqrt{2}, 7\pi/4)$

15. $(1 - 1/\sqrt{2}, \pi/4), (1 + 1/\sqrt{2}, 5\pi/4), (0, \pi/2) = (0, 0)$ **17.** $(1, 0), (-1, 0)$

19. $(0, 0), (1, \pi/2)$

Section 9.4 [pages 238–239]

1. $(-1, 0), (\sqrt{3}/2, 3/2), (1, 0), (-1, 1)$ **2.** $(2, 7\pi/4), (2, 2\pi/3), (4, 0), (\sqrt{2}, 5\pi/4), (2, 3\pi/2)$

3. $r = 2 \sec \theta$ **5.** $r = 1$ **7.** $r = \csc \theta \cot \theta$ **9.** $r = 4/(\cos \theta + 2 \sin \theta)$ **11.** $\tan \theta = 3$

13. $x^2 + y^2 = a^2$ **15.** $\sqrt{3}x - y = 0$ **17.** $x^2 + y^2 - 4x = 0$ **19.** $r = (\cos \theta - \sin \theta)/(1 + 2 \sin \theta \cos \theta)$

21. $r^2 - 2r(\sin \theta + \cos \theta) + 1 = 0$ **23.** $r^2 = \sec \theta \csc \theta$.   **25.** $(x^2 + y^2)^3 = (x^2 - y^2)^2$

27. $x^2 + y^2 - 4y - 9 + 4y^2/(x^2 + y^2) = 0$ (note that the origin is not a point of the given curve)

29. $(x^2 + y^2)(x^2 + y^2 - 1)^2 = y^2$ **31.** $y^2 = 2x + 1$ **33.** $x^2 + y^2 = 3x + 2y$

Section 9.5 [pages 243–245]

1. hyperbola, $(0, 0)$, $x = 2$, 2

3. ellipse, $(0, 0)$, $y = 2$, 2/3

5. parabola, $(0, 0)$, $y = 3$, 1

7. ellipse, $(0, 0)$, $y = 3$, 2/3

9. $r = 10/(3 + 2 \cos \theta)$

11. $r = 2/(1 + \sin \theta)$

13. $r = 25/(4 + 5 \cos \theta)$

15.

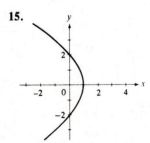

17.

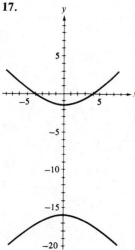

19.

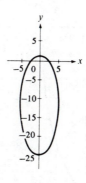

21.

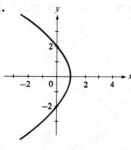

23. The shape of the conic is elliptical approaching a circle. The equation approaches $r = k$ $(k = ep)$, which is an equation of a circle.

25. $r = 184,900/(43 - 2 \cos \theta)$

27. $\cos(\alpha - \theta) = p/r$
$r \cos \theta \cos \alpha + r \sin \theta \sin \alpha = p$
$x \cos \alpha + y \sin \alpha - p = 0$

29. The line through (x_1, y_1) and parallel to the given line is $Ax + By - (Ax_1 + By_1) = 0$. Both lines can be put into the normal form by dividing by $\pm \sqrt{A^2 + B^2}$. Thus we have

$$\frac{A}{\pm \sqrt{A^2 + B^2}} x + \frac{B}{\pm \sqrt{A^2 + B^2}} y + \frac{C}{\pm \sqrt{A^2 + B^2}} = 0$$

and

$$\frac{A}{\pm \sqrt{A^2 + B^2}} x + \frac{B}{\pm \sqrt{A^2 + B^2}} y - \frac{Ax_1 + By_1}{\pm \sqrt{A^2 + B^2}} = 0$$

By choosing the same sign in both cases, the polar coordinates of Q_1 and Q_2 are

$$Q_1: \left(\frac{C}{\pm \sqrt{A^2 + B^2}}, \alpha \right), \quad Q_2: \left(-\frac{Ax_1 + By_1}{\pm \sqrt{A^2 + B^2}}, \alpha \right)$$

Thus the distance between them is

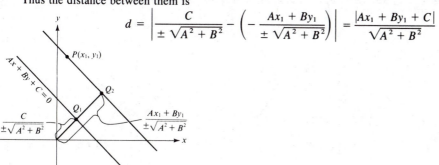

$$d = \left| \frac{C}{\pm \sqrt{A^2 + B^2}} - \left(-\frac{Ax_1 + By_1}{\pm \sqrt{A^2 + B^2}} \right) \right| = \frac{|Ax_1 + By_1 + C|}{\sqrt{A^2 + B^2}}$$

Section 9.6 [pages 251–252]

1. $x = y^2 - 2y + 2$

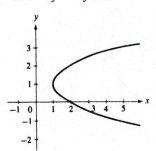

3. $y = x^2 - 1$

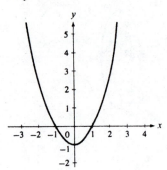

5. $y = (x + 1)^2$

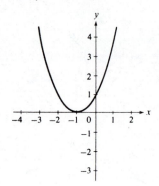

7. $x = 1 + 2r, \; y = 5 - 4r$

9. $x = 2 - 3r, \; y = 5 - 3r$

11. $x = 2 + 3r, \; y = 3$

13. $x^2 - 2xy + y^2 - 2x - 2y = 0$

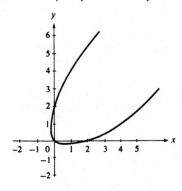

15. $y^3 = x^2$

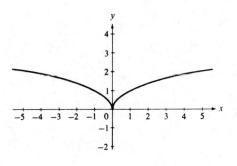

17. $(x - 2)^2 + (y + 1)^2 = 1$

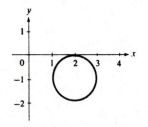

19. $(x - 3)^2 - (y - 2)^2 = 1$

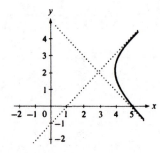

21. $y^2 = x^3$

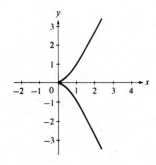

23. $x^2 + y^2 = 1$

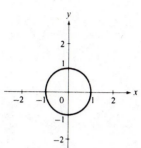

25.

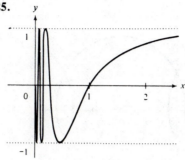

27.

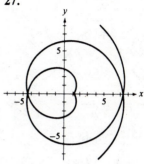

29.

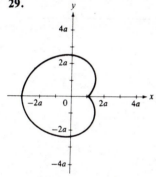

31. $x - x_1 = r(x_2 - x_1), \quad y - y_1 = r(y_2 - y_1)$

$$\frac{x - x_1}{y - y_1} = \frac{r(x_2 - x_1)}{r(y_2 - y_1)} = \frac{x_2 - x_1}{y_2 - y_1}$$

33.

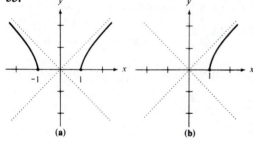

(a) (b) (c) (d)

Section 9.7 [pages 255–258]

1. $x = 1000t, \ y = -16t^2 + 1000\sqrt{3}t$ **3.** $x = 16\sqrt{3}t, \ y = -16t^2 + 16t; \ (16\sqrt{3}, 0)$

5. $x = (v_0 \cos \theta)t, \ y = -16t^2 + (v_0 \sin \theta)t$ **7.** $x = a(\theta + \sin \theta), \ y = a(1 + \cos \theta)$

9.

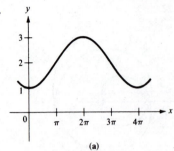

(a)   (b)

11. $r = a(1 \pm \sin \theta)/\cos \theta$ **13.** $r = (a \pm b \cos \theta)/\cos \theta$

15. $x = a \cos \theta + a\theta \sin \theta, y = a \sin \theta - a\theta \cos \theta$ **17.** $x = 5a \cos \theta - a \cos 5\theta, y = 5a \sin \theta$

Review 9 [pages 258–260]

1.   **3.**   **5.**

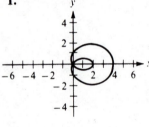   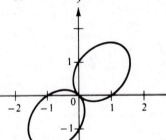   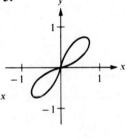

7. $(1, \pi/2), (1, 3\pi/2)$ **9.** $(-1, 3\pi/2), (0, 0) = (0, 7\pi/6)$ **11.** $(0, \pi/2), (1, 0), (-1, \pi)$

13. (a) $(0, -1), (-2, 2)$; (b) $(2\sqrt{2}, 3\pi/4), (5, \pi)$ **15.** $r = 6/(3 + 2 \sin \theta)$

17. $y = x^2 + 2x - 5$.   **19.** $x^2 - 2xy + y^2 - 2x - 2y + 1 = 0$.

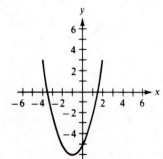

21. $x = 2r + 2, y = -8r + 3$.

23. $(x^2 + y^2 - y)^2 = x^2 + y^2$

25. $F(0, 0)$, D: $y = -2$, $e = 1$

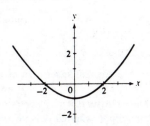

27. $y^2 = 4x^2(1 - x^2)$

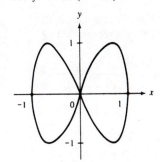

29. $x = 2a \cot \theta$, $y = a \sin \theta(1 - \cos \theta)$

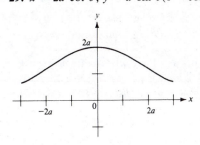

CHAPTER 10

Section 10.1 [pages 266–267]

1. $5\sqrt{2}$ **3.** 5   **5.** $\sqrt{131}$ **7.** 5   **9.** 9   **11.** $(-2, 1, 2)$ **13.** $(1, 2, 2)$ **15.** $(-9, 7, -1)$

17. $(1, 1, 5)$ **19.** $(1, 3, 3/2)$ **21.** $(8, 4, -6)$ **23.** $(-2, 4, -3)$ **25.** $5, -3$ **27.** ± 5

29. $-11, 5$

31. $r = \dfrac{\overline{P_1 P}}{\overline{P_1 P_2}} = \dfrac{\overline{Q_1 Q}}{\overline{Q_1 Q_2}} = \dfrac{x - x_1}{x_2 - x_1}$

$x - x_1 = r(x_2 - x_1)$
$x = x_1 + r(x_2 - x_1)$
By projecting onto the y axis and the z axis, we get
$y = y_1 + r(y_2 - y_1)$
$z = z_1 + r(z_2 - z_1)$

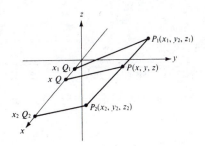

Section 10.2 [pages 272–273]

1. $-6\mathbf{i} - 2\mathbf{j} + 7\mathbf{k}$ **3.** $-3\mathbf{i} - 4\mathbf{j} + 3\mathbf{k}$ **5.** $\dfrac{4}{\sqrt{21}}\mathbf{i} + \dfrac{1}{\sqrt{21}}\mathbf{j} - \dfrac{2}{\sqrt{21}}\mathbf{k}$

7. $\dfrac{1}{\sqrt{35}}\mathbf{i} + \dfrac{5}{\sqrt{35}}\mathbf{j} - \dfrac{3}{\sqrt{35}}\mathbf{k}$ **9.** $B = (4, 0, 8)$ **11.** $A = (3, 1, 3)$

13. $A = (0, 6, -3/2)$, $B = (4, 4, -1/2)$ **15.** $60°$ **17.** $77°$

19. $3\mathbf{i} + 2\mathbf{j} + 6\mathbf{k}$, $-\mathbf{i} - 6\mathbf{j} + 4\mathbf{k}$, -1, not orthogonal   **21.** $3\mathbf{i} - 5\mathbf{j} + 5\mathbf{k}$, $\mathbf{i} + 3\mathbf{j} + 7\mathbf{k}$, 0, orthogonal

23. $\dfrac{4}{3}\mathbf{i} + \dfrac{4}{3}\mathbf{j} + \dfrac{4}{3}\mathbf{k}$ **25.** $\dfrac{7}{6}\mathbf{i} - \dfrac{7}{3}\mathbf{j} + \dfrac{7}{6}\mathbf{k}$

27. Suppose we have a vector $\mathbf{v}$ in space. Let us consider the respresentative of $\mathbf{v}$ with its tail at the origin O. The head is at $P(a, b, c)$. Let us project P onto the coordinate axes, giving points $A(a, 0, 0)$, $B(0, b, 0)$, and $C(0, 0, c)$. Since $\overrightarrow{OA}$ represents a vector of length $|a|$ that is either in the direction of $\mathbf{i}$ or in the opposite direction, depending upon whether a is positive or negative, it represents $a\mathbf{i}$. Similarly, $\overrightarrow{OB}$ represents $b\mathbf{j}$ and $\overrightarrow{OC}$ represents $c\mathbf{k}$. It is clear that $\mathbf{v} = a\mathbf{i} + b\mathbf{j} + c\mathbf{k}$.

Since the point P can be represented in rectangular coordinates by a triple (a, b, c) of numbers in one and only one way, the vector $\mathbf{v}$ has one and only one representation in component form.

29. $a_1\mathbf{i} + b_1\mathbf{j} + c_1\mathbf{k}$ is presented by $\overrightarrow{OP}$ from $(0, 0, 0)$ to (a_1, b_1, c_1) and $a_2\mathbf{i} + b_2\mathbf{j} + c_2\mathbf{k}$ by $\overrightarrow{OP_2}$ from $(0, 0, 0)$ to (a_2, b_2, c_2), or by $\overrightarrow{P_1P_3}$ from (a_1, b_1, c_1) to $(a_1 + a_2, b_1 + b_2, c_1 + c_2)$. Hence the sum is represented by $\overrightarrow{OP_3}$ or

$$(a_1\mathbf{i} + b_1\mathbf{j} + c_1\mathbf{k}) + (a_2\mathbf{i} + b_2\mathbf{j} + c_2\mathbf{k}) = (a_1 + a_2)\mathbf{i} + (b_1 + b_2)\mathbf{j} + (c_1 + c_2)\mathbf{k}.$$

Let $(a_1\mathbf{i} + b_1\mathbf{j} + c_1\mathbf{k}) - (a_2\mathbf{i} + b_2\mathbf{j} + c_2\mathbf{k}) = a\mathbf{i} + b\mathbf{j} + c\mathbf{k}$. Then
$a_1\mathbf{i} + b_1\mathbf{j} + c_1\mathbf{k} = (a_2\mathbf{i} + b_2\mathbf{j} + c_2\mathbf{k}) + (a\mathbf{i} + b\mathbf{j} + c\mathbf{k})$

$$= (a_2 + a)\mathbf{i} + (b_2 + b)\mathbf{j} + (c_2 + c)\mathbf{k}$$

$$\begin{array}{lll} a_1 = a_2 + a & b_1 = b_2 + b & c_1 = c_2 + c \\ a = a_1 - a_2 & b = b_1 - b_2 & c = c_1 - c_2 \end{array}$$

$a\mathbf{i} + b\mathbf{j} + c\mathbf{k}$ is represented by $\overrightarrow{OP}$ from $(0, 0, 0)$ to (a, b, c). Its length is
$$\sqrt{(a - 0)^2 + (b - 0)^2 + (c - 0)^2} = \sqrt{a^2 + b^2 + c^2}.$$

$\mathbf{v} = a\mathbf{i} + b\mathbf{j} + c\mathbf{k}$ is represented by $\overrightarrow{OP}$ from $(0, 0, 0)$ to (a, b, c). $\mathbf{w} = da\mathbf{i} + db\mathbf{j} + dc\mathbf{k}$ is represented by $\overrightarrow{OQ}$ from $(0, 0, 0)$ to (da, db, dc). Clearly these points lie on the same line; so $\mathbf{w}$ is in the same direction as or the opposite direction from $\mathbf{v}$, depending upon the sign of d. Furthermore
$$|\mathbf{w}| = \sqrt{d^2a^2 + d^2b^2 + d^2c^2} = \sqrt{d^2(a^2 + b^2 + c^2)} = |d||\mathbf{v}|.$$
Hence $d(a\mathbf{i} + b\mathbf{j} + c\mathbf{k}) = da\mathbf{i} + db\mathbf{j} + dc\mathbf{k}$.

31. From Theorem 10.10 and Problem 30.
$$|u + v|^2 = (u + v) \cdot (u + v) = |u|^2 + 2u \cdot v + |v|^2$$
$$= |u|^2 + 2|u| \cdot |v| \cdot \cos\theta + |v|^2 \le |u|^2 + 2|u| \cdot |v| + |v|^2$$
$$= (|u| + |v|)^2.$$
Thus $|u + v| \le |u| + |v|$.

Section 10.3 [pages 278–279]

1. $84°$, $64°$, $27°$ **3.** $103°$, $116°$, $29°$ **5.** $125°$, $125°$, $125°$ **7.** $\{-4, 2, 4\}$ **9.** $\{2, 2, -2\}$

11. $\{5, 1, -2\}$ **13.** $(3, 8, 1)$, $(4, 13, 3)$ **15.** $(3, 4, 5)$, $(5, 5, 7)$ **17.** $(5, 3, -2)$, $(9, 3, -3)$

19. x axis: $\{0°, 90°, 90°\}$, $\{1, 0, 0\}$ **21.** Parallel   **23.** Perpendicular   **25.** Coincident

27. None   **29.** Perpendicular

Section 10.4 [pages 286–287]

1. $x = 5 + 3t$, $y = 1 - 2t$, $z = 3 + 4t$; $\dfrac{x - 5}{3} = \dfrac{y - 1}{-2} = \dfrac{z - 3}{4}$

3. $x = 5 + 4t, y = -2 + t, z = 1 - 2t; \dfrac{x-5}{4} = \dfrac{y+2}{1} = \dfrac{z-1}{-2}$

5. $x = 1 + 2t, y = 1, z = 1 + t; \dfrac{x-1}{2} = \dfrac{z-1}{1}, y = 1$

7. $x = 4, y = 4, z = 1 + t; x = 4, y = 4$

9. $x = 4 + 2t, y = -3t, z = 5 + 4t; \dfrac{x-4}{2} = \dfrac{y}{-3} = \dfrac{z-5}{4}$

11. $x = 8 + 5t, y = 4 + 2t, z = 1 - t; \dfrac{x-8}{5} = \dfrac{y-4}{2} = \dfrac{'z-1}{-1}$

13. $x = 5, y = 1 + t, z = 3 + t; x = 5, y - 1 = z - 3$

15. $x = 1, y = -2 + t, z = 3; x = 1, z = 3$ **17.** Do not intersect   **19.** Do not intersect

21. The lines are identical   **23.** $(2, 1, -1)$ **25.** Perpendicular   **27.** None   **29.** Parallel

31. x axis: $\{1, 0, 0\}; y = 0, z = 0$
y axis: $\{0, 1, 0\}; x = 0, z = 0$
z axis: $\{0, 0, 1\}; x = 0, y = 0$

Section 10.5 [pages 294–296]

1. $-5\mathbf{i} + 5\mathbf{j} + 5\mathbf{k}$ **3.** $2\mathbf{i} + \mathbf{j} + 7\mathbf{k}$ **5.** $-\mathbf{i} - \mathbf{j} + 3\mathbf{k}$ **7.** $\{1, -1, 0\}$ **9.** $\{1, 1, -1\}$

11. $\{8, 5, -9\}$ **13.** $x = 3 + t, y = 2 - 3t, z = 1 - 5t$ **15.** $x = 2 + t, y = 3 - 2t, z = 1 + 4t$

17. $x = 2 + 10t, y = t, z = 5 - 8t$ **19.** $24/\sqrt{30}$ **21.** $113/\sqrt{542}$ **23.** $\sqrt{26}$

25. The area of $\triangle ABC$ is $A = 1/2\, AB \cdot BC \cdot \sin \angle BAC$. If $\mathbf{u}$ is represented by $\overrightarrow{AB}$, $\mathbf{v}$ by $\overrightarrow{BC}$ and $\theta = \angle BAC$, then $A = 1/2|\mathbf{u}| \cdot |\mathbf{v}| \sin \theta = 1/2|\mathbf{u} \times \mathbf{v}|$.

27. $\sqrt{893}/2$ **29.** $13/2$

31. $\mathbf{u} \cdot (\mathbf{v} \times \mathbf{w}) = (a_1\mathbf{i} + b_1\mathbf{j} + c_1\mathbf{k}) \cdot \begin{vmatrix} \mathbf{i} & \mathbf{j} & \mathbf{k} \\ a_2 & b_2 & c_2 \\ a_3 & b_3 & c_3 \end{vmatrix}$

$= (a_1\mathbf{i} + b_1\mathbf{j} + c_1\mathbf{k}) \cdot \left(\begin{vmatrix} b_2 & c_2 \\ b_3 & c_3 \end{vmatrix}\mathbf{i} - \begin{vmatrix} a_2 & c_2 \\ a_3 & c_3 \end{vmatrix}\mathbf{j} + \begin{vmatrix} a_2 & b_2 \\ a_3 & b_3 \end{vmatrix}\mathbf{k} \right)$

$= a_1\begin{vmatrix} b_2 & c_2 \\ b_3 & c_3 \end{vmatrix} - b_1\begin{vmatrix} a_2 & c_2 \\ a_3 & c_3 \end{vmatrix} + c_1\begin{vmatrix} a_2 & b_2 \\ a_3 & b_3 \end{vmatrix}$

$= \begin{vmatrix} a_1 & b_1 & c_1 \\ a_2 & b_2 & c_2 \\ a_3 & b_3 & c_3 \end{vmatrix}$

$(\mathbf{u} \times \mathbf{v}) \cdot \mathbf{w} = \begin{vmatrix} \mathbf{i} & \mathbf{j} & \mathbf{k} \\ a_1 & b_1 & c_1 \\ a_2 & b_2 & c_2 \end{vmatrix} \cdot (a_3\mathbf{i} + b_3\mathbf{j} + c_3\mathbf{k})$

$$= \left(\mathbf{i} \begin{vmatrix} b_1 & c_1 \\ b_2 & c_2 \end{vmatrix} - \mathbf{j} \begin{vmatrix} a_1 & c_1 \\ a_2 & c_2 \end{vmatrix} + \mathbf{k} \begin{vmatrix} a_1 & b_1 \\ a_2 & b_2 \end{vmatrix} \right) \cdot (a_3\mathbf{i} + b_3\mathbf{j} + c_3\mathbf{k})$$

$$= a_3 \begin{vmatrix} b_1 & c_1 \\ b_2 & c_2 \end{vmatrix} - b_3 \begin{vmatrix} a_1 & c_1 \\ a_2 & c_2 \end{vmatrix} + c_3 \begin{vmatrix} a_1 & b_1 \\ a_2 & b_2 \end{vmatrix}$$

$$= \begin{vmatrix} a_3 & b_3 & c_3 \\ a_1 & b_1 & c_1 \\ a_2 & b_2 & c_2 \end{vmatrix} = \begin{vmatrix} a_1 & b_1 & c_1 \\ a_2 & b_2 & c_2 \\ a_3 & b_3 & c_3 \end{vmatrix}$$

$$\mathbf{u} \cdot (\mathbf{v} \times \mathbf{w}) = (\mathbf{u} \times \mathbf{v}) \cdot \mathbf{w}$$

33. 25

Section 10.6 [pages 301–303]

1.   **3.**   **5.**

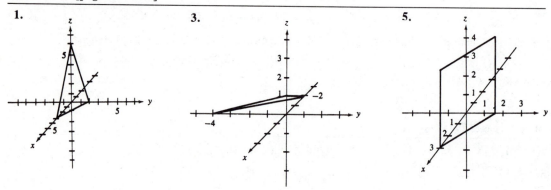

7. $3x - 4y + z + 4 = 0$ **9.** $3x - y - 2z - 17 = 0$ **11.** $3x - 4y + 2z + 9 = 0$

13. $3x + y - z - 4 = 0$ **15.** $x + y - 5 = 0$ **17.** $3x + 5y + z - 23 = 0$ **19.** $2x + y - 9 = 0$

21. $x + 5y + 3z - 26 = 0$ **23.** $3x - 7y - 5z + 22 = 0$ **25.** $x = 2 + 2t, \quad y = 5 - t, \quad z = -1 + 3t$

27. $x = 2 + t, \ y = -4 + t, \ z = 5$ **29.** Perpendicular   **31.** None   **33.** Parallel   **35.** $(1, 3, 7)$

37. $(4, 1, -1)$ **39.** Yes

Section 10.7 [pages 309–310]

1. $5/2$ **3.** 0   **5.** 0   **7.** $1/\sqrt{2}$ **9.** $2, -5/2$ **11.** $\sqrt{173}$ **13.** $5\sqrt{2}$ **15.** $\sqrt{14}/2$

17. $\sqrt{230/7}$ **19.** 7   **21.** $11/2\sqrt{26}$ **23.** $\sqrt{5}$ **25.** $52°$ **27.** $80°$ **29.** $146°$ **31.** $42°$

33. A vector perpendicular to the plane is $\mathbf{v} = A\mathbf{i} + B\mathbf{j} + C\mathbf{k}$. A, B, and C are not all 0. Suppose $A \neq 0$. Then a point in the plane is $(-D/A, \ 0, \ 0)$. The vector $\mathbf{u}$ from $(-D/A, \ 0, \ 0)$ to (x_1, y_1, z_1) is
$\mathbf{u} = (x_1 + D/A)\mathbf{i} + y_1\mathbf{j} + z_1\mathbf{k}$.
$$d = \frac{|\mathbf{u} \cdot \mathbf{v}|}{|\mathbf{v}|} = \frac{|A(x_1 + D/A) + By_1 + Cz_1|}{\sqrt{A^2 + B^2 + C^2}} = \frac{|Ax_1 + By_1 + Cz_1 + D|}{\sqrt{A^2 + B^2 + C^2}}$$

Section 10.8 [page 313]

1.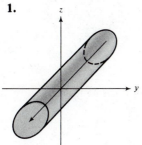

3.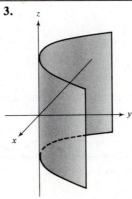

5.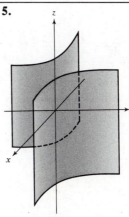

7.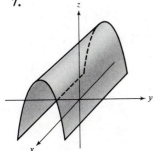

9.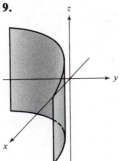

11. Sphere: $(1, 0, -2)$, 3   **13.** No locus   **15.** Sphere: $(-1/2, 3/2, -1)$, 2

17. Sphere: $(1/3, -1/3, -2/3)$, $2\sqrt{2}/3$ **19.** Point: $(1, 1/2, -2)$

21. $x^2 + y^2 + z^2 - 8x - 2y + 4z + 12 = 0$ **23.** $x^2 + y^2 + z^2 - 4x - 8y - 14z + 65 = 0$

25. $x^2 + y^2 + z^2 - 4x - 2y - 2z - 20 = 0$, $x^2 + y^2 + z^2 - 14y + 14z + 72 = 0$

27. $x^2 + y^2 + z^2 - 27x + 35y - 63z - 28 = 0$

29. If (x, y, z) is on the sphere, its distance from the center is r.
$$\sqrt{(x - h)^2 + (y - k)^2 + (z - l)^2} = r$$
$$(x - h)^2 + (y - k)^2 + (z - l)^2 = r^2$$
The above argument is reversible since r is positive.

31. $Ax^2 + Ay^2 + Az^2 + Gx + Hy + Iz + J = 0$

$$x^2 + y^2 + z^2 + \frac{G}{A}x + \frac{H}{A}y + \frac{I}{A}z + \frac{J}{A} = 0$$

$$x^2 + \frac{G}{A}x + \frac{G^2}{4A^2} + y^2 + \frac{H}{A}y + \frac{H^2}{4A^2} + z^2 + \frac{I}{A}z + \frac{I^2}{4A^2} = -\frac{J}{A} + \frac{G^2}{4A^2} + \frac{H^2}{4A^2} + \frac{I^2}{4A^2}$$

$$\left(x + \frac{G}{2A}\right)^2 + \left(y + \frac{H}{2A}\right)^2 + \left(z + \frac{I}{2A}\right)^2 = \frac{G^2 + H^2 + I^2 - 4AJ}{4A^2} = K$$

If $K > 0$, the equation represents a sphere; if $K = 0$, it represents a point; if $K < 0$, it has no locus.

Section 10.9 [pages 319–320]

1. Ellipsoid   **3.** Circular paraboloid   **5.** Circular cone

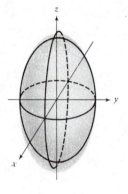

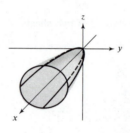

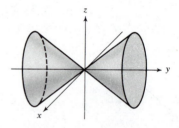

7. Circular paraboloid   **9.** Hyperboloid of two sheets

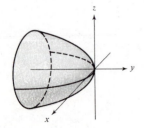

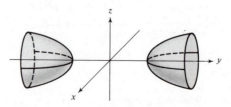

11. Hyperbolic paraboloid

13. Hyperboloid of two sheets

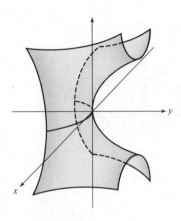

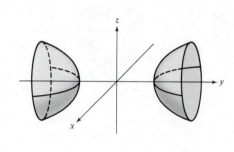

15. Circular cone

17. Hyperboloid of two sheets

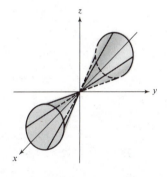

19. Hyperboloid of one sheet

21. Hyperbolic paraboloid

23. Circular cone

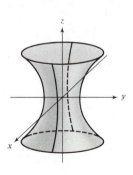

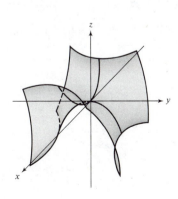

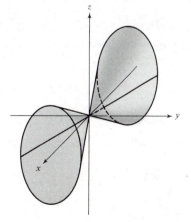

25. Hyperboloid of one sheet   **27.** $z' = x'^2 + y'^2$ **29.** $x'^2 + 4y'^2 + 9z'^2 = 36$

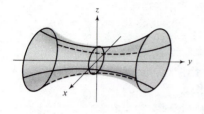
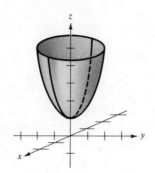
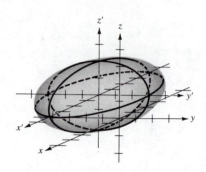

Section 10.10 [pages 324–325]

1. a. $(\sqrt{2}, \sqrt{2}, 1)$ **b.** $(-3/2, 3\sqrt{3}/2, -2)$ **2. a.** $(\sqrt{2}, 45°, 3)$ **b.** $(2, \pi/2, -2)$

3. a. $(3/2\sqrt{2}, 3/2\sqrt{2}, 3\sqrt{3}/2)$ **b.** $(0, 0, 1)$

4. a. $(2\sqrt{2}, 45°, 90°)$ **b.** $(3, \text{Arccos}\,(2/\sqrt{5}), \text{Arccos}\,(-2/3))$

5. a. $(5, 30°, \text{Arccos}\,(4/5))$ **b.** $(2\sqrt{2}, \pi/4, 3\pi/4)$ **6. a.** $(2, 45°, 2\sqrt{3})$ **b.** $(2, 2\pi/3, 0)$

7. $r = 2, \rho \sin \varphi = 2$ **9.** $r^2 = z, \rho = \csc \varphi \cot \varphi$

11. $r^2 \cos 2\theta - z^2 = 1, \rho^2(\sin^2 \varphi \cos 2\theta - \cos^2 \varphi) = 1$ **13.** $r^2 - z^2 = 1, \rho^2 = -\sec 2\varphi$

15. $z = 2xy$ **17.** $z = x^2 + y^2$ **19.** $xy = z (z \neq 0)$ **21.** $x^2 + y^2 + z^2 - x = 0$

Review 10 [pages 325–327]

1. $(-1, 2, 2)$ **3.** $(60°, 120°, 45°)$ **5.** Parallel

7. $x = 2 + 3t, y = -3 - 2t, z = 5 + t; \dfrac{x - 2}{3} = \dfrac{y + 3}{-2} = \dfrac{z - 5}{1}$ **9.** $(1, 5, -2)$

11. $x - 2y - 4 = 0$ **13.** $1/3$ **15.** $2\sqrt{3}/3$ **17.** $68°$ **19.** Point: $(1/2, 1, -3/2)$

21. Hyperbolic cylinder   **23.** Hyperbolic paraboloid

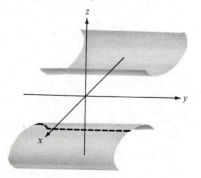
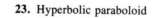
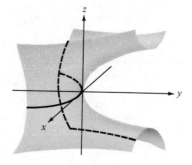

25. $z = x$ **27.** $x^2 + y^2 + z^2 - 2x + y = 0$ **29.** $A = (1/2, 9/2, -5)$, $B = (7/2, 7/2, -1)$

31. $x = 3 + 18t$, $y = 5 + 7t$, $z = -2 - 3t$ **33.** $11x - 6y - 7z - 53 = 0$ **35.** Oblate spheroid

37. Let $\mathbf{u} = \mathbf{i}$, $\mathbf{v} = \mathbf{j}$, $\mathbf{w} = \mathbf{j} + \mathbf{k}$
$\mathbf{v} \times \mathbf{w} = \mathbf{i}$, $\mathbf{u} \times (\mathbf{v} \times \mathbf{w}) = \mathbf{0}$,
$\mathbf{u} \times \mathbf{v} = \mathbf{k}$, $(\mathbf{u} \times \mathbf{v}) \times \mathbf{w} = -\mathbf{i}$,
$\mathbf{u} \times (\mathbf{v} \times \mathbf{w}) \neq (\mathbf{u} \times \mathbf{v}) \times \mathbf{w}$

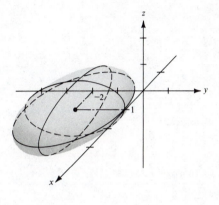

INDEX

Abscissa, 2
Absolute value of a vector, 40
Addition of ordinates, 205–206, 211
Addition of vectors, 39
Amplitude, 210
Analytic proofs, 7–8
Angle
  between two lines, 24–31, 307
  between two planes, 308
  between two vectors, 48–49, 268–269, 306
  bisector, 27–30, 78–79
  direction, 274, 276
  of incidence, 138
  of inclination, 17
  of reflection, 138
  of rotation, 168
Arcsin. *See* Inverse trigonometric functions
Asymptote
  horizontal, 185–188
  of a hyperbola, 133
  odd and even, 192–196
  slant, 204–205
  vertical, 184–185
Averages, method of, 89–91
Axes
  rectangular, 2–3, 261
  rotation of, 168–179
  translation of, 152–168, 320
Axis
  conjugate, of a hyperbola, 132
  major, of an ellipse, 124
  minor, of an ellipse, 124
  of a parabola, 118

Axis *(cont.)*
  polar, 224
  transverse, of a hyperbola, 132

Basis vectors, 41, 267

Cardioid, 227, 231, 258
Center
  of a circle, 95
  of an ellipse, 124
  of a hyperbola, 132
  of a sphere, 311
Circle(s), 95–114, 231, 244–245
  definition of, 95
  degenerate cases, 98–99
  equation of
    general form, 96
    in polar coordinates, 231, 244–245
    standard form, 95
  families of, 108–114
  involute of, 258
Cissoid of Diocles, 260
Completing the square, 96–97, 157–159
Components of a vector, 41, 268
Conchoid, 257
Cone, equation of, 317–318
Conic sections, 115–180, 203–206, 232, 239–244
  degenerate cases, 115–116, 160
  in polar coordinates, 232, 239–244
  reflection properties, 138–147
  and a right circular cone, 147–150
Conjugate axis of a hyperbola, 132
Conjugate hyperbolas, 135

Coordinate line, 1
Coordinate plane, 261
Coordinates
    cylindrical, 320–321
    left-hand, 261–262
    polar, 224–245
    rectangular, 1–3, 261–262
    relationship between rectangular and cylindrical, 321
    relationship between rectangular and polar, 235–239
    relationship between rectangular and spherical, 322–323
    right-hand, 261–262
    spherical, 322–325
Cosecant. *See* Trigonometric functions
Cosine. *See* Trigonometric functions
Cotangent. *See* Trigonometric functions
Covertex of an ellipse, 124
Cross product, 287–296
Curve fitting, 88–93
Cycloid, 254–255
Cylinder, 310–311
    directrix of, 310
    generatrix of, 310
Cylindrical coordinates, 320–321

Degenerate cases
    of a circle, 98–99
    of an ellipse, 115–116, 159–160
    of a hyperbola, 115–116, 160
    of a parabola, 115–116, 160
    of a sphere, 312
Descartes, René, 3
Difference of two vectors, 40
Directed line segment, 38–39
    equivalence, 38
    head, 38
    tail, 38
Direction angles
    of a line, 276
    of a vector, 274
Direction cosines
    of a line, 276
    of a vector, 274
Direction numbers
    of a line, 276
    of a vector, 275
Directrix
    of a cylinder, 310
    of an ellipse, 127
    of a hyperbola, 136–137
    of a parabola, 115
Distance
    between parallel lines, 76, 295

Distance *(cont.)*
    between parallel planes, 309
    between skew lines, 290–295
    between two points, 4–10, 263
    from a point to a line, 74–81, 245, 305–306
    from a point to a plane, 303–305
Domain
    of an equation, 198–203
    of a function, 31
    of parametric equations, 247
Dot product, 48–55, 269

e (base of the natural logarithm), 216
Eccentricity
    of an ellipse, 127, 240
    of a hyperbola, 136–137, 240
    of a parabola, 128, 240
Ellipse, 123–131, 155–163, 232, 239–244
    definition of, 123
    degenerate cases, 115–116, 159–160
    equation of
        general form, 156
        standard form, 124–125, 155
    in polar coordinates, 232, 239–244
Ellipsoid, 314
Elliptic cone, 317–318
Elliptic paraboloid, 316
Epicycloid, 258
Equation of a locus, 33–36, 252–258
Equations of rotation, 169, 172
Equations of translation, 153
Equilibrium of forces, 59
Equivalent directed line segments, 38
Equivalent forces, 56–60
Exponential functions, 216–219

Falling bodies, 252–255
Familes
    of lines, 81–88
    of circles, 108–114
Focus
    of an ellipse, 123
    of a hyperbola, 131
    of a parabola, 115
Forces
    equivalent, 56–60
    in equilibrium, 59
    represented by vectors, 56–60
Function, 31
    vector valued, 248–252

Generatrix of a cylinder, 310
Graph
    of an equation, 30–33
    of a number, 1

Graph *(cont.)*
  of parametric equations, 245–252
  in vector functions, 248–252
  in polar coordinates, 226–232

Head of a directed line segment, 38
Hyperbola, 131–138, 155–163
  conjugate, 135
  definition, 131
  degenerate case, 115–116, 160
  equation of
    general form, 156
    standard form, 132–134, 155
  in polar coordinates, 232, 239–244
Hyperbolic functions, 220–222
  definition of, 220
  graphs of, 220–222
Hyperbolic paraboloid, 317
Hyperboloid
  of one sheet, 315
  of two sheets, 315–316
Hypocycloid, 258

Incidence, angle of, 138
Inclination of a line, 17
Inner product, 50, 269
Intercepts
  of a curve, 183–184
  of a line, 68
  odd and even, 192–196
Invariant
  under translation, 165, 168
  under rotation, 176–179, 203–204
Inverse hyperbolic functions, 222
Inverse trigonometric functions, 212–215
Involute of a circle, 258

Latus rectum
  of an ellipse, 124–125
  of a hyperbola, 134
  of a parabola, 118
Left-hand coordinate system, 261–262
Lemniscate, 228, 232
Length of a vector, 40
Limaçon, 227, 232
Line(s)
  equation of, 62–93, 245, 249–251, 279–287
    from empirical data, 88–93
    general form, 71
    intercept form, 69
    normal form, 81, 245
    parametric, 249–251, 279–287
    point-slope form, 62
    slope-intercept form, 68
    symmetric, 284

Line(s) *(cont.)*
    two-point form, 64
    vertical, 64
  families of, 81–88
  parallel, 72–73, 278
  perpendicular, 72–73, 278
  in space, 279–287
Logarithm, 218–219
  definition, 218
  tables, 330–331

Major axis of an ellipse, 124
Midpoint formulas, 14–16, 266
Minor axis of an ellipse, 124
Multiplication of vectors
  cross product, 287
  dot product, 50, 269
  scalar multiple, 40

Nappe of a cone, 115, 147
Natural logarithm. *See* Logarithm
Normal form of a line, 81, 245
Normal line to a plane, 296–303

Oblate spheroid, 314
Octant, 261
Ordinate(s), 2
  addition of, 205–206
Origin, 1, 224, 261
Orthogonal vectors, 50, 269
Outer product, 287

Parabola, 115–122, 154–163
  definition of, 115
  degenerate cases, 115–116, 160
  equation of
    general form, 156
    standard form, 118–119, 155
  in polar coordinates, 232, 239–244
Paraboloid
  elliptic, 316
  hyperbolic, 317
Parallel lines, 21–22, 72–73, 278
Parallel planes, 301
Parameter, 82, 245
Parametric equations, 245–258
  for a line in the plane, 249–251
  for a line in space, 279–287
  of a locus, 252–258
Period, 209
Perpendicular lines, 21–22, 72–73, 278
Perpendicular planes, 301
Perpendicular vectors, 50, 269
Plane, 296–303
Point-of-division formulas, 10–16, 264–265

Point of intersection
  in polar coordinates, 232–235
  in rectangular coordinates, 31–33
Polar axis, 224
Polar coordinates, 224–245
  conic sections, 232, 239–244
  graphs, 226–232
  relationship between polar and rectangular coordinates, 235–239
Pole, 224
Prime factors, table of, 333
Product
  cross, 287–296
  dot, 48–55, 269
  inner, 50, 269
  outer, 287
  scalar, 50, 269
  vector, 287
Projectile, 252–255
Projection of one vector upon another, 51–53, 270–273
Prolate spheroid, 314

Quadrant, 3
Quadric surfaces, 314–320

Radian, 209
Radical axis, 111
Radius
  of a circle, 95
  of a sphere, 311
Range of a function, 31
Rectangular coordinates
  relationship between rectangular and cylindrical coordinates, 321
  relationship between rectangular and polar coordinates, 235–239
  relationship between rectangular and spherical coordinates, 322–323
Reflection, angle of, 138
Reflection properties of conics, 138–147
Representative of a vector, 39, 270
Residual, 89
Right-hand coordinate system, 261–262
Rose, four-leafed, etc., 228, 231
Rotation, 168–179
  equations of, 169–172

Scalar, 40
Scalar multiple, 40
Scalar product, 50, 269
Secant. See Trigonometric functions
Sector, 209

Selected points, method of, 88–89
Sine. See Trigonometric functions
Slope of a line, 17–19
Sphere, 311–313
  equation of
    general form, 312
    standard form, 311
Spherical coordinates, 322–325
Spheroid
  oblate, 314
  prolate, 314
Square roots, table of, 333
Squares, table of, 333
Strophoid, 257
Sum of two vectors, 39
Symmetry
  about the axes, 190–191, 228–232
  about the origin, 192, 228–232

Tail of a directed line segment, 38
Tangent. See Trigonometric functions
Tangent line
  to conics, 144–147
  to an ellipse, 128–131
  to a hyperbola, 138
  to a parabola, 120–122
Trace of a surface, 314
Translation, 152–168, 320
  equations of, 153, 320
Transverse axis of a hyperbola, 132
Trigonometric functions, 208–212
  tables of, 332

Vector(s), 38–61, 248–252, 267–273, 287–295
  basis, 41, 267
  definition of, 39
  difference, 40, 267
  directed along a line, 276
  functions, 248–252
  in the plane, 38–61
  product, 287–296
  proofs, 55–57
  representative of, 39, 270
  in space, 267–273, 287–295
  sum, 39, 268
Vertex
  of an ellipse, 124
  of a hyperbola, 132
  of a parabola, 118

Witch of Agnesi, 259

Zero vector, 39

STUDENT QUESTIONNAIRE

Your chance to rate Analytic Geometry, Third Edition (Riddle)

In order to keep this text responsive to your needs, it would help us to know what you, the student, thought of this text. We would appreciate it if you would answer the following questions. Then cut out the page, fold, seal, and mail it; no postage is required. Thank you for your help.

Which chapters did you cover?   (circle)   1   2   3   4   5   6   7   8   9   10   All

Does the book have enough worked-out examples?   Yes _____   No _____

enough exercises?   Yes _____   No _____

Which helped most?

Explanations _____   Examples _____   Exercises _____   All three _____

Other _____
(fill in)

Were the answers at the back of the book helpful?   Yes _____   No _____

Did the answers have any typos or misprints? If so, where?

For you, was the course elective? _____   Required? _____

Do you plan to take more mathematics courses?   Yes _____   No _____

If yes, which ones?

How much algebra did you have before this course?

Terms in high school   (circle)   1   2   3   4

Courses in college   1   2   3

If you had algebra before, how long ago?

Last 2 years _____   3–5 years ago _____   5 years or longer _____

What is your major or your career goal? _____

Your age? _____

What did you like the most about *Analytic Geometry, Third Edition?* (Riddle)

Can we quote you?   Yes _____   No _____

What did you like least about the book?

College _____ State _____

------------------------------ FOLD HERE ------------------------------

First Class
PERMIT NO. 34
Belmont Ca.

BUSINESS REPLY MAIL
No postage necessary if mailed in United States

Postage will be paid by
WADSWORTH PUBLISHING COMPANY, INC.
10 Davis Drive
Belmont, California 94002

ATTN: Rich Jones, Mathematics editor